The Rise of Critical Ani

As the scholarly and interdisciplinary study of human–animal relations becomes crucial to the urgent questions of our time, notably in relation to environmental crisis, this collection explores the inner tensions within the relatively new and broad field of animal studies. This provides a platform for the latest critical thinking on the condition and experience of animals. The volume is structured around four sections:

- engaging theory
- doing critical animal studies
- critical animal studies and anti-capitalism
- contesting the human, liberating the animal: veganism and activism.

The Rise of Critical Animal Studies demonstrates the centrality of the contribution of critical animal studies to vitally important contemporary debates and considers future directions for the field. This edited collection will be useful for students and scholars of sociology, gender studies, psychology, geography, and social work.

Nik Taylor is an Associate Professor in Sociology at Flinders University in South Australia, where she teaches and researches on human–animal relations.

Richard Twine is a sociologist and has most recently held positions at the Universities of Glasgow and Lancaster.

Routledge advances in sociology

The Rise of Critical Animal Studies

From the margins to the centre

Edited by Nik Taylor and Richard Twine

Routledge
Taylor & Francis Group

LONDON AND NEW YORK

First published 2014
by Routledge
2 Park Square, Milton Park, Abingdon, Oxfordshire OX14 4RN

and by Routledge
711 Third Avenue, New York, NY 10017

Routledge is an imprint of the Taylor and Francis Group, an informa business

First issued in paperback 2015

British Library Cataloguing in Publication Data
A catalogue record for this book is available from the British Library

Library of Congress Cataloging in Publication Data
The rise of critical animal studies : from the margins to the centre / edited
by Nik Taylor and Richard Twine.
    pages cm – (Routledge advances in sociology ; 125)
    Includes bibliographical references and index.
    1. Human-animal relationships–Research. 2. Animals–Social aspects–
Research. 3. Animal welfare–Research. I. Taylor, Nik. II. Twine,
Richard.
    QL85.R57 2014
    591.7–dc23
                                                        2013039464

ISBN 978-0-415-85857-1 (hbk)
ISBN 978-1-138-12591-9 (pbk)
ISBN 978-0-203-79763-1 (ebk)

Typeset in Times New Roman
by Wearset Ltd, Boldon, Tyne and Wear

Contents

Illustrations

Figures

Table

Contributors

Lynda Birke is Visiting Professor, Anthrozoology, University of Chester, UK. She was trained in biology (animal behaviour), but has also done interdisciplinary work, especially in human–animal studies, science studies, gender and women's studies. Her recent books include *Sacrifice: How Scientific Experiments Transform Animals and People* (Purdue University Press 2007), and *Crossing Boundaries* (Brill Academic 2012).

Matthew Cole is a sociologist teaching with The Open University, UK. His research interests include sociological analyses of how the human use of other animals is constructed as normal and acceptable, and in how vegans and veganism are represented (and often misrepresented) in public discourse. He is a former trustee of The Vegan Society and currently chairs the Vegan Society's research advisory committee. His first monograph, *Our Children and Their Animals: The Cultural Construction Of Human–Animal Relations In Childhood*, co-authored with Dr Kate Stewart, will be published by Ashgate in 2014.

Erika Cudworth is Senior Lecturer in International Politics and Sociology at the University of East London, UK. Her research focuses on the areas of environment–society relations, human–animal relations and gender, with a specific interest in concepts of intersectionality. She is author of several books including *Environment and Society* (Routledge 2003), *Developing Ecofeminist Theory: The Complexity of Difference* (Palgrave 2005), and *Social Lives with Other Animals: tales of Sex, Death and Love* (Palgrave 2011).

Jonathan L. Clark is an Assistant Professor of Sociology at Ursinus College, USA. His teaching and research focus is on animal studies, environmental studies and the intersections between these two fields. He has published in the *Journal for Critical Animal Studies*. Before earning his PhD, Jonathan practiced environmental law.

Amy J. Fitzgerald is an Associate Professor in the Department of Sociology, Anthropology and Criminology at the University of Windsor in Ontario, Canada. Her areas of interest and specialization include critical animal studies, green criminology, gender studies, and environmental sociology. More specifically, her research focuses on the perpetration of harms (criminal and otherwise) by humans against the environment and non-human animals. She has published articles and books on the coexistence of animal abuse and intimate partner violence, sport hunting culture, and harms produced by the human and "pet" food industries.

Carol S. Glasser is a sociologist and animal rights advocate. She received her PhD from the University of California, Irvine where she is a Research Fellow. She served as the Research Director at the Humane Research Council and is on the editorial team of the *Journal for Critical Animal Studies*. Carol's research focuses on critical animal studies, gender, social inequalities, and social movements. Carol is a co-founder of Progress for Science, a grassroots campaign focused on ending vivisection at the University of California Los Angeles. She resides in Los Angeles, California with a menagerie of lagomorphs, felines, canines and humans.

Jessica Gröling is a doctoral candidate and Associate Lecturer in the Department of Sociology, Philosophy and Anthropology at the University of Exeter, UK. She teaches on a number of undergraduate and postgraduate courses in sociology, anthropology and anthrozoology. Her current research is a study of broadcast and print media representations of urban foxes, particularly around the time of the ban on hunting with dogs in England and Wales. Jess has an ongoing interest in non-traditional and participatory research methods and enjoys running workshops for academics and activists on the use of the Freedom of Information Act, consensus decision-making, and participatory action research. She is also a member of the Vegan Society Research Advisory Committee and helps to coordinate the activities of the Institute for Critical Animal Studies in Europe.

Stephanie Jenkins is an Assistant Professor in the School of History, Philosophy and Religion at Oregon State University, USA. Her research and teaching interests include twentieth century continental philosophy, feminist philosophy, disability studies, critical animal studies and ethics.

Agnieszka Kowalczyk is a PhD Student at the Philosophy Institute in Poznan, Poland. Her Master's thesis concerned issues of modernity in the perspective of Bruno Latour's philosophy. She is the editor of *Theoretical Practice Philosophical Review* (www.praktykateoretyczna.pl/). Her theoretical interests include the philosophy of science, critical animal studies, materialist feminism, ecofeminism and Marxist philosophy. She is engaged in work with the Question of Boundaries Research Group at the Adam Mickiewicz University and in the organization of informal educational projects concerning economic rights and gender issues. She is the co-translator of the Polish edition of "Commonwealth" by Hardt and Negri and "Rebel Cities" by David Harvey. Finally, she is the co-editor of the anthology *Ecologies* (Łódź 2013).

Helena Pedersen, PhD in Education, is a Senior Lecturer in the Department of Child and Youth Studies, Stockholm University, Sweden. Her primary research interests include critical animal studies, critical theory, educational philosophy and posthumanism, and she is currently finalizing her research project on animal science education and critical posthumanist theory. She is author of *Animals in Schools: Processes and Strategies in Human-Animal Education* (Purdue University Press 2010), which received the Critical Animal Studies Book of the Year Award in 2010. Other recent works have appeared in the journals *International Journal of Qualitative Studies in Education, Studies in Philosophy and Education, Culture, Theory and Critique, and Policy Futures in Education*. Helena Pedersen is co-editor of the Critical Animal Studies book series (Rodopi). Together with Tobias Linné and Amelie Björck, she coordinates the research theme "Exploring 'the Animal Turn': Changing perspectives on human–animal relations in science, society and culture", funded by the Pufendorf Institute for Advanced Studies at Lund University 2013–2014.

Kay Peggs is Reader in Sociology at the University of Portsmouth, UK. She currently serves on the Board of Directors of Minding Animals International and is a member of the UK Vegan Society's Research Advisory Committee. Her current research centres on the complex inequalities associated with species-related oppressions. Her most recent book *Animals and Sociology* was published in 2012 in the Palgrave Macmillan Series on Animal Ethics. Articles related to her current interests have appeared in journals such as *Sociology, Sociological Review, Sociological Research Online* and *Society & Animals*. Her forthcoming books in the area include a monograph entitled *Experiments, Animal Bodies and Human Values*, to be published by Ashgate. She is also a research methods specialist. With her colleagues Professor Barry Smart and Dr Joseph Burridge, she has recently co-edited a major four-volume set on Observation, which was published in 2013 in the *SAGE Benchmarks in Social Research Series*.

Sara Salih is a committed vegan who teaches at the University of Toronto, Canada.

Vasile Stanescu is the Director of Speech and Debate at Mercer University, USA. He co-edits the Critical Animal Studies book series published by Rodopi Press. He serves as the Associate Editor for the *Journal for Critical Animals Studies* (JCAS). Stanescu has presented his work domestically at conferences at Harvard, Yale, Stanford and Berkeley, and, internationally, in Canada, Australia, The United Kingdom, Romania, Turkey and the Netherlands. He has reviewed texts for *Ethics and the Environment* (Indiana University Press), *Society & Animals: The Journal of Human-Animal Studies* (Brill), *Critical Sociology* (Sage Journals), and the book series *Key Themes in 20th and 21st Century Literature and Culture* (Polity Press). His research has been recognized by the Institute for Critical Animal Studies, Minding Animals International, The Woods Institute for the Environment, The Andrew Mellon Foundation, Institutul Cultural Român and the Culture and Animals

Foundation. This last June, Stanescu was one of sixteen international speakers to present at the "Zoo3OOO (Occupy species)" conference in Hamburg, Germany on the topic of Critical Animal Studies and direct action. He is currently working on a chapter for *The Year's Work in Critical and Cultural Theory* (published by Oxford University Press) on the topic of Animal Studies.

Nathan Stephens Griffin is a doctoral candidate based at Durham University, UK. His research interests include animal rights activism, veganism, queer theory, autoethnography, critical animal studies, critical pedagogy, intersectionality, biographical research methods and comics in social research. He finds the world of academia generally to be quite bizarre and alienating, but he enjoys meeting with other students/academics who share his interest in social justice and vegan junk food. Outside of studying for his PhD, Nathan plays in an acoustic punk band called "Onsind", and helps organize DIY music events in his hometown of Durham.

Nik Taylor is an Associate Professor in Sociology at Flinders University in South Australia, where she teaches and researches on human–animal relations. She has previously published *Humans, Animals and Society* (Lantern Books 2013), *Animals at Work* (with Lindsay Hamilton, Brill Academic 2013), and *Theorizing Animals* (with Tania Signal, Brill Academic 2011). Nik is the Managing Editor of *Society & Animals* and an Associate Board member of the journals *Anthrozoos*, and, *Sociology*. For more, see http://nikt6601.wordpress.com/.

Richard Twine is a sociologist and was most recently a Research Fellow at the University of Glasgow. His current research explores how to change food habits in the context of climate change. He was previously for 10 years a researcher at Lancaster University for the ESRC Centre for Economic and Social Aspects of Genomics. He is author of the book *Animals as Biotechnology – Ethics, Sustainability and Critical Animal Studies* (Routledge 2010), as well as articles on the animal-industrial complex, antibiotics, de-domestication, ecofeminism, posthumanism, bioethics and physiognomy. He is the European Director of the Institute for Critical Animal Studies (ICAS) and is on the Board of Directors of Minding Animals International. For more see www.richardtwine.com.

Acknowledgements

As always, Nik thanks the furry folk with whom she shares her life (Bailey, Loki and Squirt). "Without their constant company the writing process would be much more onerous; without their inspiration much of the rationale for the work I do, aimed at making animal lives better, would be missing".

Richard would like to thank his parents for their support over the years, his engaged group of friends both inside and outside academia who are working, intersectionally, for a better world, and Jasper, a local Lancaster ginger Tom, for his sociability and inquisitiveness.

Both Nik and Richard thank Routledge for commissioning this book, and for an excellent process.

Introduction

Locating the 'critical' in critical animal studies

Nik Taylor and Richard Twine

Introduction

We decided to compile this collection for several reasons, the main one being that we both consider critical animal studies (CAS) to be our intellectual home. However, we currently see signs of change within – and without – CAS and it is in response to these changes that the idea for this collection emerged. CAS can trace its institutional roots to 2001 and the launching of the Centre for Animal Liberation Affairs, which later became the Institute for Critical Animal Studies (ICAS) in 2007.[1] Since then, both CAS as a field of study, and ICAS as an institution, have grown considerably; the latter now has North American, Latin American, African, European and Oceanic arms, which cater to a broad range of scholars and activists. This parallels growth in human–animal studies (HAS) over the last few decades, reflected in other organisations such as the US-based Animals and Society Institute, the Australian Animal Studies Group and the global Minding Animals organisation. Nomenclature is complex in this growing field. The terms animal studies (AS), human–animal studies (HAS), and Anthrozoology, are used frequently and may mean slightly different things according to context and user. Critical Animal Studies can be added to this list, but this should not be taken to mean that CAS is another term for AS. As many authors in the current collection make clear, CAS delineates itself from the others on several key points, and for a range of good reasons. For example, AS tends to be truly interdisciplinary and while this leads to rich and burgeoning debates within the field it can also lead to tensions and boundary maintenance across the disciplines, with CAS arguably being one such tension. CAS seeks to differentiate itself from the broader AS field by having a direct focus on the circumstances and treatment of animals, or as Stanescu and Pedersen have previously put it, CAS is not only concerned with the 'question of the animal' but also with the *condition* of the animal (in Socha 2012). For example, tensions between CAS and AS often manifest themselves at conferences and may specifically focus on catering. For CAS researchers, catering practices are highly political and directly connected to the maintenance of appalling material conditions for animals. Thus, the consumption of animal products at events where 'the question of the animal' is the organising theme is seen as contradictory, unnecessary, disengaged and oppressive (see Jenkins and Twine, this volume). To a certain extent, the distinctions we are making here are partial as many who consider themselves AS scholars often dip into, or are motivated by, the political issues that CAS solely concentrates on, and attempts to point to the purity and/or presumed superior moral sentiment of one piece of work/oeuvre over another can be futile and divisive. Be that as it may, there are voices within both camps that draw pretty clear distinctions and this needs addressing here if for nothing else than to make sense of the way in which this book has been put together – we have chosen to organise the book around sections that represent some of the concerns central to CAS scholarship.

From a CAS perspective, the twenty-first century represents a pivotal period in which ecology and animal life face unprecedented threats. In this sense 'critical' expresses the urgency of our times in the context of ecological crisis. More specifically still, CAS is concerned with the nexus of activism, academia and animal suffering and maltreatment. This makes CAS more overtly political than the often wholly academic AS. CAS takes a normative stance against animal exploitation and so 'critical' also denotes a stance against an anthropocentric status quo in human–animal relations, as demonstrated in current mainstream practices and social norms. This also means that CAS is in conflict with those areas of the academy, such as animal production sciences and biomedicine, which take for granted an instrumental relationship with specific nonhuman animals. Of course, as with any field, and particularly an emergent one, mixing occurs. This can often be a good thing, leading to boundaries being pushed and ideas being stretched; the hallmark of good knowledge acquisition. However, while there are many who work across the CAS–AS divide (and calling it this may be over-stating things) the relationship between the two is increasingly fractious, with both 'sides' often locking horns over the overt political/activist, and thereby assumed ideological, stance of CAS. However, such pure boundaries are often difficult to maintain and police and we are increasingly seeing those who do not associate with ICAS as an institution as well as those who locate themselves within AS, working with the critical concepts central to CAS. This is due to, and also encourages, the legitimation of CAS as an emergent counter focus. However, this cross pollination of ideas further muddies the distinction between CAS and AS and important questions regarding how CAS scholars position themselves in terms of both AS specifically, and the academy more generally, become of paramount importance. Finally, some of the concepts that underpin CAS are increasingly on the mainstream agenda – for both the general public and those within the academy – in part due to global changes and the urgency of climate change that threatens life broadly conceived (see Warren *et al.* 2013). This means that CAS is slowly gaining ground in terms of the numbers of scholars who claim affiliation with it, but also within the walls of the academy (and elsewhere) where it is being, if not wholeheartedly accepted, at least tolerated and is now increasingly visible due to workshops and conferences; this latter in large part due to the passion and energy brought to the field by those researching in it.

It is this increasing acceptance that sparked the idea for this current collection, as it raises crucial questions. For example, is CAS becoming institutionalised and if so, is this occurring in parallel with the institutionalised acceptance of AS?[2] What do the consequences of such institutional acceptance look like? Are the messages from CAS scholars diluted – or perhaps strengthened – when they become part of the mainstream academy? How does – and should – CAS differentiate itself from AS and at what cost, what benefit? Are there concrete consequences for those who call CAS their intellectual home, for example, is it more difficult for CAS scholars – who hold minority views, research controversial topics and engage in counter-hegemonic practices such as veganism – to gain academic tenure? Are fractures between CAS and AS becoming more cemented in the field and what might the consequences of this be? Can we learn anything from the intellectual roots of CAS about this, and here we are thinking of the parallel divisions between feminism and women's studies in the late 1970s, early 1980s. To our mind, with the growth of CAS scholarship (in both output and influence), we are at a critical juncture to address these questions. Thus, the rationale for this book; we wanted to bring together established and emerging scholars in CAS to contemplate these issues, and, in doing so, begin a debate about both the content, and the positioning, of CAS. At the same time our aim is to exhibit some of the diversity of CAS and its intersection with various urgent contemporary policy debates and theoretical developments.

In doing this we do not claim to offer a comprehensive overview of CAS, nor do we claim to exhaustively cover all its distinctions from AS (this is done admirably elsewhere, see, e.g. Best *et al.* 2007; Best 2009, Gigliotti 2009; McCance 2013, Sanbonmatsu 2011; Socha 2012; Taylor 2013 and Twine 2010). What we do claim, however, is to have achieved a volume that pulls CAS scholars together across various disciplines to reflect upon the main themes integral to CAS, with a view to prompting a consideration of the *relationship* between CAS and AS, CAS and the academy, and the *future directions* of CAS. We have to note here, however, that there is a strong social science bias in this collection. This reflects both of the editor's intellectual roots in Sociology (we are both products of Sociology PhDs taken at Manchester Metropolitan University in the UK in the late 1990s) and is not to be taken as a suggestion that we favour a disciplinary hierarchy within the CAS field, although it may be instructive to think about why certain disciplines may be more likely to approach the field with an attentiveness to issues of power.

The rise of CAS signals the start of a long overdue change in contemporary academia that better reflects the importance of other animals as social beings in their own right. This entails a focus on the significance of their enmeshment with both the meanings of what it is to be 'human' as well as their presence in our everyday material lives. We do not wish to suggest that the importance of animals lies only in their contrast to humans. On the contrary, this is a legacy of humanist thinking that we, and others in CAS, strongly resist. Rather, what we mean is that animal lives are – for better or worse (usually worse) – affected by humans and as such there is a pressing need to examine how and why. As an interdisciplinary field that spans and questions the humanities, and the social and natural sciences, CAS calls for conceptual renewal, methodological innovation, theory that is relevant and engaged and, in line with many cognate influences on the academy such as feminism and environmentalism, a further softening of disciplinary boundaries.

However, CAS attempts to rupture normative understandings of academia itself, as well as having specific disciplinary critiques and, in common with much critical social science (and also with mainstream calls for academic engagement with publics and communities), wishes to stoke civil society into working toward progressive social change. The closest analogy, and one that will repeatedly surface throughout this book, is between CAS and the emergence of feminism(s) in academia, with its similarly strong links with extra-academic political activism. Both animal and feminist politics are similarly targeted against dispassionate, institutionalised scholarship based on a rationalist, liberal interpretation of (hegemonic) masculinity, and both seek to expose and overthrow the routinised and naturalised forms of practice based on oppression and abuses of power, which flow from this. It is this which makes them both explicitly *critical*. In addressing the intellectual antecedents and contexts of CAS, there is a strong affinity and intersection between feminism and animal advocacy.

Although an explicit statement of self-consciously *critical* animal studies did not emerge until the beginning of this century, reflexivity toward the human exploitation of other animals, and an emergent politicisation of the violence this entailed, began to appear alongside second wave feminism in the 1960s and 1970s. Prior to this, there were close links between the women's suffrage and anti-vivisection movements (Kean 1995). However, a significant catalyst for debate on animal ethics came from ecofeminist writings during the 1970s, 1980s and 1990s alongside, and often in tension with, the influential work of well-known animal philosophers such as Tom Regan and Peter Singer. Any contextualisation of CAS must confront the fact that, in an intellectual sense, it existed before the term was coined, and that it has since become an umbrella term for bringing together scholars who do critical research on human–animal relations. However, those who explicitly coined CAS (Best *et al.* 2007) did so with a specific sense of purpose and definition. One such vital element is the foregrounding of intersectional analysis and politics, which makes a further link with feminism and critical race studies (see e.g. Deckha 2012; Harper 2010a, 2010b; Kim 2007). This move toward intersectionality, originally pursued by ecofeminists (e.g. Adams 1994; Plumwood 1993; Merchant 2003), makes clear how the material and symbolic exploitation of animals intersects with and helps maintain dominant categories of gender, 'race' and class. In turn, this troubles the humanist premise of many extant feminist, anti-capitalist and anti-racist politics by pointing out that dominant identities and practices of gender, 'race' and class help maintain the human exploitation of animals. Essentially, CAS subverts the humanist assumptions of intersectionality theory by bringing animals and animality into these areas and also into other potentially kindred fields, such as disability studies (Nocella *et al.* 2012) and queer theory (Parry 2012; Simonsen 2012).

The affinity of CAS with feminism, expressed in attempts to understand intersections between different power relationships, becomes clearer when one appreciates the essential relationship between, on the one hand, a radical social movement and, on the other, the emergence and academic objectification of knowledge related to that movement. Both feminism and animal advocacy share a healthy scepticism of what might result in that process of objectification, formalisation and professionalization, given the collision of radical knowledge with potentially conservative disciplines and institutions. Moreover both confront an academy that has historically taken for granted naturalised readings of gender and species, necessitating confrontations with the production of scientific knowledge and the marginalisation of women and animals in interconnected *and* specific ways (e.g. Birke 1994). The CAS engagement with scientific knowledge spans, for example, animal agricultural science, biomedicine, nutrition and sustainability discourse, all of which are critiqued as sites that produce the animal as property, animal use as necessity and also produce the 'human' in various assumed and naturalised ways. We also glimpse here how CAS can be liberatory for human animals, as we contest rather fixed assumptions of human *being*, just as feminism is (counter-intuitively for some) also liberatory for men and all of us living with gender.

There is also a shared belief between feminist and CAS scholars that professionalisation should not presume a detachment from the rich diversity of civil society and protest. Critical researchers then can (and in fact, must) study such social movements, advise them and critically analyse those obdurate social institutions and practices that serve to curtail progressive social change. This process includes self-reflexive knowledge to try and guard against, for example, the re-emergence or perpetuation of oppressive theories and practices within both animal advocacy and CAS. Just as feminism has had to be attentive to the possibility of, for example, class and 'race' privilege within its theory and practice, the same reflexivity, from a CAS perspective, is similarly important within the animal advocacy movement – especially for its coalitionary bridge- building practices with other movements. While intersectionality is possibly the most important concept for analysing human–animal entanglements, it is one that underpins much of the work in the current volume and thus was too broad to use as an organising principle for the book.

Instead, we have organised this book around four themes, which simultaneously are heuristic for outlining current CAS research and for generally demarcating a difference in focus and emphasis between CAS and AS. We hope that they serve as a useful introduction for those new to the field and as points of further research and debate amongst the various overlapping communities of CAS, AS, HAS and anthrozoology.

Part I: engaging theory

CAS draws from its ecofeminist roots to articulate a stance that is explicit in its normative commitment to the removal of all forms of animal abuse, as well as an analysis of how human prerogative and exceptionalism might work through societies and cultures to reify the human being at the expense of all other beings (and this includes 'nature' broadly conceived). At the same time, however, CAS scholars are committed to *engaged* theory as opposed to 'theory for theory's sake', which some see as the hallmark of AS:

> The recipe for the 'success' of animal studies – immersion in abstraction, indulgent use of existing and new modes of jargon, pursuit of theory-for-theory's sake, avoidance of social controversy (however intellectually controversial it may often be), eschewing political involvement, and keeping a very safe distance from 'extremists' and 'radicals' agitating for animal rights – is also the formula for its failure, upon being co-opted, tamed, and neutralized by academia.
>
> (Best 2009)

While we accept that some AS work may fall into this category, we also recognise that many scholars work across both fields and suggest that the difference is not so much in the use of abstract theory but turns on the idea of *engagement*. Engaged theory is 'theory intended to support social change directly or indirectly' (Garry 2008: 99), and it is crucial to the CAS endeavour. In the CAS context, theory must be relevant to understanding and changing the material conditions of animals, and to historicising the still normative concepts that have been largely successful in shielding human–animal relations from critical scrutiny. Thus theorising will often be linked to empirical cases and events as found in media and cultural analysis. We do not think the focus on engaged theory amongst CAS scholars means eschewing *complex* theory. Complexity should not be conflated with abstraction or theory for theory's sake. Theorizing about human–animal relations has traditionally been complex, be it classic deliberations on sentience or those within feminist work on care ethics. A particularly rich vein in animal studies has emerged over the last decade from within continental philosophy (e.g. Atterton and Calarco 2004; Acampora 2006; Calarco 2008; Pick 2011; Tyler 2012; Wolfe 2003a, 2003b, 2013).

Much of this is highly cognate to CAS and the concurrent radical critique of anthropocentrism. Arguably most striking in this vein is the work of the late philosopher Jacques Derrida (2008), which constitutes both an underlining of the importance of animals to Western philosophy and a condemnation of the linguistic, symbolic and material violence that humans have enacted upon other animal species. Such work may indeed be complex, but it is also very much engaged with the lived condition of other animals as well as the destabilisation of normative constructions of the human. It thus contributes not only to our understanding of the cultural roots of animal oppressions, but gives us the theoretical and heuristic tools to begin countering it. Whilst we recognise the dangers of simplistic turns to figures like Derrida for field credibility and authorization (Fraiman 2012), and the dangers of excluding the centrality of the influence of feminism by such turns (Gaard 2011), neither do we wish to exclude this body of work from the reflections of CAS.

In common with much critical work in the humanities and social sciences, engaged theory takes an intersectional approach to power, practice and experience. However, in recognising that conceptual objects such as 'culture', the 'social' and the 'human' are anything but ontologically pure, CAS rejects a humanist frame, favouring instead an intersectional approach that includes the more-than-human. This is far from alien to other developments in the academy. For example, environmental social science has long critiqued an anthropocentric notion of society and theoretical turns towards materiality and practice have similarly placed importance on critiquing humanist ontology. CAS builds on, and borrows from, this work to focus on the impact on other-than-human-animals. Out of necessity, this involves engagement with critical work on 'nature' and 'the environment', but the linking of theory with action on behalf of other animals remains specific to CAS.

The first part of this book reflects the importance of this idea, as contributors reflect on what engaged theory – by humans in the academy, for nonhuman animals – might mean, how it might come to help the material condition of animals and how it underpins CAS as a field. Chapter 1, by sociologist Erika Cudworth, focuses on the mutual benefit inherent to an engagement between CAS and Sociology, pointing out that a focus on nonhuman species acts as a corrective to a sociological worldview which has hitherto ignored multi-species lifeworlds. As Cudworth demonstrates, a Sociology that includes species, and a CAS that draws on some of the conceptual tools of (critical) Sociology, allows an interrogation of the human domination of other species situated in an understanding of the – geographical and historical – context of human practices. Furthermore, Sociology's focus on the workings of power in society, along with its pre-occupation with how structure and agency interact to encourage, discourage, alter or confirm particular sets of practices, allows for an accounting of the discursive, symbolic *and* material position of other animals. Importantly, Cudworth points out that the historical, central, interests of Sociology – race, gender and class – give a solid grounding for intersectional approaches toward human–animal relations. In particular, Cudworth argues that given Sociology's rich history of studying the manifestation and operation of power, it is particularly well placed to study the human relations of dominance over other animals, or as she names it, 'anthroparchy'.

In Chapter 2, Kay Peggs notes that intersectionality approaches in feminism allowed a nuanced understanding of the interplay of the complexity of oppressive relationships and considers the impact this may (should) have on CAS. Like Cudworth, Peggs is a Sociologist and points out that one particularly useful aspect of critical Sociology for CAS scholars is its focus on transformation alongside its willingness to tackle tricky questions of morality and ideology in the production of knowledge. This makes it ideally placed to address several crucial issues in CAS: for example, the consequences of the hegemony of the scientific paradigm in knowledge production; how societal transformation might take place in such a way that animal interests are considered; and, in return, how including animal interests in our ontological frameworks might challenge us both in terms of being a call to activism and a fundamental challenge to the humanism that underpins the majority of (Western) knowledge systems.

In Chapter 3, Sara Salih asks why it is that for some 'knowledge is a moral burden and an ethical impetus for change, while for others it isn't'? Taking eating practices as an example, she thoughtfully considers what it might take to adopt a more ethical orientation to the world. What is the relationship between knowledge and personal and social change? She explores this within a wide-ranging paper that draws upon literary, ethological and philosophical sources. As she ponders the limits of abstract knowledge in leveraging change, Salih stresses the role of habit and desire in shaping our food practices. The themes of this chapter express well the CAS suspicion toward an over-rationalisation of the human, both generally and specifically, in the context of our relations with other animals. As an alternative means of conceptualising practices that have begun to counter the norm of everyday animal consumption, Salih turns to a notion of the 'break', or indeed break*down*, in order to explore and understand affective disruptions and turns toward vegan practice.

Part II: doing critical animal studies

In this section we bring together three scholars who consider particular methods for CAS. Again, the debt owed by CAS scholars to (some) feminist considerations regarding knowledge production becomes clear. Feminists have made it apparent that notions of objectivity in research are an impossibility, which rests upon an idealised version of the research process, itself situated within a rationalised, masculinist, liberal humanism (see, e.g. Smith 1987; Stanley and Wise 1993). This produces – academically and culturally – a hierarchy of knowledge that favours white, Western, masculine, human-centred ways of knowing the world. Appeals to scientific neutrality and objectivity effectively silence the voice of those 'researched'. In animal studies, this is particularly problematic given that our 'subjects' of research do not have a voice that is recognised as legitimate. In essence, this makes questions about the research procedure even more important and forces us to re-think not only taken for granted ways of seeing the world, but the very methods we use to investigate it. It is fitting then, that we start this section with a personal reflection from Lynda Birke (Chapter 4) regarding her own research work with horses. In a frank discussion of the challenges faced when CAS-oriented research depends on the involvement of live animals (here, horses), Birke considers the possibility of using a framework that prioritises 'interrelating', as opposed to one that focuses on the symbolic constructions of animals by humans. Seeking to discover whether research focused on relating can incorporate animal agency and the contribution that animals offer to meaning-making, Birke draws on one of her own projects, which aims to investigate 'detailed mutual engagements of horses and people'. Cautioning that research involving other animals might simply further entrench human superiority and power by noting it is produced by humans, about other animals, she calls for an ongoing reflexivity and discussion concerning the methods used by CAS scholars.

Chapter 5, by Jessica Gröling, draws deeply on the feminist roots of CAS to consider some of the challenges faced by CAS scholars who aim to study the perpetration of 'socially-sanctioned speciesist violence'. Locating the discussion around ethnographic research in slaughterhouses and laboratories, Gröling considers the role that ethnography can play in overcoming the methodological and ethical issues attached to research with those who have the ability to perpetuate dominant forms of meaning vis-à-vis human–animal relations. Crucially, she argues that traditional claims to neutrality are not only problematic inasmuch as they can never exist, but also because the preferred method for ethnographers in dealing with this is to acknowledge their own bias in their work. This, however, as she makes clear, serves only to further conceal mainstream biases. As a result, she concludes, CAS needs to move beyond reflexivity so that 'validity is no longer determined by the purported absence, or indeed the apologetically acknowledged presence, of the researcher, but rather by their committed presence and their informed and consistent attitude towards injustices' (p. 106, this volume).

In Chapter 6, Nathan Stephens Griffin considers the benefit to CAS of various 'alternative' methods – visual, reflexive and auto-ethnography/biographic. Locating his discussion within the tensions between CAS and the mainstream academy, he points out that visual (and other alternative) methods may well be a better fit with participatory action research than traditional methods, which tend to valorise the position of the researcher over that of the researched. Like others in the current collection, Stephens Griffin is clear that reflexivity must be a hallmark of CAS scholarship and that, furthermore, alternative methodological approaches allow 'a level of insight and empathy that other methods may scarcely be able to achieve' (p. 131, this volume).

Part III: critical animal studies and anti-capitalism

One of the initial points of difference stressed by those who first outlined CAS (Best *et al.* 2007) was that animal studies paid insufficient attention to the role of (capitalist) political economy in shaping human–animal relations and the exploitation of other animals. As the financial crisis in neoliberal capitalism of the last six years has made clear to more people, our current dominant economic system is intrinsically productive of social inequality. From a CAS perspective, the exploitation of human workers takes place alongside that of animals and partly through a symbolics of animalisation, wherein an implicit culture/nature dualism positions the low paid and the unpaid as 'closer to animals'. At the same time the corporate globalisation of profit-making practices (at least profitable in the short term), especially in agriculture, has played an unmistakable role in radically extending the scale of animal exploiting practices *beyond* even what might have been expected for human population increases. In 2011 for example, over 72 billion land animals[3] were killed for human consumption, a figure that rises year on year. This staggering scale of violence (and here we refer to both the ways in which these animals are forced to live, as well as their ultimate death) is bound up in the exploitation of human workers involved in slaughtering animals and processing them into 'animal products' (see Cudworth 2008; LeDuff 2003; Nibert, 2002: 109–113; Pachirat 2011). The current rise in the consumption of animals in many 'developing' countries, significantly facilitated by the exportation of a factory farm model,[4] is important for understanding this overall upward trajectory in animal consumption, as well as being a public health and climate change time bomb. This focus, for CAS, reminds animal studies scholars of the urgent political context of their work. It also underlines the inadequacy of narrowly conceived philosophical approaches to the question of the animal that valorise ethics without acknowledging the ways in which exploitative practices against animals are systemic and deeply embedded in our dominant economies, institutions, social routines and daily habits, and so are not simply amenable to better clarified ethics. This is not meant to imply either that an alternative economic system would be less exploitative of animals in any simple way, or that we should adopt a reductive understanding of human–animal relations over-focused on the economic sphere, or that ethics are unimportant. However it does mean that we favour theoretical and case study research that engages with specific human–animal relations in their economic context, work that not only can improve our understanding of these relations but more generally question critical, yet anthropocentric, analyses of contemporary capitalism. For example, what would global capitalism look like minus the exploitation of animal reproductive labour? How does that abuse intersect, in specific contexts, with that of human labour? And how can the disavowal of violence against animals illuminate, generally, theories of commodity fetishism?

Chapter 7, by Jonathan Clark, focuses on the case of lab animals (and mice and rats in particular) to take up the question of how best to think about the role of animals in the capitalist labour process. Intrigued by the discursive practice of 'human guinea-pigging' and especially its recent pharmaceutical global expansion, Clark uses this chapter to contest the humanist, Marxist conception of the labour process through an exploration of clinical labour. Clark devotes significant space to a close critical reading of Marx's faithfulness to human–animal dualism in theorising labour. Drawing upon cognitive ethology and other relevant work in CAS and AS, Clark troubles the presumptuous overly- polarised distinction in Marx's work. He then brings this discussion back to a consideration of labouring humans and animals in the context of clinical labour. The chapter concludes with a discussion of what actual benefits for animals, if any, might follow from such reconceptualisation as well as suggesting future directions in debates on labour and agency.

Chapter 8, co-authored by Amy Fitzgerald and Nik Taylor, explores the everyday naturalisation of the acceptability and assumed necessity of animal consumption. Using two types of data where meat eating is presented as normal, this chapter aims to better understand the interplay between images, identities and markets in the process of commodification. The first data source is websites for red meat exporters, aimed at companies looking to secure supplies of red meat from Australia; and the second is print advertisements published in cooking magazines in North America, aimed at individual consumers. Their analysis focuses on three emergent themes in the construction of meaning around animals as commodities: the replacement of 'realistic' animals with 'happy' animals, the romanticisation of naturalness and representations of good health and taste. Fitzgerald and Taylor then proceed to deconstruct these themes, pointing out some differences in construction between the two sources of data whilst adding to the CAS analysis of the operations and mystifications of the animal–industrial complex.

Chapter 9, by Agnieszka Kowalczyk, is deeply situated in prior ecological, ecofeminist and autonomist engagements with Marx and Marxist theory to offer a critical posthumanist re-reading of Marxist concepts in order to better construct a common site of resistance to capitalism. This leads her to discuss in detail the possibility of animal resistance and agency, arguing that resistance to power relations based on anthropocentric assumptions is both integral to, and distinct from, all other resistances to capitalism. By acknowledging the importance of other animals in the circulation of capital, Kowalczyk argues that to address the issue of struggle in contemporary global (bio)capitalism, a rethinking of Marx's definition of labour and a dichotomy of production and reproduction is necessary. In this way CAS can add a new dimension to Marxist analysis, which has marginalised the oppression of animals from consideration.

Part IV: contesting the human, liberating the animal: veganism and activism

We orientate the final section of the book around the close relationship between CAS and the broader social movements struggling for radical change in human–animal relations. It is not unusual for social movements to have their academic corollary, or for academic disciplines or fields to be historicised in terms of the overlapping evolution of social movements. CAS and the critical animal work that preceded its overt naming developed recursively with both grassroots direct action, eventually including more institutionalised forms of animal advocacy.

This section speaks to a difference in focus between CAS and AS that is notable both in the choice of research topics and the mode of academic organisation. ICAS, which has begun to successfully organise devolved CAS conferences in, for example, North America, Latin America, Europe, Africa and Australia, invites both activists and academics to attend in order to contest academic boundary-making and to attempt to create links between scholarship and activism. In common with anti-capitalist, anti-racist and feminist research, CAS scholars often assume a dual identity of activist and academic. These conferences typically also involve activist and campaign stalls, a public protest event, always ensure all vegan catering and attempt to eschew the dominant academic model of high consumption and high cost, inaccessible conferences. The assumption of professorial plenary sessions is also typically ignored in favour of a more egalitarian mode of organisation.

The CAS relation with activism is not one of conflation however. At times a critical distance is necessary in order to address the need for reflexivity and internal critique over direction and approaches. This is not an attempt to position the academic as privileged in respect of 'movement guidance'; often those working primarily as activists will have richer and deeper knowledge around grassroots activism, policy and process. However CAS, which overlaps with the interdisciplinary analysis of social movements, provides an environment for such dialogue.

This section of the book also highlights that for CAS, the examination of human–animal relations, and the embedding of this within a broader framework of intersectionality, seeks to simultaneously contest essentialisms of the 'human' and the violence inherent to the homogenisation of all nonhuman animal species under the signifier 'animal' (Derrida 2008, Chapter1; Plumwood 1993). For CAS, the promise here lies in 'liberating' people from their own assumedly fixed self-categorisations, and in 'liberating' (some) humans *and* other animals from a shared representation as animalised and, by implication, of less value than a normative white, wealthy and male model of subjecthood.[5] In subverting the taken-for-granted devaluation of the 'animal', veganism is a key practice in this endeavour, yet is only an ethico-political beginning to addressing the intercon-nected oppression of people and animals. We believe that veganism does good intersectional, oppositional work and it has rightly become partly constitutive of CAS, but it needs to be moored to a broader political vision.

Chapter 10, by Matthew Cole, offers an account of the historical formation of veganism as an ethical, rather than dietary, practice. This chapter presents a dis-course analysis of archival documents from the UK Vegan Society, focusing on the early years of *The Vegan* magazine (the quarterly journal of the Society, which began publication as *The Vegan News* in 1944). Charting the emergence of veganism as a recognisable and socially organised practice, and using mater-ials developed by vegans themselves, Cole outlines the formation of a com-munity of ethical practice that counters an often individualised or trivialised mode of representation by those outside the community. Drawing upon both Michel Foucault and Paul Rabinow's Foucault-inspired work on ethics, Cole argues that this helps us understand veganism as an ethical practice that always transcended dietary regimen, and that anticipated many of the core intersectional concerns of CAS.

Chapter 11, by Stephanie Jenkins and Richard Twine, considers some of the conceptual framings that serve to reinforce hegemonic meat/dairy-centric eating practices and in doing so brings to light key components of food norms. Jenkins and Twine focus on the concepts of autonomy, choice and privacy as acting to protect this hegemony. They simultaneously direct their critique of the depoliti-cisation of food toward AS and meat/dairy-centric societies generally. Their discussion and analysis draws upon previous work in feminist bioethics on the notion of autonomy and ecofeminist work on privacy to contribute to the critique of freedom of choice discourses as allied to individualism, neoliberalism and consumerism. They complete the chapter by developing this into a significant discussion of the potential of a posthumanist sense of autonomy, via the work of Judith Butler, arguing precisely that opening up food practices to critical inter-rogation helps engender a progressive (non-anthropocentric) notion of freedom and an unbounded ontology of the human self.

Chapter 12, by Carol Glasser, examines an important issue for animal activ-ism, namely the efficacy of illegal forms of activism. In providing empirical ana-lysis of such activism, Glasser contributes to the aforementioned reflexive discussion at the nexus of activist and academic. Specifically, in response to the paucity of empirical work in this area, Glasser draws upon primary data sets that track illegal radical activism in the United States from 1990 to 2010 and media coverage of these actions and of animal rights broadly, combined with secondary data tracking donations to animal rights groups over this same period. Although the results are inconclusive around efficacy, Glasser provides a valuable contri-bution to the analysis of radicalism and finishes her discussion with important suggestions for further empirical analysis of animal activism.

We complete the volume with a concluding discussion by Vasile Stanescu and Helena Pedersen that adeptly weaves in many of the perspectives and chapters of the volume, whilst also providing space for exploring the key con-cerns and future direction of CAS. They echo some of the concerns that drove us to compile this collection, namely that there is an 'enormous and radical' poten-tial in CAS, but that we need to guard against *studying* ourselves into an 'impasse, denoting a sense of academic detachment and passivity' (p. 271, this volume). They argue that CAS needs to move beyond traditional labels to 'post animal studies', where the practices of humans – and not other animals – are problematised.

We are immensely proud of this collection and think we have achieved pre-cisely what we set out to do – to bring scholars together to reflect on the current and future state of CAS. In large part our success is due to the rich and thought-provoking material generously offered by all our contributors. We hope you enjoy this collection as much as we did in putting it together.

Notes

1 Important figures in the early days of ICAS were Steven Best, Anthony Nocella, Richard Kahn, Carol Gigliotti and Lisa Kemmerer. Richard White was the chief editor of the associated *Journal for Critical Animal Studies* (JCAS) between 2008 and 2012. ICAS was preceded by the Center for Animal Liberation Affairs, set up in 2001 by Steven Best and Anthony Nocella.

2 We accept that the institutionalised acceptance of AS remains immature and patchy. We do not wish to suggest this has been completely achieved.

3 We calculated this figure from the UN FAOSTAT web-site (http://faostat.fao.org/ (accessed 3 December 2013)), where 2011 is the most recent year that statistics are available for. The total figure, which excludes fish and other aquatic animals and is

itself likely to be an underestimate, is 72,336,940,002 (72.3 billion). This consists of 'producing animals/slaughtered for meat (64,932,316,957), producing animals/slaughtered for eggs (6,744,343,000), and producing animals/slaughtered for milk (660,280,045).

4 See http://brightergreen.org/brightergreen.php?id=24 (accessed 3 December, 2013).

5 We place 'liberating' here in quotes in order to acknowledge both the complexity of liberating animals from conditions of domestication (though we certainly refuse the assumption that domestication fixes the fate of many animals), and the limits of a discourse of 'liberation' that might assume a possible future simplistically cleansed of all power relations.

References

Acampora, R. (2006) *Corporal Compassion: Animal Ethics and Philosophy of Body*, Pittsburgh, PA: University of Pittsburgh Press.

Adams, C. J. (1994) *Neither Man nor Beast: Feminism and the Defense of Animals*, New York: Continuum.

Atterton, P. and Calarco, M. (2004) *Animal Philosophy: Essential Readings in Continental Thought*, New York: Continuum.

Best, S. (2009) 'The Rise of critical animal studies: putting theory into action and animal liberation into higher education', *Journal for Critical Animal Studies*, 7(1): 9–53.

Best, S., Nocella, A., Kahn, R., Gigliotti, C. and Kemmerer, L. (2007) 'Introducing critical animal studies', *Journal for Critical Animal Studies*, 5(1): 4–5.

Birke, L. (1994) *Feminism, Animals and Science – The Naming of the Shrew*, Buckingham: Open University Press.

Calarco, M. (2008) *Zoographies: The Question of the Animal from Heidegger to Derrida*, New York: Columbia University Press.

Cudworth, E. (2008) '"Most farmers prefer blondes": the dynamics of anthroparchy in animals' becoming meat', *Journal for Critical Animal Studies*, 6(1): 32–45.

Deckha, M. (2012) 'Toward a postcolonial, posthumanist feminist theory: centralizing race and culture in feminist work on nonhuman animals', *Hypatia*, 27(3): 527–545.

Derrida, J. (2008) *The Animal That Therefore I Am*, New York: Fordham University Press.

Fraiman, S. (2012) 'Pussy panic versus liking animals: tracking gender in animal studies', *Cultural Inquiry*, 39: 89–115.

Gaard, G. (2011) 'Ecofeminism revisited: rejecting essentialism and re-placing species in a material feminist environmentalism', *Feminist Formations*, 23(2): 26–53.

Garry, A. (2008) 'Intersections, social change and "engaged" theories: implications of North American feminism', *Pacific and American Studies*, 8: 99–111.

Gigliotti, C. (ed.) (2009) *Leonardo's Choice – Genetic Technologies and Animals*, London: Springer.

Harper, A. B. (2010a) 'Race as a 'feeble matter' in veganism: interrogating whiteness, geopolitical privilege, and consumption philosophy of "cruelty-free" products', *Journal for Critical Animal Studies*, 8(3): 5–27.

Harper, A. B. (2010b) *Sistah Vegan: Black Female Vegans Speak on Food, Identity, Health, and Society*, New York: Lantern Books.

Kean, H. (1995) 'The "smooth cool men of science": the feminist and socialist response to vivisection', *History Workshop Journal*, 40: 16–38.

Kim, C. J. (2007) 'Multiculturalism goes imperial – immigrants, animals, and the suppression of moral dialogue', *Du Bois Review*, 4(1): 233–249.

LeDuff, C. (2003) 'At a slaughterhouse, some things never die', in C. Wolfe (ed.) *Zoontologies: The Question of the Animal*, Minneapolis: University of Minneapolis Press.

McCance, D. (2013) *Critical Animal Studies – An Introduction*, Albany, NY: SUNY Press.

Merchant, C. (2003) *Reinventing Eden: The Fate of Nature in Western Culture*, London: Routledge.

Nibert, D. (2002) *Animal Rights/Human Rights: Entanglements of Oppression and Liberation*, Lanham, MD: Rowman and Littlefield.

Nocella, A., Bentley, J. and Duncan, J. (2012) *Earth, Animal, and Disability Liberation: The Rise of the Eco-Ability Movement*, New York: Peter Lang.

Pachirat, T. (2011) *Every Twelve Seconds – Industrialized Slaughter and the Politics of Sight*, New Haven: Yale University Press.

Parry, J. (2012) 'From beastly perversions to the zoological closet: animals, nature, and homosex', *Journal for Critical Animal Studies*, 10(3): 7–25.

Pick, A. (2011) *Creaturely Poetics – Animality and Vulnerability in Literature and Film*, New York: Columbia University Press.

Plumwood, V. (1993) *Feminism and the Mastery of Nature*, London: Routledge.

Sanbonmatsu, J. (ed.) (2011) *Critical Theory and Animal Liberation*, Lanham, MD: Rowman & Littlefield.

Simonsen, R. R. (2012) 'A queer vegan manifesto', *Journal for Critical Animal Studies*, 10(3): 51–80.

Smith, D. (1987) *The Everyday World as Problematic: A Feminist Sociology*, Boston: Northeastern University Press.

Socha, K. (2012) *Women, Destruction, and the Avant Garde – A Paradigm for Animal Liberation*, New York: Rodopi.

Stanley, L. and Wise, S. (1993) *Breaking Out Again: Feminist Ontology and Epistemology*, London: Routledge.

Taylor, N. (2013) *Humans, Animals and Society: An Introduction to Human-Animal Studies*, New York: Lantern Books.

Twine, R. (2010) *Animals as Biotechnology – Ethics, Sustainability and Critical Animal Studies*, London: Earthscan/Routledge.

Tyler, T. (2012) *CIFERAE: A Bestiary in Five Fingers*, Minneapolis: University of Minnesota Press.

Warren, R., VanDerWal, J., Price, J., Welbergen, J. A., Atkinson, I., Ramirez-Villegas, J. and Osborn, T. J. (2013) 'Quantifying the benefit of early climate change mitigation in avoiding biodiversity loss', *Nature Climate Change*, 3: 678–682.

Wolfe, C. (2003a) *Zoontologies: The Question of the Animal*, Minneapolis, MI: University of Minneapolis Press.

Wolfe, C. (2003b) *Animal Rites: American Culture, the Discourse of Species and Posthumanist Theory*, Chicago: University of Chicago Press.

Wolfe, C. (2013) *Before the Law – Humans and other Animals in a Biopolitical Frame*, Chicago: University of Chicago Press.

Part I
Engaging theory

1 Beyond speciesism

Intersectionality, critical sociology and the human domination of other animals

Erika Cudworth

Introduction

Talking about non-human animals and the profound difference of 'species' has proven difficult for sociology, whose disciplinary boundaries were historically constituted around the designation of an arena – 'the social' – which was defined as exclusively human. Whilst sociology has broadened its subjects, objects and processes of study, it has held fairly fast to the conception of the social as centred on the human. This chapter argues for a sociology that is not human-exclusive, and acknowledges the way species shapes the human and non-human lifeworld as part of the condition of life. In this sense, sociology needs Animal Studies as a corrective to its limitations and to paint a more convincing picture of social lives, which are those of multi-species. Sociology cannot continue to produce work on the body, on work or on the 'family' for example, which assumes that all bodies or workers are exclusively human and that we dwell in single-species households. The conventional trilogy of social domination, of class, 'race' and gender that has been the focus of critical sociological approaches to matters of oppression and exclusion, has been challenged by new concerns with other differences – of place and location, age and generation, sexuality and various forms of embodied difference. Despite these important developments, most sociology stops short at the difference of species. It is time for sociologists of a more critical persuasion to properly consider the difference of species in social relations.

In turn, Animal Studies might benefit from the insights of a critical sociological framework. The concept of 'speciesism' has been of great significance in raising political questions about the use and treatment of non-human animals. It is a foundational concept for more critical approaches in Animal Studies and has introduced an analysis of power and domination into discussions of human–animal relations. Whilst speciesism has been used across the multiple disciplines of Animal Studies (in particular, in philosophy and political theory), it does not translate well into contemporary sociological discussions around inequality and difference. In the discipline of sociology, critical perspectives have moved away from the terminology of 'isms' and 'phobias' (sexism, racism, homophobia, for example) and have developed concepts that raise questions about the patterns of social relations and the norms associated with them (in terms of gender, ethnicity and sexuality, and so on). Thus gender for example, is used to discuss the multiple and complex ways in which the difference of sex is constituted and reproduced. This chapter will suggest that 'speciesism' is not adequate for capturing the full range and forms of our social relations with non-human animals.

A critical and sociological analysis of human relations with non-human animals will understand contemporary societies as structured in terms of particular configurations of species relations. It can provide us with the tools for the theorisation of species in terms of human domination, exploitation and oppression so important in Critical Animal Studies, whilst remaining sensitive to differences in the kind and degree of human practices which are geographically and historically situated. A critical sociology of species must account for both the discursive and material placement of non-human animals, interrogating institutional contexts and related practices, and considering the extent to which these change and/or reconstitute themselves over time. Being critically sociological, it should be underpinned by the conception that the oppressions of human and non-human animals are intersected. Finally, a critical sociology of species will draw political inspiration from Critical Animal Studies. Thus it must be an engaged sociology, a call to action, which grounds its attempts to theorise, document and explain the world in the context of political struggles to change it.

This chapter maps the territory on which a more satisfactory sociological approach to thinking about non-human animals might be grounded. The argument advanced here focuses particularly on those species whose lives human beings have urgently shaped – domesticates. With respect to these non-human animals, I contend that species difference means human domination, and the paper ends with a brief discussion of human domination as a complex system of social relations that privileges the human but within which, different degrees of domination of non-human animal species might be found, dependent on social context.

Sociology and animal studies

Sociology has generally held to the conceptions of those foundational figures such as Simmel and Durkheim, of a society that emerges 'through the symbolically constituted and linguistically mediated encounters and interactions through which meanings and representations are communicated from one mind to another in the course of human association' (Scott 2010: 16–17). It is only fairly recently that influential voices have been heard to argue for the radical configuration of the discipline. For Latour, whose sustained intervention here has been significant, sociology must fully embrace the world of non-human beings, objects and things and the ways our lives are constituted with them (Latour 2010: 75–78). Critiques such as this have helped to open up the discipline to new areas of concern, such as Animal Studies, which has emerged as a diverse and expanding transdisciplinary field.

Whilst bringing animals 'in' to the humanities and social sciences has encouraged inter- and post-disciplinarity, much of the work remains cast within disciplinary paradigms. It is interesting to note however, that its impact has been far greater in arts (particularly in history, cultural geography, philosophy, literature and film) than social science disciplines. Sociology has had limited engagements with Animal Studies, certainly in the UK. Where it has, an interest in non-human animals has often been brought into view through more established sub-areas such as science and technology studies (Twine 2010), food and eating (Stewart and Cole 2009), the family (Charles and Aull-Davies 2008) and rural studies (Wilkie 2010). As Burawoy (2005: 268) has argued, it is critical sociologies that often draw attention to the omissions of the disciplinary mainstream and identify new subjects and objects for study. Twine (2010: 8) has suggested that sociological Animal Studies might be understood in Burawoy's terms as a 'critical sociology' (Twine 2010: 8). The 'bringing in' of animals as new subjects of sociological study has been via both mainstream and critical routes. On the one hand, we have a sociology that includes non-human animals and can certainly be considered critical in terms of Burawoy's framework. On the other hand we have sociological Animal Studies, which raise questions about the exploitation and oppression of 'Other' animals, and this, I would suggest, is more reflective of critical traditions in sociological enquiry. Parallels here might be drawn with the influence of feminism in sociology. Whilst more liberal scholarship considered that 'adding' the concern with gender to existing frameworks and approaches might be sufficient, more critical approaches contended that approaching social life with the 'lenses' of gender radically alters not only the subjects and objects of enquiry, but the nature of enquiry itself (see Peterson and Runyan 2010; also Peggs, Chapter 2 this volume).

A new sociological subject?

Whilst the notion of species suggests taxonomic classification of kinds, types or varieties, it is also a social assignation. Human social relations also shape the biology and sociality of other species as they are incorporated into and co-constituted with, social institutions and practices. Interrogating naturalised categories and practices has been an important sociological preoccupation, and sociology lends itself to problematising the human–non-human animal binary and the ways this is played out in social formations. In addition, species is constituted by and through 'human' hierarchies – ideas of animality and of 'nature' are vitally entangled in the constitution of 'race', gender, class and other 'human' differences with which critical sociologies have very well established preoccupations.

However, as Alger points out, there is a 'hard line that sociology has always drawn between humans and other species' (2003: 69); reflective of the 'species apartheid' that characterises the social sciences more broadly (Twine 2010: 2). Arluke (2004) has noted there are very real barriers to the acceptance of Animal Studies as a research area within the discipline. Grant capture and publishing attest to the continued anthropocentrism of the subjects and objects of sociological enquiry. In addition, even critical sociologies are resistant to the study of non-human animals, shaped by the belief that studying non-human animals lessens or undermines the notion of oppression. Anthropocentrism can be seen as foundational for the discipline, which emerged with the mission of countering biological explanations in social life, and is thus wary of engagements with the non-human animate lifeworld. In this sense, studying animals and our relations with them is to redefine the 'disciplinary matrix' of sociology, and undermine its dualistic organising categories (Benton 1994: 29). Some sociologists have undertaken grounded studies that disrupt the common conflation of 'society' with the human use of language. Irvine for example, critiques the anthropocentrism of Mead and his view that non-human animals sit outside sociology due to their inability to perceive, imagine and speak, which renders them incapable of socially meaningful behaviour (2004: 121). Rather, she suggests that some humans and some animals share both meanings and communication and therefore, are the proper subjects of sociological enquiry (Irvine 2007; Sanders and Arluke 1993).

The debate is raised, of course, about how we deal with these tricky categories in producing social theory. Do we interrogate them, refine them and continue to deploy them with their imperfections, or do we abandon them? Such debates have persisted over the last two decades with respect to the tension between amorphous collective concepts such as 'gender' and 'ethnicity' and localised and situated differences in social form, identity and the possibilities for agency. Various philosophical interventions in Animal Studies have sought to problematize and rework the classical binary formulations of the human–animal distinction in Western thought (Derrida 2002: 135; 2008: 153–160). More radical philosophical positions push further, arguing for the abolition of the 'guardrails of the human-animal distinction' (Calarco 2008: 149). I think however, this is a step too far. In the social world, species is a powerful and persistent discursive and material distinction; and some sociologists have usefully made these distinctions the focus of their attempts to conceptualise human relations with non-human animals.

Ted Benton (1993) has sought to both retain and problematise the catch-all concept of non-human animals by using a notion of species as 'differentiation'. Drawing on ethological work, he argues that many species have overlapping forms of 'species life' with humans, with certain needs, forms of sociality and ecological dependency. He challenges the presumption of human separateness from 'Other' animals, arguing that we should think about 'differentiations' rather than differences between animal species (ibid.: 45–57). Differentiations of species, and particular social, economic and ecological contexts give rise to different categories of human animal relationship. This leads Benton to a sociological categorisation of 'Other' animals in terms of their form of relation to humans. Certain non-human animals may be labourers of various kinds (from guarding, carrying and pulling to guiding visually impaired humans); some species will be food and resources (for human clothing and shelter needs); a limited number may be companions; and many are 'wild' (that is, outside incorporation into human social practices, or in conditions of limited incorporation). In addition,

Benton categorises animals as human entertainment (in hunting, shooting, fishing and fighting, for example), as cultural symbols and as human edification (for example in 'wildlife' documentaries) (ibid.: 2–8). Benton uses these categories in arguing that humans and animals stand in social relationships to each other, that animals are constitutive of human societies and that these relationships are incredibly varied across time and space. These relationships are fundamental to the structuring of human societies (ibid.: 68–69). Although this social categorisation of animals is useful, it elides some important forms of relationship between human and non-human animals and underestimates the contingency of animals' social location (see Stewart and Cole 2009: 461).

In these attempts to categorise non-human animals in terms of social relations, human power is foregrounded. Whilst the abandonment of categories might seem attractive by virtue of being radically transgressive, under current social arrangements the categories into which animals are placed are a description of the material world that animals inhabit. In the web of social practices and institutions which non-human animals, particularly those we have 'domesticated', are very much caught, the difference of species structures material practice and has real effects on the lives and deaths of non-human animals. For a critical sociology of non-human animals, we need the highly problematic human–animal distinction as the theoretical basis of a politics that contests humanocentrism.

Species relations in modernity

One important way of drawing species into sociology has been in terms of understanding the process of historical change through formations of human relations with non-human animals. Historical accounts have used modernity as a framework for theorising human relations with non-human animals, mapping changing attitudes towards animals accompanying the dramatic changes of transition in European modernity, from relations of dependency, contingency, and religious inspired anthropocentricity to those of distance, sentimentality and ambivalence in more secular times (Thomas 1983: 166–167).

Least controversial for the disciplinary mainstream is a position that considers human relations with non-human animals to be revealing about human beings themselves. Here, non-human animals are seen as sociologically relevant and an apparently neutral position is usually articulated on the quality of species relations. Tester (1992) for example, concentrates on the imposition of social relationships through regulation of human relations with other animals. He draws on the work of Elias in suggesting that the development of anti-cruelty legislation was part of the 'civilizing process' to discipline the working class (ibid.: 68–88). Tester argues that how we think about animals does not tell us about the ontological condition of animals, but about ourselves. So for example, animal rights 'is not a morality founded on the reality of animals, it is a morality about what it is to be an individual human who lives a social life' (ibid.: 16). Animal rights has nothing to do with any concern for sufferings that humans may inflict upon animals, but is about humans making themselves feel 'good' as moral agents, arguing for those who cannot argue for themselves (ibid.: 78). Tester offers us an account of human–animal relations that is clearly sociological but highly anthropocentric and of little help in considering the specificity of human relations with non-human animals; for animals, in this analysis, could be replaced with any other marginalised social group.

A less anthropocentric account is provided by Franklin (1999), who contends that we have recently seen significant qualitative changes in species relations as the categorical boundary between human and other animal species has been challenged with 'postmodernisation'. Modernity defined humans as rational, capable of self-improvement and potential goodness, and established clear boundaries between humans and animals. From the seventeenth to the twentieth centuries, animals were treated primarily as a resource for human improvement, so that meat eating and the use of animals in research became standard practices. As we move towards postmodernity however, 'misanthropy' has become a feature of social life, as we collectively reflect on our destruction of the natural world.

Animals are also associated with a sense of 'risk', which can be seen in food scares, or concerns about the preservation of 'wildlife'. Finally, individuals suffer 'ontological insecurity' due to the depletion of family ties and sense of community and neighbourhood, with changes in domestic relations (increased divorce rates and re-marriage) and patterns of employment (with 'flexible' labour markets, higher unemployment and less job security). Consequently, people look to relationships with pets, for example, to provide stability and a sense of permanence in their lives (ibid.: 36). Thus Franklin suggests that we are developing 'increasingly empathetic and decentred relationships' with other species evidenced across a range of sites of human–animal relations – from entertainment to food, pet keeping to hunting (ibid.: 35).

However, there are a number of difficulties with such an account. First, it makes some significant and empirically unsubstantiated sociological assumptions; for example, that certain social changes (such as those in the structure of the family) have led to certain practices (like more people keeping pets), in order to provide security. Second, whilst human relations with animals have undoubtedly changed, there have been different and competing conceptions of how humans can relate to other animals and both continuity and change in material practices. A model of increasing sentimentality ignores the contradictions embedded in our relations with animals and the different kinds of relations that humans have with specific species. In addition to these particular problems, it is not sufficient for sociologists merely to say that animals are co-constitutive of human social arrangements, and that these relations change over time. We need to consider the power relations articulated through such relationships. Since the 1970s, such theorising has been largely philosophical, and early work in Animal Studies articulated human relationships with 'Other' animals in terms of the political language of interests and rights.

The concept of speciesism

Singer is a name much associated with the use of the terminology of 'liberation' and 'oppression' to describe human relations with animals. The key concept underpinning Singer's contributions is 'speciesism', which, in short, is discrimination based upon species membership. Nibert has suggested that in order to develop a sociological account of speciesism, we need to consider the way it is manifested in social institutions and practices, and in social relations (Nibert 2002: 195). I am not at all convinced, however, that the notion of speciesism itself, linked to discriminatory practice, is best suited to the development of a sociological approach, or a critical account of human relations with other animals.

In questioning the practices of speciesism, philosophical Animal Studies has been preoccupied with the drawing and re-drawing of species boundaries in deciding 'which' other animals count as ethical subjects. Debates have considered vertebrate sentiency, interests and the importance of a central nervous system, mammals being 'subjects of a life' and a critique of the anthropocentrism of the focus on similar minds. There is a tendency to measure the extent to which animals do or do not approximate to human capabilities, failing to understand that animals have their own ways of life and being in the world (Whiten *et al.* 1999: 682). In turn, there has been a difficulty in extending humanist concepts of rights and interests to the incredible array of different kinds and types of being that fall under the label 'non-human animal'. The difficulties with humanism are also present in deploying the language of political extension. This language of rights, interests and discrimination in animal rights theory means that we cannot get away from our humanocentrism, as similarities between some humans and some species are usually foregrounded in advancing rights and interest claims.

The undoubted strength of theorising about animal rights and interests however, has been to set an agenda in which the lives and well-being of non-human animals is analytically foregrounded. To consider 'species' as a problematic, socially constituted and oppressive category has been a fundamentally important innovation, problematising the certainties and the qualities of human power. Decades have passed since arguments were made for the sentience of animals and the irrationality of the ways in which humans treat them, yet fundamental changes in human relations with non-human animals have been negligible. As Benton argues, the difficulty with the rights and interests discourse is its inability to take account of the prevailing social structures and relations apparent in certain places and at certain historical junctures (1993: 210). Animals are in and of human societies, and we need fundamental changes in human social practices before we will see any shift in the treatment of animals.

Intersectionality and the discourse of species

Some work in Animal Studies has stressed the operationalisation of speciesism as a discourse of power rather than a form of discrimination. These accounts are

far more attentive to the ways in which the discourse of species is often constituted with overlapping and intersected discourses around human difference and
domination. For Wolfe, for example, speciesism is seen as a set of discourses
embedded in a range of texts of popular culture, and occasionally also challenged therein. The discourse of species understood through such texts 'in turn
reproduces the *institution* of speci*esism*' (2003a: 2, original emphasis). Whilst
Wolfe is clear that 'the violent effects of the discourse of speciesism fall overwhelmingly in institutional terms, on nonhuman animals' (ibid.: 6), he is attentive to the ways in which the 'discourse of animality [has] historically served as
a crucial strategy in the oppression of humans by other humans' (Wolfe 2003b:
xx); an insight with a long legacy in feminist and postcolonial theory and, in particular, in ecofeminist work.

From the early 1970s, ecofeminists suggested that cultural discourses carry
binary normalisations that construct a dichotomy between women and 'nature',
including the multifarious species of non-human animal, and male dominated,
Western, human culture. The arguments presented often drew on a form of
standpoint epistemology: gender roles constituted through such discourses (such
as social practices of care) render women in closer material proximity and relation to the environment and non-human animals (Salamone 1982). Additionally,
it was contended that women may empathise with the sufferings of animals, as
they have some common experiences, for example female domestic animals are
most likely to be 'oppressed' via control of their sexuality and reproductive
powers (Benny 1983: 142). Others examined the speciesism of linguistic practices and the links between this and our gendered and racialised use of language
(Dunayer 1995; Adams 2003); or looked at the interrelations between gender
and the environmental and species impact of colonial practices (Lee Sanchez
1993). Such writing has been influential in alerting us to the intersectional qualities of oppression. However, there is often a tendency in this literature to conflation – the use of an all-encompassing theory of gender relations to explain
intersected oppressions. Carol Adams and Josephine Donovan (1995: 3) for
example, have contended that patriarchy is 'prototypical for many other forms of
abuse', and Suzanne Kappeler has asserted that patriarchy is 'the pivot of all
speciesism, racism, ethnicism, and nationalism' (1995: 348; Collard 1988; Gaard
1993).

Val Plumwood provides a more satisfactory conceptualisation of gender,
nature, race, colonialism and class as interfacing in a 'network' of oppressive
'dualisms' (1993: 2). For Plumwood, these exist as separate entities but are also
mutually reinforcing in a 'web' of complex relations (ibid.: 194). This does not
mean different forms of oppression are indistinguishable; they are distinct yet
related. Both Wolfe and Plumwood have an analysis of intersected discourses,
but a crucial move is absent. Their understanding is ideational – we do not see
how these ideas of separation, of human uniqueness and the animal as 'Other',
are articulated in located contexts and inform what sociologists would understand as social institutions and related practices. This is a gap, but it is not one
I am necessarily criticising Wolf or Plumwood for failing to fill. It is time for

sociology to step up to the task of outlining the social institutions in which the discourse of species is embedded and to provide an analysis in terms of social relations.

Material practice: capitalism and the oppression of animals

Sociology has made a most useful contribution in the theorising of human relations with non-human animals in terms of Marxian influenced analyses. These approaches are not without their difficulties however, and in critiquing them here I will draw together the arguments made so far in outlining the elements for developing a critical sociology of species.

David Nibert (2002: 7) explicitly uses the concept of oppression in relation to the historical development of human relations with non-human animals. He argues that social institutions are foundational for the oppression of animals – not individual attitudes and moral deficiencies. Nibert isolates three elements in his model of non-human animal oppression. First, we have economic exploitation where animals are exploited for human interests and tastes; second, power inequalities coded in law leave animals open to exploitation; and third, this is legitimated by an ideology which naturalises the oppression of animals in its many forms. Nibert retains the concept of speciesism, but redefines it to mean 'an ideology, a belief system that legitimates and inspires prejudice and discrimination' (ibid.: 17).

In turn this 'ideology' emerges from economic institutions and practices. Contemporary cultural processes and institutional arenas though which animals are exploited and oppressed – zoos, the breeding and keeping of pets, the 'use' of animals in research, hunting and food production – are explained in terms of profit creation, corporate interest and the generation and sustaining of false commodity needs. Nibert appears to endorse a multiple systems model of interacting systems of oppression: '*the arrangements that lead to various forms of oppression are integrated in such a way that the exploitation of one group frequently augments and compounds the mistreatment of others*'. (Nibert ibid.: 4, original italics). Disappointingly however, the overriding thesis is that the human oppression of other animals is caused and reproduced by relations of capitalism (ibid.: 3). The same criticism might be made of Bob Torres (2007), who applies Nibert's model to the case of highly industrialised capital-intensive agriculture in the global north. Animals are largely understood as labourers, who labour by eating and breeding in producing commodities such as milk and eggs in dull, barren and stressful conditions (for further discussion of animals as labourers in co-constituted contexts with humans, see Clark, Chapter 7 this volume). Animals are also property, which enables their transformation into embodied commodities such as meat and leather (Torres 2007: 36–58). Torres allows that the histories of exploitative systems are different and that the oppression of animals can exist before and beyond capitalism (ibid.: 156), yet capitalism is his key analytic tool.

In considering the work of those such as Plumwood and Wolfe, I argued that their position is idealist – i.e. focusing analysis on cultural discourses – albeit attentive to questions of interacting social difference. Torres and Nibert on the other hand, provide us with an overwhelmingly materialist analysis that elides the oppression of animals with the systemic imperatives of capitalism. What is required is a full analysis of social intersectionality: an analysis of social difference, inequality and domination in terms of various kinds of relational systems of power: species, gender, 'race', class and so on. We also need an analysis of the social practices and institutions that constitute, reproduce and rearticulate the relations of species specifically. To this end, the remainder of this chapter provides an outline of a theory of the human domination of other animals.

Human domination in species relations

Human relations with other species are constituted by and through social institutions and processes and these can be seen as sets of relations of power and domination, which are consequential of normative practice. These interrelate to form a social system of human domination that I refer to as 'anthroparchy' (Cudworth 2005). In anthroparchal society, humans have socially formed relational power over other species. The social and ecological effects of species as a system of relational power are co-constituted with other kinds of complex forms of domination (such as patriarchy, capitalism and orientalism) and assume specific spatialised and historical formations. Such systems of social domination exist within a relational *matrix* – intermeshing and coalescing in a particular pattern, articulated in different ways, in different times, places and spaces. I will begin by outlining the elements of a social system of species domination, which I call anthroparchy, and proceed to discuss the relational matrix in more depth.

The domination of non-human animals

Human relations with non-human animals need to be considered as a system of social institutions and practices that have their own conceptual repertoire. Non-human beings are dominated by humans – a set of systemic relations I call anthroparchy. This involves different degrees of domination of other animals by humans, and I use the concepts of oppression, exploitation and marginalisation in order to describe this.

Iris Marion Young has used the notion of 'oppression' as an overarching concept to describe 'systematic institutional processes', which prevent people developing their potential through exploitation, marginalisation, powerlessness and violence (1990: 38). It is her understanding of oppression that Nibert (2002) deploys in relation to non-human animals. I prefer to use 'oppression' more restrictedly and chose the term 'domination' as a general descriptor for systemic relations of power that inhibit the potential of an individual organism, group, micro or macro landscape, to 'flourish'. I use 'oppression' to describe a harsh degree of relations of dominatory power and its application is species specific – some species can be oppressed (such as farmed animals) and others cannot (such as intestinal flora). Exploitation refers to the use of something as a resource, and this can apply broadly, including for example, the exploitation of farmed animals for labour, skin, fur, flesh and other products, or the use of animals in guarding and herding. Marginalisation is the rendering of something as relatively insignificant, and decentred, and has a similar meaning to 'anthropocentrism'.

The political economy of animal agriculture exemplifies all three levels at which anthroparchal relations operate. Marginalisation is involved in the definition of certain species of animal as a resource and as a human food. The oppression and exploitation of animals can be seen in the denial of species' specific behaviours, incarceration, physical harm and ultimately of course, killing. Animals are exploited as a set of resources, for example in terms of the utilisation, modification and magnification of their reproductive capacity. There is some diversity in the levels of operation of anthroparchal practice. For example, intensive forms of production in animal agriculture can be seen as extreme or strongly oppressive institutional sites.

The domination of non-human animals in contemporary Western societies, for example, might be understood as constituted through groups of social relations, which can be found in particular arenas – such as production, domestication, violence, polity and culture. For example, animal agriculture is an institutional system and a set of production relations endemic to human domination. The condition of domestication may involve physical confinement, the appropriation of labour and fertility and incarceration, which are foundational for the farming of animals. Institutionalised violence is also systemic, and for species with greater levels of sentiency, operates in similar ways to violence affecting humans. For example, animals raised and killed for food may experience pain and fear. The lives, deaths and dismemberments of animals for 'meat' articulates a range of forms and degrees of physical violence and in some cases, psychological harms, and such violence reflects the complex intersections of relations of social power (Cudworth, 2008; see also Gröling, Chapter 5 this volume). Huge numbers of a limited range of species are essential forms of property and/or labour. Animals are a specific form of embodied property however, and it is the distinction of human from non-human life that is a priori for such commoditisation. Animals produce commodities in terms of offspring, milk and eggs, which become human food. These social relations are framed by law, culturally mediated and politically supported or contested.

Anthroparchal cultures are exclusively humanist, and at best, marginalise non-human animals or presents them in ways that are framed by human interests and re-inscribe the norms of human domination. If we again refer to the example of 'meat', the hierarchical ordering of the Western diet is reproduced in the popular culture of cooking and eating in which the eating of animals is constituted as normative. This process of reproducing food cultural norms is also shaped by various kinds of intra-human difference in addition to the distinction of species that enables 'meat' to be eaten (Adams 1990; Cudworth 2010; Fitzgerald and Taylor, Chapter 8 this volume). Animal foodways have been relatively stable, despite significant social change, and in the West our representative regime of animals-as-meat continues to be framed by intersected discourses of difference and power, in particular those constitutive of formations of gender, sexuality and nation.

It is unlikely that all animals 'used' by humans experience domination in the same way, although there are strong similarities in the ways in which processes of domestication affect both companion and farmed animals. The oppressive experiences of farmed animals may be very different from that of prized 'working animals', such as those providing assistance for humans who are blind, deaf, ill or aging, for example. The lives of animals kept as 'pets' are often very different from those of farmed animals, but there is much evidence of cruelty in the ill-treatment, neglect and abandonment of animals by their human 'companions' (Cudworth 2011a). The industries that have emerged around pet keeping in the West involve intensive breeding and also strong genetic selection for the reproduction of desirable breed (and other) traits.

In this social system of species relations centred on human domination, non-human animals have limited agency. A sociological account of agency requires that agency is not understood as a capacity or property that humans and/or non-human animals possess, but as socially structured – i.e. options for actors are shaped by social relations (Carter and Charles 2011). The lives of most farmed animals are so tightly constrained by structures of oppressive power that they cannot exercise agency. Human companions (of dogs, for example) may live with animals, which they understand as agentic beings, and a limited agency of a fundamentally unequal but co-constituted kind may be possible. Whilst ultimately, companion animals live precarious lives that are usually determined by the well-being and disposition of human 'owners', there are also significant differences in the degree to which different kinds of domesticates, in different kinds of social contexts, might be said to be 'oppressed'. These different forms and degrees of human domination of specific kinds of non-human animals need always to be considered.

Intersectional domination

Whilst I have argued for an understanding of human domination as socially constituted and systemic, there are differences in the specific forms that species relations assume geographically and culturally. The category 'human' is also a catch-all category and the domination of both human and non-human animals is intersectional. Whilst I concur with much of the analysis provided by those such as Nibert on the oppression and exploitation of domesticates in animal agriculture, an explanation based on an analysis of capitalism does not capture the range of interlinked processes involved. We must also consider the ways in which, for example, the intersection of colonialist and patriarchal relations is particularly marked in the farming of animals for food.

In the contemporary West, the meat industry is patriarchally constituted. Farmed animals are disproportionately female and are usually feminised in terms of their treatment by predominantly male human agricultural workers. Farmers disproportionately breed female animals so they can maximise profit via the manipulation of reproduction. Female animals that have been used for breeding can be seen to incur the most severe physical violence within the animal food system, particularly at slaughter (Cudworth 2008). Female and feminised animals are bred, incarcerated, raped, killed and cut into pieces, in gargantuan numbers, by men who are often themselves subjected to highly exploitative working conditions (Eisenitz 1997: 85). These working conditions are structured by the gendered division of labour and also characterised by a culture of machismo (Cudworth 2008).

Furthermore, operations of local, regional and global networks of relations shaped the development of animal food production, and the production and consumption of animals as meat was an historical process in which systemic relations of species are constituted with and through relations of colonialism. In the eighteenth and nineteenth centuries, European countries established the global international system of meat production. Britain and Germany in particular invested heavily in land and also, later, in factories in South America, primarily in Argentina in the eighteenth century, and in Brazil in the nineteenth (Rifkin 1994: 145–147; Velten 2007: 153). The colonial model of meat production was further enabled by the development of refrigerated shipping, which made it possible to ship 'fresh' meat to Europe from the USA, South America and Australasia (Franklin 1999: 130). This enabled Europeans to consume greater quantities of meat, but in order to make best use of the potential market in Europe the price had to be minimised by intensifying production and saving labour costs through increased mechanisation, processes which led to the development of intensive agriculture in Europe and the USA, models of production now spread across the globe with corporate interventions in Asia, Africa and the Caribbean (Cudworth 2011b).

Finally, as social and natural systems are co-constituted, we must also consider the impact of farmed animal agriculture on the worlds of other species and things. As is becoming increasingly recognised, industrialised animal agriculture is claimed to be a driving force behind all of the contemporary and pressing environmental problems that we face – deforestation, water scarcity, air and water pollution, climate change and loss of biodiversity (CIWF 2002; Steinfeld *et al.* 2006; Worldwatch Institute 2004; World Bank 2001). Thus, whilst farmed animal agriculture is an integral element of a social system of species relations in which domesticates are oppressed, it is also constituted by relations of capital, colonialism and patriarchy and shaped in important ways by intra-human difference.

Conclusion: disciplinarity, interdisciplinarity and criticality

The emergence of sociological animal studies has been a very positive development, yet the discipline of sociology generally remains entrenched in its anthropocentric humanism; and this holds as fast in supposedly critical Marxist influenced sociologies as it does in the mainstream of liberal scholarship. Other disciplines (such as philosophy and political theory or human and cultural geography) have been far more inclusive of the non-human than has sociology to date. However, I have argued in this chapter that in its analysis of the processes of historical change, in its attention to both the power of ideas and beliefs and the analysis of concrete social practices, the discipline of sociology has much to offer those who seek to understand our relationships with non-human animals. Sociology has much to learn from Animal Studies. It might reflect on the human-centred nature of its humanism and the human exclusivity of its emancipatory projects. Thinking about 'Other' animals does not undermine the emancipatory impulse of critical sociological approaches to difference and inequality; rather, it broadens and deepens them. I have suggested that a critical sociology of species will understand species relations as another important formation of difference that, like gender or 'race' or 'ability', is socially constituted. It will see current formations of species as organised in terms of power and hierarchy and as constituted in cross-cutting ways with other social formations of intra-human difference and domination with which critical sociologies have been so much concerned. As such, it will be part of the project of Critical Animal Studies.

A critical sociology of species will also make a distinct disciplinary contribution to the way Critical Animal Studies can approach questions of human power and problematise our relationships with non-human animals. It will enable us to move beyond a core terminology based on the idea of discriminatory practice (the conceptual repertoire of speciesism) to one of social relations that privilege the human. The interdisciplinary reach of Animal Studies, more broadly, is a strength, whilst the attention in Critical Animal Studies to the overlapping and co-constituted oppression(s) of all creatures in various ways is a key distinguishing characteristic. We live in a complex world of multiple social relations and an important contribution of sociology has been the increasingly sophisticated mapping of the way these interact. These are of utmost relevance to understanding the social forms which our relationships with non-human animals take and can add depth to our approaches to intersectionality in Critical Animal Studies.

The human domination of non-human animals exists in a milieu of multiple systems of social domination. Whilst we who inhabit the category 'human' are implicated in a system of species domination, being human is insufficient to describe who we are. 'We' humans have incredibly different relations to non-human animals depending on our geographical and cultural location, and our gender, ethnicity and relationship to wealth. We also, of course, have different relations according to our politics. A critical sociology *for* non-human animals must be a politicised sociology and it must take account of, and present a challenge to, the intersected dominations of all the beings on this planet we inhabit.

References

Adams, C. J. (1990) *The Sexual Politics of Meat*, Cambridge: Polity.

Adams, C. J. (2003) *The Pornography of Meat*, London: Continuum.

Adams, C. J. and Donovan, J. (1995) 'Introduction', in C. J. Adams and J. Donovan (eds) *Animals and Women: Feminist Theoretical Explorations*, Durham, N. Carolina: Duke University Press.

Adams, C. J. and Proctor-Smith, M. (1993) 'Taking life or taking on life? Table talk and animals', in C. J. Adams (ed.) *Ecofeminism and the Sacred*, New York: Continuum.

Alger, J. M. (2003) 'Drawing the line', *International Journal of Sociology and Social Policy*, 23(3): 69–93.

Arluke, A. (2004) 'A sociology of sociological animal studies', *Society & Animals*, 10(4): 369–374.

Benny, N. (1983) 'All one flesh: the rights of animals', in L. Caldecott and S. Leyland (eds) *Reclaim The Earth*, London: Women's Press.

Benton, T. (1993) *Natural Relations: Ecology, Animal Rights and Social Justice*, London: Verso.

Benton, T. (1994) 'Biology and social theory in the environment debate', in M. Redclift and T. Benton (eds.) *Social Theory and the Global Environment*, London: Routledge.

Burawoy, M. (2005) 'For a public sociology', *American Sociological Review*, 70(1): 4–28.

Calarco, M. (2008) *Zoographies: The Question of the Animal from Heidegger to Derrida*, New York: Columbia University Press.

Carter, B. and Charles, N. (2011) 'Human-animal connections: an introduction', in B. Carter and N. Charles (eds) *Human and Other Animals: Critical Perspectives*, Basingstoke: Palgrave.

Charles, N. and Aull-Davies, C. (2008) 'My family and other animals: pets as kin', *Sociological Research Online*, 13(5), available at www.socresonline.org.uk/13/5/4.html (accessed 14 December 2013).

Collard, A. with Contrucci, J. (1988) *Rape of the Wild: Man's Violence Against Animals and the Earth*, London: The Women's Press.

Compassion in World Farming (CIWF) (2002) *Detrimental Impacts of Industrial Animal Agriculture*, Godalming, Surrey: CIWF.

Cudworth, E. (2005) *Developing Ecofeminist Theory: The Complexity of Difference*, Basingstoke: Palgrave.

Cudworth, E. (2008) ' "Most farmers prefer blondes" ' – dynamics of anthroparchy in animals' becoming meat', *The Journal for Critical Animal Studies*, 6(1): 32–45.

Cudworth, E. (2010) ' "The recipe for love" '? Continuities and changes in the sexual politics of meat', *The Journal for Critical Animal Studies*, 8(4): 78–99.

Cudworth, E. (2011a) 'Walking the dog: explorations and negotiations of species difference', *Philosophy, Activism, Nature*, 8: 14–22.

Cudworth, E. (2011b) 'Climate change, industrial animal agriculture and complex inequalities', *The International Journal of Science in Society*, 2(3): 323–334.

Cuomo, C. (1998) *Feminism and Ecological Communities: An Ethic of Flourishing*, London: Routledge.

Derrida, J. (2002) 'The animal that therefore I am (more to follow)', trans. D. Wills, *Critical Enquiry*, 28: 369–418.

Derrida, J. (2008) 'I don't know why we are doing this', in M.-L. Mallet (ed.) *The Animal That Therefore I Am*, trans. D. Wills, New York: Fordham University Press.

Dunayer, J. (1995) 'Sexist words, speciesist roots', in C. J. Adams and J. Donovan (eds) *Animals and Women: Feminist Theoretical Explanations*. Durham, N. Carolina: Duke University Press.

Eisenitz, G. (1997) *Slaughterhouse: The Shocking Tales of Greed, Neglect and Inhumane Treatment inside the U.S. Meat Industry*, New York: Prometheus Books.

Franklin, A. (1999) *Animals and Modern Cultures: A Sociology of Human-Animal Relations in Modernity*, London: Sage.

Franklin, A. (2007) 'Human-nonhuman animal relationships in Australia: an overview of results from the first national survey and follow-up case studies 2000–2004', *Society & Animals*, 15(1): 7–27.

Gaard, G. (1993) 'Living interconnections with animals and nature', in G. Gaard (ed.) *Ecofeminism: Women, Animals, Nature*, Philadelphia: Temple University Press.

Irvine, L. (2004) *If You Tame Me: Understanding Our Connections with Animals*, Philadelphia: Temple University Press.

Irvine, L. (2007) 'The question of animal selves: implications for sociological knowledge and practice', *Qualitiative Sociology Review*, 3(1): 5–22.

Kappeler, S. (1995) 'Speciesism, racism, nationalism ... or the power of scientific subjectivity', in C. J Adams and J. Donovan (eds) *Animals and Women: Feminist Theoretical Explanations*, Durham, N. Carolina: Duke University Press.

Latour, B. (2010) 'A plea for earthly sciences', in J. Burnett, S. Jeffers, and G. Thomas (eds) *New Social Connections: Sociology's Subjects and Objects*, Basingstoke: Palgrave.

Lee Sanchez, C. (1993) 'Animal, vegetable, mineral', in C. J. Adams (ed.) *Ecofeminism and the Sacred*, New York: Continuum.

Nibert, D. (2002) *Animal Rights/Human Rights: Entanglements of Oppression and Liberation*, Lanham, MA: Rowman and Littlefield.

Peterson, V. S. and Sisson Runyan, A. (2010) *Global Gender Issues in the New Millennium*, Boulder, CO: Westview Press.

Plumwood, V. (1993) *Feminism and the Mastery of Nature*, London: Routledge.

Rifkin, J. (1994) *Beyond Beef: the Rise and Fall of Cattle Culture*, London: Thorsons.

Rollin, B. (1981) *Animal Rights and Human Morality*, Buffalo, NY: Prometheus Books.

Salamone, C. (1982) 'The prevalence of natural law within women: women and animal rights', in P. McAllister (ed.) *Reweaving the Web of Life: Feminism and Non-violence*, San Francisco: New Society.

Sanders, C. R. and Arluke, A. (1993) 'If lions could speak: investigating the animal-human relationship and the perspectives of non-human others', *Sociological Quarterly*, 34: 377–390.

Scott, J. (2010) 'Sociology and the sociological imagination: reflections on interdisciplinarity and intellectual specialisation', in J. Burnett, S. Jeffers and G. Thomas (eds) *New Social Connections: Sociology's Subjects and Objects*, Basingstoke: Palgrave.

Steinfeld, H., Gerber, P., Wassemaar, T., Castel, V., Rosales, M. and de Haan, C. (2006) *Livestock's Long Shadow: Environmental Issues and Options*, Rome: United Nations Food and Agriculture Organisation.

Stewart, K. and Cole, M. (2009) 'The conceptual separation of food and animals in childhood', *Food, Culture and Society*, 12(4): 457–476.

Tester, K. (1991) *Animals and Society: The Humanity of Animal Rights*, London: Routledge.

Thomas, K. (1983) *Man and the Natural World: Changing Attitudes in England 1500–1800*, London: Allen Lane.

Torres, B. (2007) *Making a Killing: The Political Economy of Animal Rights*, Oakland, CA: AK Press.

Twine, R. (2010) *Animals as Biotechnology: Ethics, Sustainability and Critical Animal Studies*, London: Earthscan.

Velten, H. (2007) *Cow*, London: Reaktion Books.

Whiten, A., Goodall, J., McCrew, W. C., Nishida, T., Reynolds, V. and Sugiyama, Y. (1999) 'Culture in Chimpanzees', *Science*, 399, 17 June.

Wilkie, R. (2010) *Livestock/Deadstock: Working with Farm Animals from Birth to Slaughter*, Philadelphia: Temple University Press.

Wolfe, C. (2003a) *Animal Rites: American Culture, the Discourse of Species and Posthumanist Theory*, Chicago: Chicago University Press.

Wolfe, C. (2003b) 'Introduction', in C. Wolfe (ed.) *Zoontologies: the Question of the Animal*, Minneapolis: University of Minnesota Press.

Wolfe, C. (2003c) 'In the shadow of Wittgenstein's lion: language, ethics and the question of the animal', in C. Wolfe (ed.) *Zoontologies: the Question of the Animal*, Minneapolis: University of Minnesota Press.

World Bank (2001) *Livestock Development: Implications for Rural Poverty, the Environment and Global Food Security*, Washington, DC: World Bank.

Worldwatch Institute (2004) *State of the World Report: Consumer Society*, Washington, DC: Worldwatch Institute.

Young, I. M. (1990) *Justice and the Politics of Difference*, Princeton, NJ: Princeton University Press.

2 From centre to margins and back again
Critical animal studies and the reflexive human self

Kay Peggs

Introduction

Some thirty years ago, in her cutting edge work *Feminist Theory, from Margins to Center*, bell hooks observed of feminism that 'the dearth of material by and about black women' (1984: x) signified that it had not acknowledged or addressed the most victimized of women. In consequence, she argued, those 'who are most victimized by sexist oppression; women who are daily beaten down, mentally, physically, and spiritually – women who are powerless to change their condition in life' (hooks 1984: 1) had little voice in the feminist movement at the time. hooks was not just describing the situation as she saw it; she was arguing that such marginalized voices should be brought to the centre of feminism as 'White women who dominate feminist theory have little or no understanding of white supremacy' (hooks 1984: 4). By employing the innovative concept of 'intersectionality' (which she used to explore and to understand how some women, though disadvantaged in terms of patriarchy, can also hold privileged identities in terms of, for example, 'race' and class), hooks sought to develop an understanding of the complexity of oppressive relationships. So that this understanding could be developed, she urged mainstream feminists to reflect on their own oppressive positions and to bring marginalized groups from the margins to the centre of feminism (hooks 1984: 3). In this chapter I want to think about some of the similarities and differences between hooks's call to feminists and the position in Critical Animal Studies (CAS), which seeks to offer a critique of more traditional forms of Animal Studies through the development of an understanding of the material experiences of nonhuman animals in their relationships with humans (Twine 2010: 8; Best 2009). In this regard CAS seeks to compel Animal Studies to 'confront the reality of the suffering of the actual animal herself' (Stanescu and Twine 2012: 4).

I have set myself a considerable task and, of course, I will not be able to cover all the complexities in a book chapter. My thinking is grounded in my reflections on the challenge of CAS to Animal Studies seen not least in the aspiration of bringing CAS from the margin to the centre of Animal Studies and to the centre of academic work more generally, in the bringing of the material consequences for nonhuman animals in their relations with humans from the margin to the centre, and in the need for the deconstructed and reflexive human to move from the margin to the centre in such studies. In this regard I aim to pick up on hooks's idea that an understanding of the complexity of oppressive relationships necessitates a commitment to self-reflection because it is only through reflection on our own oppressive positions that we can bring the marginalized from the margins to the centre (hooks 1984: 3). I argue that a critical engagement with the human oppression of nonhuman animals requires reflexivity; a different form of human being towards nonhuman animals, reflexive humans who no longer take themselves for granted in relation to nonhuman animals.

In short, to my way of thinking, CAS entails a move of the 'human' from centre to margins and back again, a displacement or decentring of the unreflexive and oppressively self-interested human by an exposed, self-reflexive human who is critically engaged.

In order to achieve my aim, the discussion is divided into three main sections corresponding to three key aspects of hooks's book that I feel speak to CAS in relation to the complexities of marginalizing and centre-staging; these are: (1) CAS and the understanding of the perspectives of the most victimized, that is the lives of nonhuman animals, (2) CAS and political engagement that centres on the oppression of nonhuman animals and the place of advocacy of the oppressed in academic work, and (3) reflexivity and the role of reflection and self reflection in CAS. At times, in order to think about these aspects, I use sociology as a case study. This is because sociology is the area I know best and because sociology has been rather slower than other areas of study to explore human–nonhuman animal relations (Arluke 2002; Kruse 2002). In this regard I hope the discussion will also serve to demonstrate the relevance of sociology to CAS and of CAS to sociology.

From margin to centre: critical animal studies and the awareness of the lives of nonhuman animals

In her *Introduction to Critical Animal Studies*, Dawn McCance comments that 'particularly since the seventeenth century, modern Western ways of knowing nonhuman animals, [are] inseparable from violent techniques practised on them, [which has] turned animals into "stone," that is, into inert objects, useful and disposable things' (McCance 2013). A critique of this human way of 'knowing' nonhuman animals is at the centre of CAS which, initially as a specialism in the field of analytic philosophy, emerged forty years ago 'to expose, and to offer ethical responses to, today's unprecedented subjection and exploitation of animals' (McCance 2013). The critical stance adopted by CAS to this way of knowing centres on, to use hooks's words in reference to feminism, 'the absence of choices' (hooks 1984: 5) that humans bestow on billions of nonhuman animals every year. This absence of choices is the 'primary point of contact between the oppressed and the oppressor' (hooks 1984: 5).

The absence of choices for nonhuman animals in their relationships with humans most often centres on, and results in, terrible abuse,[1] and the accounts of these abuses are hard to bear. Ted Benton suggests that the most extensive and most systematically organized of these abuses occur in intensive factory farms and in research laboratories (1998: 171). The abuses that Benton is referring to in these contexts are the authorized abuses that are endorsed in the interests of humans and, in consequence, are outside the reach of the criminal justice system (Flynn 2008: 170). Although such institutionalized violence accounts for the majority of the violence humans execute against nonhuman animals, it is shielded from the everyday experience of most humans (Emel and Wolch 1998: xi) with the result that the farm, the slaughterhouse, and the scientific laboratory have 'a whole hidden geography' (Philo and Wilbert 2000).

Writers and activists in CAS make it their business to reveal the terrible realities of the lives of these nonhuman animals, not least because public perceptions often do not match the realities.

At the dinner table 'we are brought into direct touch with the most extensive exploitation of other species that has ever existed' (Singer 1990 [orig. 1976]). Public perceptions of rural agriculture often centre on a rural idyll in which 'calves nuzzle their mothers in a shady meadow, pigs loaf in the mudhole, and chickens scratch in the barnyard' (Mason and Finelli 2006: 104). This might be the human mental picture but, as Jim Mason and Mary Finelli (2006) make clear, this is a pole away from the reality.

The majority of farms are industrial, intensive farms in which billions of nonhuman animals live and die in the most terrible of ways. For example, 54 per cent (by weight)[2] of the ten billion land animals (fish are not included in the figure) in the United States who are slaughtered every year for meat, milk, or eggs live in just five per cent of the country's industrial farms (Humane Society of the United States 2002: 7). Before their deaths, these nonhuman animals live their foreshortened, miserable lives in the most awful conditions, often without grazing land, natural light, and room to move. Their lives are 'utterly mechanized, standardized, and commodified' (Donaldson and Kymlicka 2011: 76). If they do not die on the way to the slaughterhouse they often experience agonizing and atrocious deaths at the hands of poorly paid, ill-trained, and often emotionally-detached humans (Stibbe 2001; Wilkie 2010).

The vast majority of nonhuman animals intended for the meat industry 'suffer every waking minute they are alive. Physically, they are sick, plagued by chronic, debilitating diseases. Psychologically, they are ill, weighed down by the cumulative effects of disorientation and depression' (Regan 2004: 89–80). The cows raised for the insatiable human diet of milk and milk-based products fare little better, if better at all. In order to produce the amount of milk that humans require, cows must give birth every twelve or fourteen months and each cow must produce around 14 litres of milk a day (Blaney 2002: 2). They suffer from appalling exhaustion and many experience dreadful health problems associated with incessant birthing and milking (Blaney 2002: 2). In the interests of the overproduction of eggs for human consumption chickens, who are raised to overproduce eggs in factory farms, lead appalling lives. Their miserable and cramped lives are devoted to managing the inevitable battles that accompany overcrowding and coping with the filthy and cramped conditions in which they are forced to live. After about eighteen months their abused bodies are considered useless and they are mashed for processed food (McCance 2013).

The farming of fish is equally cruel. Increasing numbers of fish live out their lives in factory farms (Rowlands 2002: 175), crammed into small tanks where injury and disease are rife. They are often fed on a diet laced with chemicals and those who survive these appalling conditions end up suffocating or being bludgeoned to death in preparation for the plate. In sum, nonhuman animals live lives of squalor and misery in factory farms (Rowlands 2002) since the focus is on maximizing economic returns rather than on their welfare (Cassuto 2007: 59). Biotechnical techniques are being used in the global farming industry to provide even greater potential for raising nonhuman animals for human consumption (Donaldson and Kymlicka 2011; Twine 2010). Thus, for Bernard Rollin, 'The animal is now an inexpensive cog in a machine, part of a factory, and the cheapest part at that, and thus totally expendable' (Rollin 2008: 11). For Sue Donaldson and Will Kymlicka (2011) nonhuman animals have been reduced to 'widgets'.

Intensive, industrial type farms have displaced family farms in the 'production of animals' (McCance 2013) but this is not to say that nonhuman animals intended for human food who live outside of factory farms are much better off. They might live their often foreshortened lives in superior conditions to those lived by their miserable counterparts who live in factory farms, but they are still intended for human consumption. For example, free-living fish are caught with hooks and are left to suffocate to death in nets (Francione 2004: 109). Compassionate or 'happy meat' might not be sliced from the bodies of nonhuman animals who are raised in gestation crates or on intensive farms, but still these nonhuman animals live their lives as 'mere instruments for our ends' (Torres 2007: 26).

Nonhuman animals raised in farms that use traditional farming techniques are often abused and neglected, are always exploited, and are usually killed (Donaldson and Kymlicka 2011: 76).

Many of the deaths and painful practices meted out by humans to nonhuman animals in laboratories violate anticruelty laws (Beirne 1999, 2002). Annually an average of 115 million living, vertebrate, nonhuman animals are used as subjects in experiments worldwide (Taylor, *et al.* 2008: 327) and this figure does not include the vast number of invertebrate nonhuman animals used (Peggs 2010). This staggering figure can only be estimated as the majority of countries (79 per cent) do not publish figures (ibid.). However, using statistical calculations, Katy Taylor and her colleagues estimate that the USA, Japan, China, Australia, France, Canada, UK, Germany, Taiwan, and Brazil use the highest number of nonhuman animals in experiments (2008: 327).

In 2005 in the European Community alone, 12.2 million living vertebrate nonhuman animals were used in experiments (Commission of the European Communities 2007) and in 2010, 3.6 million nonhuman animals were used in experiments in the UK, representing a 3 per cent increase on the figure for the previous year (Home Office 2011: 8). Such experiments cover a range of painful and life threatening-procedures including scalding, breaking of bones, and blinding, and many of these maltreatments are inflicted without anaesthesia (Rowlands 2002: 124). Moreover, despite the welfarist emphasis on the 3Rs (Replacement, Refinement, and Reduction of the use of nonhuman animals), developments in biotechnology and genetics have increased the utility of nonhuman animals for use in laboratories (Knight 2011; Brown and Michael 2001). Only the millions of nonhuman animals concerned and the relatively small group of humans involved experience this gargantuan scale of suffering, distress, and exploitation. These 'sites of harm', to use Glen Elder, Jennifer Wolch and Jody Emel's phrase, keep 'mass, mechanized and industrialized violence toward animals "out of sight" [which] is necessary to legitimize suffering on the vast scale required by the mass market's demand for meat and medicine' (1998: 85).

John Berger (2009) sees our present relations with nonhuman animals as based in spatial separation. Because humans are so spatially separated from living nonhuman animals (though, as we have seen, humans are often spatially close to dead nonhuman animals who are on the plate (Singer 1990 [orig. 1976]); Peggs, 2012), Berger (2009) argues that we can no longer imagine them, and thus we are reduced to making representations of them in the forms of toys and pictures. However, these imaginings do not centre on the suffering of inflicted nonhumans – rather they focus on romantic representations. Such romantic representations serve to hide rather than elucidate the reality of human–nonhuman animal relations. Steve Baker suggests that it is only by reflecting on the ways in which nonhuman animals figure in our lives, which includes reflecting on what romantic representations of nonhuman animals reveal about our attitudes to them, that we are able to examine the consequences of our attitudes (1993: 3).

Writers in the CAS tradition have an acute awareness of the consequences of our attitudes to nonhuman animals and to the effects of these attitudes on the lives of nonhuman animals. Those enslaved in sites of harm such as farms and laboratories and others that I have not mentioned (such as nonhuman animals who are imprisoned in zoos or are utilized in sport, who are used as trophies or exploited for entertainment) lack choice and often have their needs unmet. Such nonhuman animals are at the centre of CAS, which seeks to speak on behalf of nonhuman animals whose oppressed lives are marginalized. As we can see, this is unlike the feminist theory hooks referred to, which she felt rarely included 'knowledge and awareness of the lives of women and men who live in the margin' (hooks 1984: x). Knowledge and awareness of the lives of nonhuman animals who live in the margin are the central focus of CAS. As with feminism, the focus of CAS is on oppression, which places struggle in a radical political framework (hooks 1984: 5), and it is to efforts by CAS to move the radical political framework from the margin to the centre that I turn to next.

From margin to centre: critical animal studies and political engagement about the oppression of nonhuman animals

From its inception in analytical philosophy, the scope of CAS has widened to encompass a range of disciplines and areas that seek to engage the broad field of Animal Studies in a critical way (McCance 2013). This critical stance centres on a re-evaluation of 'the role and presence of nonhuman animals' (Twine 2010: 1) which, for McCance (2013), involves: engaging the 'critical' in a critical questioning of the Cartesian and religious legacy about the 'human' and the 'animal'; a critical exploration of the side effects of today's subjection of nonhuman animals; and contemplation on the 'better way' that might result from this critical engagement. This critical approach of CAS differs from animal studies because CAS is critical of Animal Studies itself (Best 2009) as it seeks to take Animal Studies further by confronting the problems associated with a lack of challenge to anthropocentric perspectives found in Animal Studies and the concerns expressed in Animal Studies about taking a standpoint on the human oppression of nonhuman animals. Part of the problem for Animal Studies is the standpoint taken for animals.

As an area of study, Animal Studies has sought to move from the academic margin to the academic centre. In 2002, Kenneth Shapiro announced 'modest gains' in the attempts of Animal Studies to 'secure the place of animals other than humans in the "moral landscape,"' (2002: 332). In this regard Shapiro argued that Animal Studies was located 'solid, at the margin' (2002: 331). In the years since Shapiro wrote this piece, Steve Best (2009) perceives a more substantial move of traditional forms of Animal Studies from the margin to the academic mainstream, which is paradoxically, he argues, 'both laudable and lamentable'. For Best (2009), in its quest to be accepted, Animal Studies is in danger of being contained, for example by trying to avoid controversy through shunning political movements that are pro animal rights. In sum, he argues that being embraced by the mainstream is 'a formula for its failure, upon being co-opted, tamed, and neutralized by academia' (Best, 2009). Nevertheless, the movement from the margin to the centre for Animal Studies means that the study of human–nonhuman animal relations has become more central to a range of academic areas and this is an enormous achievement. A good example of this movement is found in sociology, which has been traditionally more resistant to exploring human–nonhuman animal relations than areas like philosophy and anthropology (Arluke 2002; Kruse 2002). The call made by Clifton Bryant (1979) in the late 1970s for a 'zoological connection' in sociology has been slowly, but surely, taken up. Animal Studies is now a burgeoning area of study in sociology, demonstrated by study groups devoted to it in learned societies such as the American Sociological Association and The British Sociological Association. Although to my mind it would be premature to say that Animal Studies has entered mainstream academic sociology, it is certainly on its way there. However, CAS is more firmly placed at the margins and, indeed, is located at the margin of Animal Studies itself.

CAS encompasses both academic pursuits and activist objectives (McCance 2013) and it is its political engagement with activism that academic pursuits resist. As we have seen, Best (2009) makes clear the paradox here and Shapiro (2002) pointed to this paradox some years before. Shapiro pointed out that,

> Politics outside the academy also is a critical contributor to the well-being of HAS [Human-Animal Studies][3] as the twin emergence of the contemporary animal rights movement (ARM) and HAS historically were, and no doubt will continue to be, intertwined. There are assets and liabilities in this association. For HAS, the association with ARM gives a supplementary institutional infra-structure and audience outside the academy. It also gives it relevance, cachet, and a compelling set of practical applications and policy implications. On the liability side, the undeserved charge of violence and terrorism, with all that term currently carries, readily spills over to HAS. More insidiously, HAS is vulnerable to the charge that an ARM agenda biases its investigations and scholarship.
>
> (2002: 336)

In the light of the oppression of billions upon billions of nonhuman animals, how can the study of 'human'–'animal' relations fail to be politically engaged? Can academic pursuit really subscribe to the notion that humans can study their oppressive relations with nonhuman animals in an 'objective' and detached or disengaged way? After all, academics are humans and benefit from the privileges associated with being human (Nibert 2003; Peggs 2013). Feminist thinking deconstructs notions of objectivity by centring on the 'situated knowledge' that, in making its partiality clear, can be held accountable (Haraway 1991: 188–190).

This is plain in hooks's (1984) work. She notes that 'throughout the work my thoughts have been shaped by the conviction that feminism must become a mass based political movement if it is to have a revolutionary, transformative impact on society' (hooks 1984: x). CAS is similar in that it seeks to make a 'better way' (McCance 2013): for CAS writers the themes of academia, advocacy and activism are inseparable. However, academia is not always welcoming to advocacy-oriented approaches, which hinders CAS's movement to the centre. Drawing on sociology as a case study, we can compare and contrast feminism and CAS in terms of mainstream acceptance of advocacy-oriented study.

It is useful to centre on sociology as, for some sociologists, a central aim of their sociology is the transformative effects it can have. For example, with reference to his early reading of Albert Camus's *The Rebel*, Zygmunt Bauman reflects that:

> I suppose it was from Gramsci's Prison Notebooks which I read a year or two after absorbing Camus's cogito 'I rebel, therefore I am', that I learned how to rebel armed with sociological tools and how to make sociological vocation into a life of rebellion.
>
> (2008: 233)

Michael Burawoy (2005) seems to have some sympathy with this rebellious approach. Burawoy (2005) sees the political direction of sociology as giving room to make a better world. This making of a better world is a characteristic of Burawoy's conceptualization of sociology as embracing four ideal-typical forms; that of professional sociology that is based in accumulated bodies of knowledge, policy-oriented sociology that is in service of a goal defined by a client, public sociology that is in conversation with publics, and critical sociology that promotes new research areas and examines professional sociology with a view to making it aware of its biases and silences (2005: 271).

This critical approach can become more radical, in the CAS sense, as we will see below. However, before we get there, it is significant that there is no consensus in sociology about the adoption of this more critical approach to sociology. In this regard Martyn Hammersley offers a distinction between the functional purpose of sociology (what purpose does it serve to the world) and the moral purpose of sociology (what purpose should it serve) (1998: 13). In response to Alvin Gouldner's notion that sociology should embrace questions of morality (1975: 25) and provide the basis for 'right living' (1970), Hammersley warns that this would lead the sociologist to depart from a 'sociological explanatory mode' into 'rational talk about the mission of sociology' (1998: 15). Hammersley suggests that what is envisaged by Gouldner is an engagement between sociology and the transformation of society, which would be 'facilitated by a strong relationship between radical sociologists and political activities; albeit with some autonomy preserved on both sides' (1998: 33). Hammersley cautions that this could expose sociology to bias, because it could lead the sociologist to prioritize her or his values over what he sees as truth (1998: 41). Here there are clear reverberations between Hammersley's thinking and the charge of 'agenda biases' that Shapiro feels could be levelled at a critical, politically engaged Animal Studies – that is CAS.

Gouldner's general support for a close relationship between sociology and values led to the development of a critical sociology that has become more radical (e.g. see Hollands and Stanley 2009) in that it is often associated with advocacy-oriented sociology. Moreover, Liz Stanley observes, advocacy-oriented sociology (such as feminist sociology) has become mainstream (2000: 60–61). An example I have used elsewhere, which I rehearse below, offers a good case in point (Peggs, in press). In his mainstream social research methods textbook, Alan Bryman (2008) points out that women's standpoint is put into practice in feminist research. He notes that, for feminist researchers,

> to do research on women in an objective, value-neutral way would be undesirable (as well as being difficult to achieve) because it would be incompatible with the values of feminism. Instead, many feminist researchers advocate a stance that extols the virtues of a commitment to women and to exposing the conditions of their disadvantage in a male-dominated society.
>
> (2008: 103)

This is not questioned; it is embraced. But it has taken time.

As with feminism in sociology it is likely that CAS needs more time to move to the mainstream of sociology and other academic pursuits. Sociology has a relatively small, but growing engagement with critical work about human– animal relations. Ted Benton's (1993) early critical work has been followed by work by sociologists such as David Nibert (2002, 2003), Leslie Irvine (2007), and Richard Twine (2010). All are explicit in their advocacy for nonhuman animals.

Nevertheless, sociology has traditionally failed to report on, much less take a stand on, the oppression of nonhuman animals (Peggs, in press) to the extent that some sociologists remain resistant to the inclusion of more traditional forms of Animal Studies in sociology (for discussion see Kruse 2002; Peggs 2012; Taylor and Signal 2008). It is this taking seriously and centre-staging of more traditional forms of Animal Studies that can help with the transition from the margin to the centre of CAS providing this more traditional form does not marginalize CAS itself. Moreover, support for CAS from other advocacy-oriented approaches can help. However, this is not always forthcoming. Returning to feminism, Josephine Donovan and Carol J. Adams cite resistance from some feminists, who argue that focusing on nonhuman animals removes the focus away from more pressing human needs such as violence against women (1995: 3). In this way the link between feminist ethics and a moral concern for nonhuman animals is denied. For example Kathryn Paxton George argues that concerns for nonhuman animals assigns 'arbitrary moral burdens' on women (1994: 408). Donovan and Adams disagree with Paxton George. Using an approach that draws on hooks' (1984) notion of intersectionality, Donovan and Adams have sought to bring nonhuman animals to the centre of feminist analyses of oppression. In their view:

> Women must not deny their historical linkages with animals but rather remain faithful to them, bonded as we are not just by centuries of similar abuse but by the knowledge that they – like us, often objectified as Other – are subjects worthy of the care, the respect, even the reverence that the sacredness of consciousness deserves. Such an assertion of subjectivity is necessarily subversive of domination in all its forms.
>
> (Donovan and Adams 1995: 7)

Of course, the oppression of nonhuman animals is based on structures that derive from and find expression in the behaviour of human individuals. The ethic of care outlined and promoted by Donovan and Adams (1995) requires human self-reflexivity and I turn to this in the final section.

From margin to centre: the role of self-reflection in critical animal studies

Reflexivity, according to Margaret S. Archer (2010) 'has the self referential characteristic of "bending back" some thought upon the self'; reflexivity is grounded in 'internal conversation' and subsequent opinion formation. Although Archer seems to suggest that this is a peculiarly human activity (such an assumption is problematic for many writers in a CAS tradition),[4] I want to focus on this idea of human engagement with reflexivity in CAS and, in doing so, I want to explore the idea that reflexivity brings the human back to share centre-stage with nonhuman animals. CAS does this by veering away from taking the human for granted and, as a result, seeks to render the human visible. So, for me, CAS entails a move of the 'human' from the centre to the margins and back again of the 'human', exemplified by replacement of the centrally-placed, taken-for-granted human with the centrally-placed, exposed, and self-reflexive human who is critically engaged. The concept of reflexivity is relevant and helpful here and though I want to draw on sociology for this discussion, of course, reflexivity is central to a number of academic pursuits (Archer 2010).

In keeping with the flow of this chapter, I explore thoughts on reflexivity in terms of gender and the male standpoint in sociology. This is a useful focus as Dorothy Smith argues that claims of objectivity in sociology come from the male standpoint which, although 'appearing to view the world from no place, in fact operates from the standpoint of the patriarchal relations of ruling' (1987: 221). This resonates with the aim of CAS to challenge the taken-for-granted, anthropocentric standpoint. Such challenges involve self-reflection on our own viewing of the world. This is, of course, a central element of hooks's thesis as, using Rita Mae Brown's observations on class, hooks argues that participants in struggle must look at oppressive behavioural patterns in themselves since 'it is these behavioral patterns which must be recognized, understood, and changed' (hooks 1984: 3).

This 'internal conversation', as Archer would call it, is central to reflexive sociology, which seeks to make transparent the opaque working practices and assumptions involved in producing knowledge claims. Gouldner advocates a reflexive sociology that deepens 'our understanding of our own sociological selves and of our position in the world' (1970: 512) by encouraging us to offer critiques that are based in awareness of the subjective and personal circumstances of ourselves as intellectuals (Lemert and Piccone 1982: 733). Thus, a reflexive approach invites us to accept and make clear that the knowledge we produce is rooted in our own perspectives. We can see such an approach in Bauman's work as he views the work of the sociologist as entailing 'Piercing through the "curtain of prejudgements" to set in motion the endless labour of reinterpretation, opening for scrutiny the human-made and human-making world' (Jacobsen and Tester 2011: 315). Leaving aside the anthropocentric focus in Bauman's work, it is clear that such a reflexive approach is essential to CAS. For example, in relation to sociology and animals, David Nibert suggests that:

Members of the discipline [sociology], who like most other humans in society partake in the privileges derived from entangled oppressions – such as eating and drinking substances derived from the bodies of 'others', wearing their skin and hair, and enjoying the entertainment value their exploitation provides – can do so only by accepting the self-interested real-ities crafted by powerful agribusiness, pharmaceutical and other industries that rely on public acquiescence in oppressive social arrangements.

(2003: 20–21)

CAS writers seek to challenge rather than partake in these entangled oppressions and it is the reflexive approach that can turn us to CAS. This can be seen in Erica Cudworth's autobiographical piece in her book *Social Lives with Other Animals: Tales of Sex, Death and Love* (2011). In this piece, she considers the implica-tions of her relationships with nonhuman animals in her childhood and how these relationships were formative in her critical approach. For example, she reveals that:

> My Grandpa refused to eat meat from when he was a young child.... Everyone else I knew ate meat, and most of my family and family friends liked, or lived with, dogs. Who do we love, and what do we eat? This was the first time the boundaries of species presented themselves as a problem-atic question for me, as I chewed incessantly on the glob of ultra fibrous flesh in my mouth.

(Cudworth 2011: 4–5)

This reflexive approach, that is central to writing for animals, brings the lives of the writers of CAS to centre-stage (for example in disclosures about personal veganism, animal advocacy and activism) and in this way CAS writers are not indistinct figures; rather, through autobiographical disclosures, they move from the margins to the centre of their work.

Concluding remarks

In this chapter – by drawing on hooks' critique of feminism – I have sought to consider the complexity of notions of centrality and marginality in respect of CAS. CAS seeks to bring to the centre the lived experiences of billions of oppressed nonhuman animals. CAS has done much to reveal the material experi-ences of nonhuman animals that result from the systematic imbalances in power relations between humans and nonhuman animals, which disadvantages non-human animals. Those enslaved in 'sites of harm' and others who are kept as 'pets' or for sport, are at the centre of CAS. It is CAS that has centred on this marginalized oppression. As a branch of Animal Studies, CAS has been less suc-cessful than its more traditional counterpart in moving from the margin to the centre in more general academic pursuits. Although this means that CAS is still at the margins, it has retained its commitment to activism and advocacy that is central to its work. Nevertheless, CAS has sought to compel Animal Studies to engage with a critical approach that draws on political movement. Should this prove successful, the steps forward that Animal Studies is making in the main-stream could provide an opportunity for CAS.

I have argued that in tandem with the bringing of the material consequences for nonhuman animals in their relations with humans from the margin to the centre, there is a need for the deconstructed and reflexive human to move from the margin to the centre in such studies. CAS is a reflexive approach, but it cannot rest easy. As McCance (2013) suggests, the critical in CAS also requires a self-critical approach. This is important if CAS is to progress. Let me take two examples. First, in what we might see as a critical approach to CAS, Stanescu and Twine argue that CAS should move towards ending the distinction between 'the 'so called' human and the 'so called animal', as this distinction is anthropocentric (2012: 19). Among other things, this entails confronting the application of hierarchy to difference, the homogenization of a multitude of species under the term 'animal', and the target-ing of the notion of the 'human' as distinct from other species.[5] To my mind this is essential and something I try to do myself (e.g. Peggs 2009).

However, there seems to be no consensus among CAS writers about who is an animal. If McCance (2013) is right in her suggestion that 'not all critical animal theorists would concede animal status to a primitive fish', then before an end to the distinction between 'human' and 'animal' can be sought CAS should be clear about what it considers an 'animal' to be. If 'animal' is being used in an exclusionary way by some CAS theorists, then serious implications for scholar-ship, advocacy, and activism follow. An end to the distinctions of 'human' and 'animal' could serve to remove the focus on the definition of 'animal' and thus obscure, rather than illuminate, exclusions in conceptualizations of 'animal'. Second, CAS writers often reveal their ethical commitments by, for example, making clear that they take the ethical and political stance of vegan. I feel com-pelled to say this myself, but I am mindful of hooks' (1984) suggestion that identifying ourselves in such ways can hinder rather than help political commit-ment. In relation to feminism, hooks suggests that, in order to emphasize engage-ment with political commitment we:

> could avoid using the phrase 'I am a feminist' (linguistic structure designed to refer to some personal aspect of identity and self-determination) and could state 'I advocate feminism'... because.... The shift in expression from 'I am a feminist' to 'I advocate feminism' could serve as a useful strategy for eliminating the focus on identity and lifestyle.
>
> (1984: 29–30)

This focus on the difference between lifestyle and political engagement in fem-inism resonates with current questions about lifestyle and ethical approaches to veganism – perhaps hooks provides the basis for a way forward.

The positioning of CAS and its focus on the material conditions of oppressed nonhuman animals is a complex one when it is considered in relation to marginality and centrality. It seems to be a moot point about whether it is desirable for CAS to move from the margin to the centre, as there are concerns that such a move could result in the loss of the critical from the approach. This is because the critical is vital to the field of CAS. This critical stance means that nonhuman animals and the oppression of nonhuman animals will remain central to the field. However, it also means that the other animals, us humans, must come back to the centre as well – not as assumed and invisible creatures but as exposed reflexive creatures, who focus critically on our own oppressive relations with nonhuman animals. This is because CAS is central to the challenge of the marginalization of the oppressive human.

Notes

1 Of course, such terrible abuse is not an obvious feature of all human–nonhuman animal relationships. For example, nonhuman animals who are treated as companions and who are loved and cared for do not seem to fit the relations of oppressor–oppressed. Nevertheless, cared for nonhuman animals like these are 'owned' and do not have the autonomy of free-living nonhuman animals. For discussion see, for example, Arluke and Sanders (1996).
2 This figure, based on weight, is a problematic and revealing way of expressing a proportion that is about nonhuman animals who are individuals.
3 This has been the preferred term for Animal Studies of the Animals and Society Institute.
4 For example, Marc Bekoff (2002) speaks of 'minding animals' in the sense that we should embrace an ethics of care for nonhuman animals and that we should appreciate the agency, emotionality, etc. of nonhuman animals.
5 Editor's communication.

References

Archer, M. S. (2010) 'Introduction: the reflexive return', in M. S. Archer, *Conversations About Reflexivity: Ontological Explorations*, Abingdon, Oxon: Routledge.

Arluke, A. (2002) 'A sociology of sociological animal studies,' *Society & Animals*, 10(4): 369–374.

Arluke, A. and Sanders, C. (1996) *Regarding Animals*, Philadelphia: Temple University.

Baker, S. (1993) *Picturing the Beast: Animals, Identity and Representation*, Champaign: University of Illinois Press.

Bauman, Z. (2008) 'Bauman on Bauman – pro domo sua', in M. Hviid Jacobsen and P. Poder (eds), *The Sociology of Zygmunt Bauman: Challenges and Critique*, Aldershot: Ashgate.

Beirne, P. (1999) 'For a nonspeceisist criminology: animal abuse as an object of study', *Criminology*, 37(1): 117–147.

Beirne, P. (2002) 'Criminology and animal studies: a sociological view', *Society & Animals*, (10)4: 381–386.

Bekoff, M. (2002) *Minding Animals: Awareness, Emotion and Heart*, Oxford: Oxford University Press.

Benton, T. (1993) *Natural Relations, Ecology, Social Justice and Animal Rights*, London: Verso.

Benton, T. (1998) 'Rights and justice on a shared planet: more rights or new relations?', *Theoretical Criminology*, 2(2): 149–175.

Berger, J. (2009) *Why Look at Animals*, London: Penguin.

Best, S. (2009) 'The rise of critical animal studies: putting theory into action and animal liberation into higher education', *State of Nature*, not paginated.

Blaney, D. P. (2002) *The Changing Landscape of U.S. Milk Production. Statistical Bulletin Number 978*, Washington, DC: US Department of Agriculture.

Brown, N. and Michael, M. (2001) 'Switching between science and culture in transpecies transplantation', *Science, Technology and Human Value, 1* (Winter 2001): 3–22.

Bryant, C. D. (1979) 'The zoological connection: animal-related human behavior', *Social Forces*, 58(2): 399–421.

Bryman, A. (2008) *Social Research Methods*, 3rd edn, Oxford: Oxford University Press.

Burawoy, M. (2005) 'For public sociology', *American Sociological Review*, 70(1): 4–28.

Carter, B. and Charles, N. (2011) 'Power, agency and a different future', in B. Carter and N. Charles (eds), *Human and Other Animals: Critical Perspectives*, Houndsmills, Basingstoke: Palgrave Macmillan.

Cassuto, D. N. (2007) 'Bred meat: the cultural foundation of the factory farm,' *Law and Contemporary Problems*, 70(1): 59–87.

Commission of the European Communities (2007) *Fifth Statistical Report on Laboratory Animals*, Brussels: Council of European Communities Publications Office.

Cudworth, E. (2011) *Social Lives with Other Animals: Tales of Sex, Death and Love*, Basingstoke: Palgrave Macmillan.

Donaldson, S. and Kymlicka, W. (2011) *Zoopolis: A Political Theory of Animal Rights*, Oxford: Oxford University Press.

Donovan, J. and Adams, C. (1995) 'Introduction', in J. Donovan and C. J. Adams (eds), *Animals and Women: Feminist Theorectical Exploration*, Durham, North Carolina: Duke University Press.

Elder, G. W. (1998) 'Race, place, and the bounds of humanity', *Society & Animals*, 6(2): 183–202.

Emel, J. and Wolch, J. (1998) 'Preface', in J. Wolch and J. Emel (eds), *Animal Geographies: Place, Politics, and Identity in the Nature-Culture Borderlands*, London: Verso.

Flynn, C. P. (2008) 'A sociological analysis of animal abuse', in F. R. Ascione (ed.), *The International Handbook on Animal Abuse and Cruelty: Theory, Research and Application*, Indiana: Purdue University Press.

Francione, G. L. (2004) 'Animals – Property or Persons?', in C. R. Sustein (ed.), *Animal Rights*, Oxford: Oxford University Press.

Gouldner, A. (1970) *The Coming Crisis of Western Sociology*, New York: Basic Books.

Gouldner, A. W. (1975) *For Sociology: Renewal and Critique in Sociology Today*, Harmondsworth: Penguin.

Hammersley, M. (1998) 'Sociology, what's it for? a critique of Gouldner,' *Sociological Research Online*, 4(4), available at: www.socresonline.org.uk/socresonline/4/3hammersley.html. No pagination (last accessed 3 January 2014).

Haraway, D. (1991) *Simians, Cyborgs, and Women: The Reinvention of Nature*, London: Free Association Books.

Hollands, R. and Stanley, L. (2009) 'Rethinking 'current crisis' arguments: Gouldner and the legacy of critical sociology', *Sociological Research Online*, 14(1), available at: www.socresonline.org.uk/14/1/1.html, No pagination (last accessed 3 January 2014).

Home Office (2011) *Statistics of Scientific Procedures on Living Animals: Great Britain 2010*, London: The Stationary Office.

hooks, b. (1984) *Feminist Theory from Margin to Centre*, Boston: South End Press.

Humane Society of the United States (2002) *Factory Farming In America: The True Cost of the Animal Agribusiness for Rural Communities, Public Health, Families, Farmers, the Environment, and Animals.* Washington, DC: HSUS.

Irvine, L. (2007) 'The question of animal selves: implications for sociological knowledge and practice', *Qualitative Sociology Review*, 3(1): 5–22.

Jacobsen, M. H. and Tester, K. (2011) 'Poet-intellectual and public sociologist: an interview with Zygmunt Bauman', *Academic Quarter*, 2 (Spring): 303–318.

Knight, A. (2011) *The Costs and Benefits of Animal Experiments*, Basingstoke: Palgrave.

Kruse, C. R. (2002) 'Social animals: animal studies and sociology', *Society & Animals*, 10(4): 375–379.

Lemert, C. and Piccone, P. (1982) 'Gouldner's theoretical method and reflexive sociology', *Theory and Society*, 11(6): 733–757.

McCance, D. (2013) *Critical Animal Studies*, Albany: SUNY Press.

Mason, J. and Finelli, M. (2006) 'Brave New Farm?', in P. Singer (ed.), *The Defense of Animals: The Second Wave*, Malden, Oxford: Carlton Blackwell.

Meyer, J. (2009) 'Representing the experimental animal: competing voices in Victorian culture', in S. C. McFarland (ed.), *Animals and Agency*, Leiden, the Netherlands: Brill.

Nibert, D. (2002) *Animal Rights/Human Rights: Entanglement of Oppression and Liberation*, Plymouth: Rowman and Littlefield.

Nibert, D. (2003) 'Humans and other animals: sociology's moral and intellectual challenge', *International Journal of Sociology and Social Policy*, 23(3): 5–25.

Paxton George, K. (1994) 'Should feminists be vegetarians?', *Signs*, 19(2): 405–434.

Peggs, K. (2009) 'A hostile world for nonhuman animals: human identification and the oppression of nonhuman animals for human good', *Sociology*, 43(1): 85–102.

Peggs, K. (2012) *Animals and Sociology*, Basingstoke: Palgrave Macmillan.

Peggs, K. (2010) 'Nonhuman animal experiments in the European Community: human values and rational choice', *Society and Animals*, 18(1): 1–20.

Peggs, K. (2013) 'The "animal-advocacy agenda": exploring sociology for nonhuman animals', *The Sociological Review*, 61(3): 592–606.

Philo, C. and Wilbert, C. (2000) 'Animal spaces, beasty places: an introduction', in C. Philo and C. Wilbert (eds), *Animal Spaces, Beasty Places: New Geographies of Human Animal Relations*, London: Routledge.

Regan, T. (2004) *Empty Cages: Facing the Challenge of Animal Rights*, Lanham, MD: Rowman and Littlefield.

Rollin, B. (2008) 'The ethics of agriculture: the end of true husbandry', in M. S. Dawkins (ed.), *The Future of Animal Farming: Renewing the Ancient Contract*. Oxford: Blackwell, pp. 7–19.

Rowlands, M. (2002) *Animals Like Us*, London: Verso.

Shapiro, K. (2002) 'Editor's introduction: the state of human-animal studies: solid, at the margin!', *Society & Animals*, 10(4): 331–337.

Singer, P. (1990 [orig. 1976]) *Animal Liberation*, 2nd edn, New York: New York Review of Books.

Smith, D. E. (1987) *The Everyday World As Problematic*, Boston: Northeastern University Press.

Stanescu, V. and Twine, R. (2012) 'Post animal studies: the future(s) of critical animal studies', *Journal for Critical Animal Studies*, 10(4): 4–19.

Stanley, L. (2000) 'For sociology, Gouldner's and ours', in J. Elderidge (ed.), *For Sociology: Legacies and Prospects*, Durham: Sociology Press.

Stibbe, A. (2001) 'Language, power and the social construction of animals', *Society & Animals*, 9(2): 145–161.

Taylor, N. and Signal, T. (2008) 'Throwing the Baby out with the Bathwater: Towards a Sociology of the Human-Animal Abuse 'Link'?', *Sociological Research Online*, 13(1), available at: www.socresonline.org.uk/13/1/2.html (last accessed 3 January 2014).

Taylor, K., Gordon, N., Langley, G., and Higgins, W. (2008) 'Estimates of worldwide laboratory animal use in 2005', *Alternto Labouratory Animals*, 36: 327–342

Torres, B. (2007) *Making a Killing: The Political Economy of Animal Rights*, Oakland, CA: AK Press.

Twine, R. (2010) *Animals as Biotechnology – Ethics, Sustainability and Critical Animal Studies*, London: Earthscan.

Wilkie, R. M. (2010) *Livestock/Deadstock: Food Animals, Ambiguous Relations, and Productive Contexts: Working with Farm Animals from Birth to Slaughter*, Philadelphia: Temple University Press.

3   Vegans on the verge of a nervous breakdown

Sarah Salih

I

'We cannot know his legendary head' begins the first stanza of Rilke's 'Archaic Torso of Apollo.' The 'eyes like ripening fruit' are missing from the statue, but the poem ends with the extraordinary warning that you are seen by every place – in other words, every*thing* is seeing you. You cannot know, and yet you must change your life.

This is an essay about what you need to know, or think you need to know, in order to do this, the type of knowledge that's needed for change. An unwieldy question surfaces at once: What is knowledge? Is it, as Rilke's poem implies, the result of seeing, the physical perception of an actual object or person – I see, therefore I know?[1] If you imagine what's absent (here, the legendary head with its ripening eyes) are you 'knowing' in a sense? How do you know when you're knowing? And if you don't change your life after knowing, does it mean you haven't really known, that the knowledge wasn't 'true' in some sense? Again, how do you know you haven't known, or that you've known unknowingly?

Like Rilke, I'm interested in pairs of absent eyes, in bodies we can't 'know' because we have limited access to them physically and ontologically. I want to think about why some people feel themselves to be accountable to these bodies all the same while others do not, even though this latter group might well be in possession of various forms of knowledge – images, statistics, narratives of pain, suffering and death. It would seem that for some, knowledge is a moral burden and an ethical impetus for change, while for others it isn't, and I've spent a long time wondering and worrying about why this is so. No doubt my concern is linked to a long-standing doubt about what it is that those of us who dwell and work in the academy are doing as knowledge-producers, particularly when the subjects (objects?) out of whose existences we generate our books, articles and university courses live unimaginably horrible lives and die painful, undignified deaths.[2] It's easy for us to be dazzled by our subjects or 'texts' as we might think of them, dazzling in the ways we write about or present them, and we may call this 'activism' and think it's sufficient. That's why I always feel winded when I read those last lines of Rilke's 'Archaic Torso.' You believe you're just an innocent bystander admiring a gorgeous old statue and then the poem rounds on you (perhaps it too, is one of those 'places' that's seeing?) and issues the order for you to change your life.

In part this is a matter of 'what we do with what we know', a question to which ethologist Marc Bekoff dedicates a whole chapter in his book about animal emotion. Bekoff is someone who believes that knowledge makes a difference: he assumes that when you've learned certain things about nonhumans – for example, the particular ways in which they suffer and experience pain – you'll be obliged to adapt your behaviour accordingly.[3] '[O]nce we agree that animal emotions exist and that they matter ... then what?' asks Bekoff. 'Then we must consider ethics. We must look to our actions and see if they are consistent with our knowledge and beliefs' (Bekoff 2007: 5). This, he says, is a form of Socratic ethics, which requires us to ask how we ought to live, to reappraise who we are, what we do and what we want to be (ibid.: 135).

I will take up this ethical paradigm shortly, but first let's think about where Bekoff's confident 'must' comes from and what sort of power – if any – it might have with regard to his readers' actual practice. To state the obvious, Rilke's and Bekoff's resounding 'musts' are rhetorical: they don't emanate from legal statute and thus have no performative power at all. In the present day, eating animals who have been raised and killed in industrial facilities is entirely legal, and it's not recognized as a crime by most people, although many – including Coetzee's Elizabeth Costello – might believe that it is one. If you're Elizabeth Costello, you may feel you're under a moral obligation to change once you know that an ethical (as opposed to juridical) crime of unimaginable proportions is being committed every day and everywhere in the industrialized West, but it's also the case that such knowledge doesn't in fact oblige you in any straightforward sense to change.[4] Accordingly many people 'know' but don't alter their practices one whit: I think for example of my brother-in-law living in a household of proselytizing vegetarians but staunchly eating meat for lunch at work every day; or my friend the Philosophy professor, who accepts that the Levinasian injunction 'Thou shalt not kill' should be extended to nonhuman animals, but declares himself unable to give up the pleasures of steak and cheese. This is just random anecdotal 'evidence' of course, and yet I'm sure there are scores of good people who are moved when they read Peter Singer's *Animal Liberation* or Jonathan Safran Foer's *Eating Animals* (2009), but somehow find that this information makes little or no lasting difference to their habits of consumption. You might look, you might know, you might feel you need to change your life, and yet still you fail to do so.

If we knew, would we change?, asks Safran Foer towards the end of his stark exposé of the meat industry. His conclusion is gloomy: 'There are some things ... we don't need labels to know,' he says, indeed, we *already* know, which means we're the generation of whom it will fairly be asked: What did you do when you learned the truth about eating animals? (Safran Foer 2009: 193, 252). Re-reading Safran Foer's prediction, I'm reminded of the World War I poster that used to hang on the wall of the history classroom in my secondary school.

A little girl sits on her father's lap as she points at a picture in an open book; a boy plays with tin soldiers and some kind of ancient army vehicle on the floor. The girl is turning towards her father, and whatever she's saying has made him look troubled and uncomfortable. *'Daddy, what did YOU do in the Great War?'* reads the caption beneath this image of cowardice and family disharmony. We are to assume that daddy did nothing for the war effort, or else why would he wear such a guilty expression?

The poster was produced in 1915 and, like Safran Foer's comment, it asks us (in a rather manipulative fashion, it has to be said) imaginatively to anticipate how we'll feel when the next generation holds us accountable for what we did or didn't do in the face of our knowledge.[5] You *knew* a war was going on, the little girl's open book and her accusing finger imply. Surely you did something about it? How could you not – indeed, why would I assume otherwise? (note that the caption says *'What* did you do?' rather than 'Did you do anything daddy?'). We're waging war on animals, Safran Foer declares after citing how many billions of birds are killed by the US industrial meat machine every year. The question Safran Foer wishes us to ask (and it is a flawed question, for all sorts of reasons, not least that it assumes future generations will have a clearer ethical perspective than we do) is: how will you answer when your son or daughter demands to know what you did about it? (ibid.: 33).[6]

It's telling that the Great War poster and Safran Foer's book deploy the same rhetorical technique. By evoking the viewer/reader's future sense of moral burden – guilt, really, and perhaps also shame – they emphatically draw our attention to a wrong that's being committed in the present. You might not regard your current actions as morally deficient, these texts suggest, but hindsight will show you the degree to which you were knowingly or unknowingly engaged in an enormous collective immorality. Apparently, it isn't easy to discern right and wrong in the present, or else why would we need to be so proleptic about 'what it is right to do', to hold ourselves accountable even if only in our fantasies, to our accusing, ultra-moral children?[7] And even if you do have a clear, in-the-moment Socratic or Kantian sense of what's right to do – fight in the Great War, stop eating meat – you might continue to act in a way that runs counter to your moral sense. This is again to state the obvious perhaps, and yet much of the animal studies discourse produced in the last decade or so seems to be predicated on the notion that to arm the reader with knowledge in the form of information is an effective step towards necessary change. Accordingly, books from Peter Singer's *Animal Liberation* to the aforementioned *Eating Animals* provide uncompromising accounts of precisely what it is that we do to nonhumans in industrial facilities, and these books also draw on scientific studies to detail what nonhumans think and feel about being used in this way. The 'conscientious omnivore', Michael Pollan, read *Animal Liberation* while eating a steak in a steakhouse, and even he was put off his meal (although apparently he managed to finish it). When confronted with the kind of information Singer presents, Pollan writes in the 2002 *New York Times* article in which he recounts the experience, you can look away or stop eating animals – or like him, you can adopt the practice of a 'humanocarnivore', which involves making the effort to meet your meat, as the PETA (People for the Ethical treatment of Animals) film has it, in order to ensure that the animal whose flesh you're eating or whose milk you're drinking has been 'humanely' handled (Pollan 2002: 111).[8]

Pollan's thoughtfulness – indeed, his conscientiousness – are laudable, although I think he's mistaken in the conclusions he ultimately reaches as regards his own (and others') consumption practices.[9] Still, he does make one rather throw-away comment that rings troublingly true: 'We certainly won't philosophize ourselves to an answer,' he writes at the end of a description of a farm where 'humane slaughter' is practised. At this point, the question being posed is unclear, but I'm struck by Pollan's suggestion that philosophizing will only take you so far as regards what you do. If we assume for a moment that 'the animal' is a problem, then you could say that it isn't really a problem of philosophy or epistemology (philosophy after all, means 'love of knowledge'), since as I've already suggested, there's no necessary connection between knowledge, or even love, and ethical practice. If the animal is a problem, it would seem that it's primarily a problem of how we understand and perceive the world and all the beings in it. Accordingly, I want to spend some time thinking about how breaks in perception – perhaps even break-*downs* – are more likely than abstract knowledge to precipitate lasting physical and psychological re-orientations in relation to 'the more than human' world (Abram 1997: 22).

II

Carnivores vastly outnumber vegetarians – there seem to be roughly 97 meateaters to every three vegetarians in the US and 98 to two in the UK – which means, of course, that there are many different varieties of meat eater.[10] Among these, there are two types with whom I'm most accustomed to arguing. First, there are those who claim that animals are 'lower' than humans on some kind of scale or great chain of being. For this group, animals are insentient, unintelligent, with no sense of futurity, in short, more 'thing' than 'person'. As a consequence, it's fine for humans to use animals in whatever ways they see fit, whether as food, in experiments, as labourers and so on, and indeed, some of these 'Type A' carnivores as we might dub them, will cite scripture to 'prove' to you that God created animals for this very purpose. A second, perhaps these days more numerous, group (call them 'Type B'), accepts that animals are sentient, suffer, feel pain, form bonds and would undoubtedly rather not give up their lives prematurely just because humans have decided they should. Nonetheless, Type B enlightened carnivores still continue to eat meat, dairy and eggs, mainly because they enjoy doing so but also they say, because humans are biologically, evolutionarily designed for such a diet and will fall ill from lack of protein, iron, calcium and other nutrients if they adopt a herbivorous diet.[11] Type B carnivores are more likely than Type As to engage in Pollan's conscientious omnivorism: they'll often seek out 'ethical' meat and dairy products, wherever possible avoiding those which have been intensively farmed – although if push came to shove and they found themselves in a steakhouse confronted by a piece of 'meat' of uncertain provenance, like Pollan they'd probably finish their meal.

What I've loosely called 'animal studies discourse' – by which I mean the body of ethically-engaged, animal-oriented writing produced in the three-and-a-half decades since the appearance of Singer's *Animal Liberation* – has largely assumed its rhetorical shape from these two types of carnivorism and their accompanying justifications. This makes sense if you reflect that animal studies discourse in general and Critical Animal Studies discourse in particular, is dialectical in its very nature. Much of it is responsive to pre-existing, predominant discourses, and although Critical Animal Studies is not of necessity activist, many of the thinkers and writers who locate their work within this field are seeking to bring about some kind of change through the circulation of ideas: most of the time, we are not engaging in ethical reflection for the pure philosophical joy of it ('love of knowledge' again), but in order to bring about tangible change in our readers' attitudes and practices. Again, there are many types of animal studies discourse, but two are likely to predominate given the most common justifications for carnivorism. There are those texts which explicitly counter what Singer would call the speciesist assumptions of Type A carnivores (i.e. the 'animals are inferior' argument), by carefully setting out to demonstrate the myriad ways in which such notions are false. Usually this involves lining up scientific as well as anecdotal evidence of animal sentience, cognition, emotion, sociality, morality and so on, and the literature in this area is growing as cognitive ethology becomes more sophisticated in its technologies and techniques. Additionally, and quite often in the same text, many animal studies authors set out to horrify Type B carnivores into changing their consumption practices by presenting them with evidence of the collective and massive harms in which they're engaged when they consume animal products and by-products. This evidence usually takes the form of statistics (numbers of animals killed per day, per hour, per minute, in the time it took you to read this sentence, and so on), along with graphic accounts and/or images of what is done to animals in intensive 'farm' installations or CAFOs (concentrated animal feeding operations).[12]

Both these loosely-defined approaches provide readers/viewers with knowledge so that they'll simultaneously change their minds and their consumption practices, with the implicit underlying hope that readers who are exposed to new, more 'truthful' knowledge about what it means to use animals for food, clothes, soap etc., will be prompted into a radical reshaping of their attitudes and lives. Many carnivores will need to be reassured that it's possible to deviate from the current norm (i.e. meat-eating) without suffering from protein deficiency or dying of malnutrition, and with Type A carnivores, it's frequently necessary to start at an even more basic level than that, laying the groundwork by pointing out that our consumption practices are not in fact neutral or 'normal' – they're ethical choices. So, for example, Peter Singer and Jim Mason begin their book *The Ethics of What We Eat* with the observation that although until now people haven't typically thought of their food choices as ethical, an increasing number of consumers are adopting alternative modes of consumption as an explicitly political action (Singer and Mason 2006: 3, 5).[13] The authors also point out that the US food industry spends $11 billion annually in its attempts to induce consumers to crave their products, whereas said industry has a concomitant vested interest in keeping consumers ignorant about the ethical components of their food choices (ibid.: 7). In the US especially, you're not supposed to inquire too closely into the visible and invisible components of your food, because if you did, you probably wouldn't want to eat it. This is Safran Foer's wager, and it's also the thrust of Marion Nestle's uncompromising account of the powerful stranglehold that corporate US food producers maintain over the information that's currently disseminated about food and its relation to 'health'.[14] As far as I know, Nestle is not a vegetarian, yet throughout *The Politics of Food* she repeatedly emphasizes that advice about the health benefits of a plant-based diet has not changed in more than 50 years and is consistently supported by scientific research. In spite of this, Nestle asserts, Americans are increasingly confused about their food choices because the aggressive tactics of food-producing corporations dominate and distort what consumers are allowed to know about nutrition (Nestle 2007: xiii, 6, 7, 366, xviii). As Nestle herself states it, the central thesis of her book is that diet is a political issue, whether or not consumers realize the extent to which they're manipulated by corporations that work hard to get people to eat more of the wrong foods. Market competition is indeed fierce, since the US food industry produces twice as many calories as its citizens need to eat per day. '[O]ne of the great unspoken secrets about the American food system: overabundance', writes Nestle, whereas the 'eat less' message of those who are truly concerned about the nation's health is almost entirely submerged in the flood of misinformation and marketing (Nestle 2007: 1, 13).[15]

'The industry' as it's portrayed in Nestle's book, has such a powerful grip on the nutritional information available to US citizens that one is left wondering how on earth it's possible to make an ethical and informed choice regarding what to eat. Indeed, Nestle concludes, you might believe you're making informed choices about food, but actually it's impossible to do so if you remain oblivious of the ways in which the food industry subtly and less subtly influences your decisions (Nestle 2007: 360). Like Singer and Mason, Nestle believes your food choices will be more ethical (as well as healthier) only if you seek out knowledge about the ways you're currently manipulated by the food industry: indeed, she claims, the overabundant food system and the flourishing economy which give rise to the industry's strong-arm tactics, also mean that most of us are able to make a political statement every time we eat (ibid.: 372).[16]

Whereas Nestle focuses on the nefarious practices adopted by the corporate US food industry to ensure that the nutritional reputation of their product (meat, dairy, eggs and so on) remains untarnished by unwelcome scientific data and/or health warnings, Singer and Mason are concerned with ethics as much as with health.[17] To know about the life that was led by the factory-farmed animal and to continue to consume factory-farmed meat is, they imply, a failure to lead an eth-ically good life. Once you *know*, you can't plead ignorance, you can't un-know; and yet, as I've already suggested, it's clear that for many people the mere fact of knowledge doesn't necessarily lead to a lasting shift in perception or practice. Knowledge on its own is not enough. It's possible to forget, to put what you know aside, to compartmentalize. So you might read Singer's account of the miserable, degraded lives led by intensively farmed dairy cows, but then you're confronted with a tempting hunk of cheese and somehow Peter Singer and his bovines fade to the background for now. As you're savouring that piece of cheese, you won't be thinking about the tiny stall in which the cow was con-fined, her painful life of repeated artificial insemination and continuous lactation, her abrupt separation from her calf, the degraded death she'll eventually be made to die. You'll just be thinking about cheese, and if you're a Type B carnivore you might promise yourself that you'll fail better next time.

I'm not suggesting that knowledge can't prompt us to change our lives – indeed, I think in many cases it does – but I wonder whether the peculiarly complex and powerful nature of eating necessitates a different approach or set of approaches to the ones we've been adopting thus far (i.e. presenting carnivores of both 'types' with scientific data, statistics, horrible photos and so on). Eating is bound up with habit, with who we think we are, our place in the world. Our culture, gender, sexual orientation, social status, familial positioning – all, to a degree, hinge on what we eat, how and with whom, and what we eat also hinges on these things. For many humans, eating is also a matter of desire – I *want* this piece of cheese, regardless of what I know about its history – so that to try to persuade you to eat otherwise is to come up against an element of your libido which, unlike sexual desire, has remained largely unregulated and unchallenged, unthought, even.[18] An apparently innate sense of disgust might prevent you from eating certain kinds of flesh (e.g. that of your recently-deceased grandfather or your dog), but as far as I know, there are no laws preventing you from indulging your tastes should you feel like having baked dog for dinner.[19] In the face of such strong urges, your ability to recall the data you've come across might well be rather feeble. Yes, you know the cow suffered, but your tongue, your saliva, your stomach, are all telling you to bite into that piece of intensively-farmed cheese. How then, do you give up on your desire, to reverse Lacan's famous dictum?[20] How do you narrow the gap between what you know is right (if you're Type B) and what you think you want? Is it pos-sible to effect a synaptic shift so that if I offered you a hunk of cheese, you'd recoil – just as if I'd offered you a hunk of my beloved golden retriever, who died last week?[21] Clearly, what's needed is a break in perception, yet most of us are so attached to our tastes and our eating habits that we don't believe it's possible for this to occur. Perhaps this is why vegans are accustomed to hearing carnivores say 'But I *couldn't* live without cheese/milk/eggs/chicken, etc.; I just love them too much!' Little attention has been paid to the shifts in perception that occur when one moves from a carnivorous to a vegan way of life, and so I want to consider some of the perceptual processes and 'breaks', which are likely to occur during the process of changing your life in this particular way.

III

Leave your culture, gender, social status, orientation and so on to one side for now, and think carefully about your gastronomic leanings, the types of food you think you want to eat. Taste, and the pleasure that often accompanies taste, can be difficult to pin down. 'What part of us feels pleasure?' asks Pascal in his *Pensées*: 'Is it our hand, our arm, our flesh, our blood? It must obviously be something immaterial' (Pascal 1995: 108). Since as Pascal hints, pleasure and taste aren't obviously located anywhere on or in the body, one might indeed be led to conclude that they're all in the mind. Malcolm Gladwell suggests as much in his popular *Blink* (2005), a study of 'the power of thinking without thinking'. Using our sense of taste as an example, Gladwell demonstrates the startling extent to which we've lost our ability to know our minds when it comes to food and other preferences (e.g. what kind of furniture or music we like). From Gladwell's account, it emerges that our tastes in food are indeed as much psychological as physiological, and also that they're highly unstable. For one thing, rather than experiencing taste directly, we're susceptible to so-called 'sensation transference', a groundbreaking marketing insight discovered and pioneered by Ukrainian-born Louis Cheskin in the mid-twentieth century. Cheskin found that people transfer sensations or impressions about a product's packaging to the product itself. As Gladwell summarizes it: 'Cheskin believed that most of us don't make a distinction – on an unconscious level – between the package and the product. The product is the package and the product combined.' So, for example, if product designers add an image of a tiny sprig of parsley to the logo on a tin of meat, consumers are more likely to believe that the meat is 'fresh' – and presumably to taste it as such (Gladwell 2005: 160, 164). We deposit our unconscious associations onto the food we buy and eat (especially if it's packaged), so that, as Gladwell puts it, in the real world, there's no such thing as blind tasting – or you might say, there's no such thing as 'pure taste'. When you're drinking (for example) Coca-Cola, you automatically load onto your sensation of drinking all the unconscious associations you have of the brand, the image, the can and the colour of the logo. Your taste buds and your ideas are sutured to one another to such a degree that it's difficult to discern how you're *actually* experiencing the fizzy substance in the can (ibid.: 166). Only 'expert tasters' – those who are professionally trained to account for their food preferences – can reliably account for their reactions to food: the rest of us have lost this ability, and our reactions are accordingly shallow and unstable. Change the colour of the logo, modify the image, and most of us will think we prefer the taste of *this* product over *that* one (ibid.: 176, 181, 184).

Gladwell seems to believe that we have 'true preferences', and that marketing and the injunction to account for what we find delicious have messed up our underlying sense of what we really like to eat and drink. On the other hand, you could argue that from infancy onwards, taste is *never* separate from context, whether what you're eating has been packaged and marketed or not (perhaps mummy's milk tastes nicer because she smiled at you the first time she breast-fed; and you might have unconscious associations about the 'healthiness' of brown rice, even though you buy it in bulk and have never seen an advert for brown rice on the television). Presumably, even expert tasters arrive at their dinner plates with unconscious associations and cultural assumptions about what food should taste like, even though they might be better than other people at focusing on the sensations occurring in their mouths. Although the notion of true preference could certainly be deployed by carnivores, who claim to have a 'real' taste for meat, again, you might draw the converse conclusion from *Blink*. If, as Gladwell suggests, we've 'lost the fundamental ability to know our mind' when it comes to food, then how do we know that our tastes are *ever* in fact our tastes? Do you really like cheese, or are you merely habituated to thinking you do for a variety of personal and less personal reasons (mummy's milk again, perhaps)? Our tastes are, Gladwell suggests, highly associative and unstable; you might be relishing the meat stew I give you at a dinner party until I inform you that what you've just eaten contains the flesh of my late golden retriever (Joy 2010: 11). In that case, your pleasure will rapidly be supplanted by disgust (unless you come from a dog-eating culture), which suggests that your *idea* of food to a large degree conditions how you taste it, that you can never entirely separate 'preference' from 'idea' and that pleasure may very easily be supplanted by aversion.

I'm sure the drift of my argument is clear enough. Michael Pollan would have enjoyed his steak much more if he hadn't brought along *Animal Liberation* as a dinner companion. On reading Singer's treatise, Pollan's idea of 'meat' began to shift, as did the experience of eating a food he'd hitherto enjoyed.[22] Eating is connected to cognition and, as Pollan suggests, people in the industrialized West currently suffer from a form of cognitive dissonance when it comes to food. This, says Pollan, is likely to render us subject to 'the compound madness of an impeccable industrial logic' – i.e. the lunacy of intensive 'farming' (Pollan 2011: 58, 62). Most of us don't read *Animal Liberation* while eating steak – although I wonder if this would be a good thing to advocate, along with raising and killing the cow yourself – so the circumstances of Pollan's conversion to conscientious omnivorism are unusual. Nor does Pollan assert that he loses the taste for meat once he knows how it's produced. He finishes his steak, after all, and sets out to find more 'moral' sources of flesh so he can satisfy his gastronomic desires and his conscience at the same time. Nonetheless, I think Pollan's paradigm is instructive, because it suggests that we're in possession of what philosophers such as David Hume and Adam Smith would call 'moral tastes'. These tend to cut two ways: first, our tastes are at least to some degree contingent upon what we think of as 'good' (can you enjoy that cup of coffee quite as much if I tell you the beans were harvested by underpaid, undernourished 10-year-olds?), and second, it may be possible to align our desires so that we only enjoy what's 'good' – in this case, food that's been produced with a minimum amount of harm.[23]

In other words, I'm suggesting that it's possible to experience a form of *moral* sensation transference, whereby your notions of what's good and bad attach themselves to what you eat and drink in much the same way as the colour of the Coca Cola logo influences how you taste the liquid in the can. You might think you like beef, cheese, etc., but if your tastes are as unsteady and shallow as Gladwell suggests, and if I sit you down and tell you the truth about eating animals as Safran Foer puts it, then maybe your preferences will begin to shift as one set of associations replaces another (images of happy cows frolicking in meadows *à la* butter adverts vs. the grim realities of intensive dairy production as graphically documented by Sue Coe and many others).

Let's return once more to Pollan in his steakhouse: what he experienced was a break in his perception because he had Singer's book on the table right next to his plate of meat. For whatever reason, Pollan was open to suspending what he thought he knew, to seeing the world in a different way – indeed, to *seeing* what he had previously overlooked. What results is, quite literally, a break with his former self. Who is Pollan now that he's understood something of what happens behind the closed doors of America's CAFOs and no longer wishes to consume the products of such cruelty? 'When you look, you begin to sound like one of those animal people', he remarks (Pollan: 62), as though the suspension of one identity (unthinking carnivore) necessitates the immediate taking up of another (animal people). Not favouring that particular ontology, Pollan instead creates another, reconstituting himself as a humanocarnivore or conscientious omnivore. The rapid substitution of one identity category for another is noteworthy in itself, but in spite of his brisk and rather patronizing rejection of animal rights discourse, one has to respect Pollan's willingness to break with himself, to make his life more difficult by acting upon his *sense* of what it is right to do and good to eat (presumably, he can no longer dine in any old steakhouse now). In other words, Pollan allows his feelings about intensive farming to move him into a different relation to animals, eating, humans and so on. Here, it is attention (Pollan's attention to Singer, to the meaning of the steak), which precipitates the break. Could it be that 97 or 98 out of 100 people simply fail to attend, or refuse to allow their attention to be moved in a particular direction? Of course, it's difficult to say with any certainty, but I *am* suggesting that vegans exercise a different form of attention to the majority of people, and as a result, perceive the world differently. Furthermore, unless they've been brought up to avoid meat from birth, most vegans will have undergone a break-through or break-down similar to the one Pollan describes. To be a vegan is to refuse to accept the end-product as it is presented to you (e.g. a piece of cake), but also to repeatedly break with other humans (refusing the cake) because you have broken products down into their constituent parts (cake=flour, milk, eggs; eggs=chickens; chickens=battery 'farms', battery farms=cruel; cruel=I don't want to eat the cake). Certainly, it may be exhausting to pay one's universe such scrupulous and unremitting attention, and the initial moment of break-down, i.e. the first move towards ethical consumption of whatever kind, is likely to be followed by many others. J. M. Coetzee captures this with characteristic starkness in his portrayal of Elizabeth Costello's acute cognitive dissonance and dismay. After a visit to Appleton College, where she has given a series of talks about the lives of animals, Costello breaks down as her son John is driving her to the airport. In tears, she articulates her sense of dislocation from her fellow-humans:

I seem to move around perfectly easily among people, to have perfectly normal relations with them. Is it possible, I ask myself, that all of them are participants in a crime of stupefying proportions? Am I fantasizing it all? I must be mad! Yet every day I see the evidences. The very people I suspect produce the evidence, exhibit it, offer it to me. Corpses, fragments of corpses that they have bought for money.

It is as if I were to visit friends, and to make some polite remark about the lamp in their living-room, and they were to say, 'Yes, it's nice isn't it? Polish-Jewish skin it's made of, we find that's best, the skins of young Polish-Jewish virgins.' And then I go to the bathroom and the soap-wrapper says, 'Treblinka – 100% human stearate.' Am I dreaming? I say to myself? What kind of house is this?

(Coetzee 1999: 69)

I've come to think of this as 'the Elizabeth Costello moment' – an interlude of break-down in self-certainty and ease with one's universe, which can occur at any moment (cycling behind a meat van, for example). In those moments, you seem to be at utter variance with other humans and you're indeed likely to ask yourself whether your perceptions are mad and fantastical, since everyone else seems to be entirely comfortable with what you regard as horrifying – criminal even.[24] It's telling that the scene in which Costello breaks down and admits her cognitive disorientation is placed at the end of *The Lives of Animals* (it concludes a section in *Elizabeth Costello*), since it leaves a perceptual chasm yawning at the text's close. Coetzee's reader must ponder for herself why it is that where others see 'food', Costello perceives evidence of a massive crime – corpses, fragments of corpses which have been purchased, in all good conscience she believes; these people don't seem to be evil, they are full of human kindness. Why does Costello view the world in this way? Why don't you?

For one thing, you might object to the direction in which Costello's fantasy escalates: indeed, Coetzee pre-empts his reader's outrage by having Abraham Stern, a fictional professor, refuse to break bread with Costello because of the analogy she draws between the slaughter of Jews in World War II and the slaughter of animals in the present (Coetzee 1999: 94). All the same, there's something compelling about the levels of specificity compressed into Costello's brief outburst, revealing as they do the degree to which her views are precisely that – a matter of perspective. What makes Costello's speech so powerful – and for many, so unacceptable – is that she moves rapidly from an allusion to a generalized 'crime', to the corpses this crime produced, before ruthlessly invoking Polish-Jewish victims, then Polish-Jewish virgins and, finally, Treblinka. This, 'the dreaded comparison' as one critic has called it in a different context, is in essence a way of knowing and perceiving the animal and the humans who consume them, and it's structurally similar to the parenthetical cake=bad break-down I gave above.[25] That Costello resorts to analogy – indeed, to *this kind* of analogy – is in itself noteworthy, for it conveys how far humans might be from regarding their food, their soap, their shoes, etc. as 'corpses, fragments of corpses.' In other words, there's the suggestion that for Costello, 'meat' itself is a kind of analogy or euphemism, which covers over its truth. And of course, Costello's comparison also reminds the reader that there are other ways of looking, seeing, imagining and representing, which depart quite radically from the norm.

To perceive the world as Costello does is to leave the norm behind, to marginalize yourself by insisting on looking at 'the evidence' in a way that most people don't.[26] Here, the question of the animal is indeed a question of perception as I've suggested. Costello doesn't merely know in the abstract about the wrongs done to animals in stupefying proportions; clearly she *feels* this, and her altered perceptions mean she must replace one set of symbols or signifiers (e.g. 'meat') for another ('corpse'). Costello refuses to accept a pre-given interpretation of her world, even though such a refusal leaves her isolated and confused; it is also a kind of perception that involves repeated efforts of imagination because one is not merely responding to what's immediate. You don't see the eggs in the cake, or the chicken who laid the eggs (or the Polish-Jewish virgins for that matter), so you must bring them to mind. Obviously, for Costello the author, imagination is a crucial mode of perceiving and being in the world. '[T]here is no limit to the extent to which we can think ourselves into the being of another', she declares in her first Appleton College lecture: 'there are no bounds to the sympathetic imagination' (Coetzee 1999: 80). Again, to engage in this kind of limitless sympathetic imagination you have to be willing to suspend your sense of self along with your assumptions about the world, to accept as in Rilke's poem, that what's absent ('the legendary head', which you cannot know) may nonetheless be exactly the thing that precipitates you into change.

IV

Breaking-down and breaking-with as I'm characterizing them, could be said to constitute a more ethical orientation towards the world, a form of knowing in unknowing. The psychoanalyst Jonathan Lear similarly sees in what he calls 'break', the possibility for new possibilities (Lear 1998: 118). Describing break as disruptions of primary process, a mental activity, which may be more or less meaningful, Lear also rejects 'the ethical' as a limit to the kinds of possibilities psychoanalysis seeks to open up. Instead, psychoanalytic practice might take advantage of what Lear calls 'lucky breaks' – those moments when the self-disrupting mind breaks down and opens itself to new possibilities. 'There is a rip in the world itself', writes Lear, 'and this makes room for a different kind of analytic moment' (ibid.: 119, 130, 89). Analysis begins where virtue ends, he writes; it requires a departure from established paths of virtue, a willingness to live without a principle. '[A]nalytic mindedness … is, as it were, an existential Sabbath from ethical life' (ibid.: 165, 128). Obviously, it would be strange to characterize vegans as people who live without a principle (although the principle for them is a pretty broad one, i.e. that of least harm), and yet I find there's something useful in Lear's concept of 'break', those moments in which as Lear says, it's possible to experience vividly how things *don't* fit together (ibid.: 126). In the contexts I've been discussing, this is a matter of suspending what you think you know in order to see and know otherwise, from an oblique angle and from the margins. In order to change your life, you may have to break down and give up – to give up some of your habits certainly, but also to give up trying to know, as Rilke suggests, or believing you have to 'know' in an abstract way. At least, if you allow yourself to break down even just a little, as Pollan did in his steakhouse, you may find yourself perceiving and therefore knowing differently – although it seems that Pollan was only able to push his knowledge so far. In other words, it's not a matter of replacing one moral system with another (and it's not exactly an ethical Sabbath either), but of opening your perceptions to an expanded kind of knowledge even if it means discarding the comfortable notions with which you've habitually protected yourself.

But I haven't accounted for why it is that some people are able or liable to 'break' while others are not. Why did Pollan take *Animal Liberation* into the steakhouse with him? Bravado perhaps and, presumably, in part because he knew it would make good copy, yet there's no reason to believe the changes he says he made to his life afterwards were merely a gimmick. Are Type B carnivores impercipient (Type As are simply ignorant, I think), unaware of the horrors to which the Elizabeth Costellos of this world can't help but turn their attention? Why do some people perceive what others do not? Am I constitutionally different from you? Was I born with some kind of ethical superpower, which provides me with X-Ray vision when I look at pieces of cake and Kentucky Fried Chicken? Obviously not. It's difficult to say why some people exercise this kind of 'vision' while others can't or don't or won't. I suspect there's a degree of choice in it, but I think there's also something more inchoate, some kind of predisposition to break and be broken – to allow yourself to be knocked over by the injunction to change.

I've been suggesting that there are different ways of 'knowing' and that abstract knowledge by itself won't precipitate you into a different mode of being. Especially when it comes to animals, it's a matter of *how*, rather than *what* you know. I might tell you and tell you about the horrors of industrial farming, but if something doesn't break in Lear's sense of the word, you'll eat the piece of cake next time it's offered to you. Have I managed to activate some kind of synapse in my brain, which means my response to eggy cake is different to yours? Again, I don't know, but what I do know is that my responses to eggy cake or animal flesh are not purely intellectual, which is why, at the risk of sounding mystical and essentialist, I'm suggesting that veganism is a matter of how I perceive and respond to a world that often seems weird and incomprehensible. When I see a piece of cake or meat or cheese, I also see what is not there to see: the animal whose life went into it. This is a form of sympathetic imagination, perhaps even of identification, and I know it's possible to develop this capacity because I haven't always had it. I used to eat pieces of cake and chunks of cheese without giving them a second thought or glance, and I drifted into vegetarianism out of a vague sense that it was 'healthier' and more moral. Then one day, more than a decade ago, I was rounded upon in much the same way as I've suggested Rilke's poem rounds on its reader. When I went into my local cheese shop and asked for vegetarian free-range organic cheese, the owner of the cheese shop himself threw up his hands and bluntly informed me that I shouldn't bother with cheese if I cared that much, because the dairy industry and the meat industry were hand-in-glove. What they do to cows is appalling, he said: You must change your life and give up dairy products altogether. At the time, I loved milk, cheese and yoghurt, yet in that moment I knew, not just at an abstract level, but somewhere in my body, that I couldn't eat them any more (and I had no reason to doubt what the cheese shop man was telling me: he owned a cheese shop, after all. Why would he lie?) I can't tell you what led up to and went into that moment of break as I've come to think of it. It happened too long ago.

Perhaps I was already on the verge; perhaps I had a notion of the depredations of the dairy industry but had suppressed the knowledge and/or chosen not to consider the matter too deeply; perhaps I didn't like cheese all that much anyway and was looking for a pretext to give it up; maybe the cheese shop man reminded me of my father, or some other authority figure; maybe I had a bad experience as an infant and unconsciously loathed dairy products forever after. You can psychoanalyze it all you like, but I don't think it really matters. I knew he was right, and as a consequence, from that moment onwards, I could no longer see or taste the world as I had before.

Contrary to what many Type As and Bs think, veganism is not a religion, nor is it about constructing a rigid moral code and sticking to it come what may. There are no 'rules', only perceived wrongs, felt injustices and an ongoing capacity for break.

Notes

1 For discussions of the connections between knowledge and perception see for example Martin Jay's *Downcast Eyes, The Denigration of Vision in Twentieth-Century French Thought* (1993) and David Michael Levin (ed.) *Modernity and the Hegemony of Vision* (1993).

2 In *On the Psychotheology of Everyday Life*, Eric Santner describes how Franz Rosenzweig turned away from academia and embraced a 'mundane' everyday life because he saw scholarship as a vampire, draining the scholar of humanity (Santner 2001: 16–18).

3 This is the central question of Barbara MacDonald's article, ' "Once you know something, you can't not know it." An empirical look at becoming vegan' (2000).

4 Coetzee (2004: 114).

5 Safran Foer writes that his son's impending birth was the immediate impetus for him to give up meat and write *Eating Animals* (Safran Foer 2009: 5–6).

6 For a critique of this kind of moral discourse which is predicated on the figure of the child, see Lee Edelman, *No Future. Queer Theory and the Death Drive* (2004).

7 Children, it is implied by this reasoning, have an innate ability to distinguish right from wrong. For a classic study of childhood morality, see Jean Piaget's *The Moral Judgement of the Child* (1968).

8 The steakhouse episode is also included in *The Omnivore's Dilemma* (2006). For a trenchant response to Pollan's omnivorism, see B. R. Myers' article in *The Atlantic* (2007).

9 For a critique of Pollan's arguments and those of locavorism more generally, see Vasile Stănescu, '"Green" Eggs and Ham? The Myth of Sustainable Meat and the Danger of the Local' (2010).

10 This is the conclusion of a 2003 Harris poll: See www.vrg.org/journal/vj2003issue3/vj2003issue3poll.htm (accessed on 14 December 2013). A July 2012 Gallup poll estimated that 5 per cent of Americans 'consider themselves vegetarians' while 2 per cent say they are vegans (www.gallup.com/poll/156215/consider-themselves-vegetarians.aspx) (accessed on 14 December 2013). In Britain, the Vegetarian Society estimated that in 2012, 2 per cent of adults and children reported that they were vegetarian while less than 1 per cent reported following a vegan diet. In France (where I wrote this article) the figure is similar. See the European Vegetarian Union for worldwide statistics www.evana.org/index.php?id=70650 (accessed on 14 December 2013). Perhaps unsurprisingly, the best place to be vegetarian is India, where 40 per cent identified themselves as such in a 2006 poll.

11 This belief is widespread, and it dovetails with the carb-phobia of the last decade or so. A well-known example of this is the paleo diet, which has been around since the 1970s but has recently been popularized by Loren Cordain's books. See also Lierre Keith's *The Vegetarian Myth* (2009). For a critique of these 'hunter-gatherer' beliefs, see Milton (2000).

12 Peter Singer's *Animal Liberation* is the most obvious example of this kind of rhetoric, and Jonathan Safran Foer's *Eating Animals* adopts a similar approach. Another representative example is David J. Wolfson and Mariann Sullivan's 'Foxes in the Hen House. Animals, Agribusiness and the Law. A Modern American Fable' (in Sunstein and Nussbaum 2004).

13 Cf. Adams (2003: 5).

14 See Pollan (2002: 110).

15 See Nestle (2007: 130, 22, 17).

16 Singer and Mason also claim that the adoption of more ethical, politically-motivated eating practices is not confined to one socio-economic class of people (Singer and Mason 2006: 5).

17 For example, Nestle reports that when she worked for the US Public Health Service, she was not allowed to advocate eating less meat as a way to reduce the intake of saturated fat (Nestle 2007: 3).

18 Safran Foer asks why taste is exempt from the moral rules that govern other senses (Safran Foer 2009: 93).

19 The aversion we feel towards certain kinds of meat seems to be largely cultural, since different cultures find it acceptable to eat different kinds of flesh. See Paul Bloom's discussion of the Meiwes 'cannibal' case in Germany and what it illustrates about the pleasures of taste (Bloom 2010: 24–27). See also Safran Foer's modest proposal concerning the uses to which dead dogs might be put (Safran Foer 2009: 24–29).

20 This is the widely cited 'ne pas céder sur son désir' from Lacan's Seminar VII. Marc de Kesel points out that Lacan has been misquoted and taken out of context: the phrase 'ne pas céder sur son désir' nowhere appears in the text, but it has been extrapolated as an injunction (de Kesel 2009: 262).

21 See Joy (2010: 11).

22 Pollan says this is similar to reading *Uncle Tom's Cabin* while watching slaves labouring on a plantation (Pollan 2011: 58).

23 This may well be a version of Epicureanism, the notion that individuals can be happy and take pleasure only from what is morally 'good'.

24 The intersections between vegan ontological 'breakdown' and recent queer theoretical discussions of opacity, responsibility and the ethical subject are compelling. See for example Judith Butler's *Giving An Account of Oneself* (2005). My thanks to Nik Taylor and Richard Twine for pointing out this connection.

25 See Spiegel (1996). For an excellent discussion of intersectionality, disgust and the

tensions between feminist and critical animal studies discourses, see Richard Twine, 'Intersectional disgust? Animals and (eco)feminism'.

26 Carol Adams describes veganism as a minority position (Adams 2003: 28). See also Coetzee (2004: 88), where Elizabeth Costello observes that Gandhi's vegetarianism was far from an exercise of power; rather, '[i]t condemned him to the margins of society'.

References

Abram, D. (1997) *The Spell of the Sensuous*, New York: Vintage.

Adams, C. J. (2003) *The Sexual Politics of Meat: A Feminist-Vegetarian Critical Theory*, New York: Continuum.

Bekoff, M. (2007) *The Emotional Lives of Animals*, California: New World Library.

Bloom, P. (2010) *How Pleasure Works: The New Science of Why We Like What We Like*, New York: W.W. Norton & Co.

Butler, J. (2005) *Giving an Account of Oneself*, New York: Fordham University Press.

Coetzee, J. M. (1999) *The Lives of Animals*, Princeton: Princeton University Press.

Coetzee, J. M. (2004) *Elizabeth Costello*, New York: Vintage.

Cordain, L. (2010) *The Paleo Diet. Lose Weight and Get Healthy by Eating the Foods You Were Designed to Eat*, New York: Wiley.

De Kesel, M. (2009) *Eros and Ethics: Reading Jacques Lacan's Seminar VII*, trans. Sigi Jottkandt, New York: SUNY Press.

Doidge, N. (2007) *The Brain That Changes Itself*, New York: Penguin.

Edelman, L. (2004) *No Future: Queer Theory and the Death Drive*, Durham: Duke University Press.

'Fact Sheets', (2011) The Vegetarian Society, available at: www.vegsoc.org/page. aspx?pid=750 August 12 (accessed 14 December 2013).

Gladwell, M. (2005) *Blink: The Power of Thinking Without Thinking*, New York: Little Brown and Company.

'How many vegetarians are there?' *Vegetarian Journal* (2003) available at: www.vrg.org/ journal/vj2003issue3/vj2003issue3poll.htm (accessed 11 August 2012).

Jay, M. (1993) *Downcast Eyes: The Denigration of Vision in Twentieth-Century French Thought*, Berkeley: California University Press.

Joy, M. (2010) *Why We Love Dogs, Eat Pigs and Wear Cows: An Introduction to Carnism, the Belief System that Enables Us to Eat Some Animals and Not Others*, San Francisco: Conari Press.

Keith, L. (2009) *The Vegetarian Myth: Food, Justice and Sustainability*, New York: P.M. Press.

Lear, J. (1998) *Open Minded: Working Out the Logic of the Soul*, Cambridge: Harvard University Press.

Levin, D. M. (ed.) (1993) *Modernity and the Hegemony of Vision*, Berkeley: California University Press.

McDonald, B. (2000) ' "Once you know something, you can't not know": An empirical look at becoming vegan', *Society and Animals*, 8(1): 1–23. Accessed 11 January 2012.

Milton, K. (2000) 'Hunter-gatherer diets – a different perspective', *American Journal of Clinical Nutrition*, 71: 665–667. Accessed 11 August 2012.

Myers, B. R. (2007) 'Hard to swallow', *The Atlantic* (n.p.). Accessed 5 August 2012.

Nestle, M. (2007) *Food Politics: How the Food Industry Influences Nutrition and Health*, Berkeley: California University Press.

Pascal, B. (trans. A. J. Krailsheimer) (1995) *Pensées*, London: Penguin.

Piaget, J. (trans. Marjorie Gabain) (1968) *The Moral Judgement of the Child*, London: Routledge.

Pollan, M. (2002) 'An animal's place', *New York Times Magazine*: 58–111. Accessed 12 June 2011.

Pollan, M. (2006) *The Omnivore's Dilemma: A Natural History of Four Meals*, New York: Penguin.

Pollan, M. (2011) *The Omnivore's Dilemma. A Natural History of Four Meals*, New York: Penguin.

Rilke, R. M. (trans. Stephen Mitchell) (1982) *The Selected Poetry of Rainer Maria Rilke*, New York: Random House.

Safran Foer, J. (2009) *Eating Animals*, New York: Little Brown & Company.

Santner, E. (2001) *On the Psychotheology of Everyday Life. Reflections on Freud and Rosenzweig*, Chicago: Chicago University Press.

Singer, P. (2002) *Aniaml Liberation*, New York: HarperCollins.

Singer, P. and Mason, J. (2006) *The Ethics of What We Eat. Why Our Food Choices Matter*, London: Rodale.

Spiegel, M. (1996) *The Dreaded Comparison: Human and Animal Slavery*, New York: Mirror Books.

Stănescu, V. (2010) ' "Green" Eggs and Ham? The Myth of Sustainable Meat and the Danger of the Local', *Journal for Critical Animal Studies*, 8: 8–32.

Sullivan, M. and Wolfson, D. J. (2004) 'Foxes in the hen house: animals, agribusiness and the law: a modern American fable', in C. Sunstein and M. Nussbaum (eds) *Animal Rights: Current Debates and New Directions*, Oxford: Oxford University Press.

C. Sunstein and M. Nussbaum (eds) (2004) *Animal Rights: Current Debates and New Directions*, Oxford: Oxford University Press.

Taylor, C. (1989) *Sources of the Self: The Making of Modern Identity*, Cambridge: Harvard University Press.

Twine, R. (2010) 'Intersectional disgust? Animals and (eco)feminism', *Feminism and Psychology*, 20(3): 397–406.

Varela, F. J. (1999) *Ethical Know-How: Action, Wisdom and Cognition*, Stanford: Stanford University Press.

Varela, F. J., Thompson, E. and Rosch, E. (1993) *The Embodied Mind: Cognitive Science and Human Experience*, Cambridge Mass: MIT Press.

Part II
Doing critical animal studies

4 Listening to voices

On the pleasures and problems of studying human–animal relationships

Lynda Birke

> "Lots of people talk to animals", said Pooh.
> "Maybe, but..."
> "Not very many *listen*, though", he said.
> "That's the problem", he added.
> (From *The Tao of Pooh*, Benjamin Hoff
> 1982: 29)

It's often tricky, being an academic with activist interests. Because I am passionate about animals and interspecies relationships, I want to figure out ways of studying and thinking about them. But that's where the problems start: studying interspecies relationships is fraught with difficulties, both methodological and political. How can we study our relationships to other animals in ways that truly include them, or that give them voice? Or, more importantly, in ways that enable us to *listen*, as Winnie-the-Pooh reflected? And how easy is it to devise methods that fit with the ethos of Critical Animal Studies?

Reading much of the animal studies literature, and especially that from Critical Animal Studies (CAS) perspectives, I have been struck by two things. The first is a reason for celebration, in my eyes, in that there is still a strong commitment to radical politics evident in some of the writing.[1] To be sure, there are many research studies which seem overly academic, apparently eschewing any reference to political issues. But there are also many studies that do shed light on the very complex ways in which humans interact with, and subordinate, animals. I hope that efforts to maintain the connection between scholarship and political action (represented by this volume) continue.

The second is that this scholarship is very wide-ranging, and we have seen some excellent investigations of a vast array of human–animal engagements. However explicit (or not) the politics, the breadth of these studies builds a strong base to better understand the myriad entanglements between and across species. But, and to me this is a big but, *actual animals* do not often seem to figure: rather, what predominates, it seems to me, are studies of how we humans represent nonhuman others, or how we build infrastructures around them (in farms, zoos, labs, the wilderness). Where actual animals enter the fray seems to be limited to more behavioural studies, looking at (say) human impacts on animals.

Are these, though, merely another form of animal experimentation? Such studies may involve an artificial situation, staged for the purpose, or may involve physically moving an animal from place to place.[2] In the sense that the researcher is looking at how animals respond, and may stipulate as part of the study how and why subjects are moved, and that human and animal do particular tasks, then the answer to the question is yes. "Experimentation", however, conjures horrifying images; I am certain that every contributor to this volume shares an abhorrence of some of the things that can go on in biomedical laboratories. But *is* what we do in animal studies "experimentation" in this awful way? Or are some forms of manipulation for research purposes ever acceptable?

Pondering on these questions, I think about how important to us are the writings of cognitive ethologists, which emphasize the sentience and intellectual abilities of other species. I do not condone research that imposes any kind of suffering on animals – but to judge that, we need to know how that species experiences its world, not simply extrapolate from ourselves. Most Animal Studies scholars are interested to know more about the sentience and cognitive abilities of nonhumans, and readily cite ethological studies. Yet many of these are based on "experiments" in the sense of manipulating situations that frame what the animal does. It is from these studies that we learn more about how different species think, and about what their choices are (Dawkins 2006; Bekoff 2002) – knowledge which is surely central to understanding animals' lives, and to making their voices heard.

Here, I explore some of these issues, which continue to trouble my own research practice and thinking about how we humans interrelate with other species. To begin, I will consider the practical questions inherent in my own work, which is concerned with specific relationships between humans and horses – how might we do research that focuses on *interrelating*, rather than on one or other participant? How can we pay attention to the animal's presence and activity? How can the problems of achieving interdisciplinarity across species be overcome? To illustrate this, I use as a case study a project I am currently involved in, on "horse–human relationships". I, and other researchers involved in this study, have struggled to think about how to study the relationship, both in terms of appropriate methodology, as well as in terms of how to consider simultaneously *both* horse and human as constructors of the relationship. But we can also ask critical questions of this work – does it (can it ever?) really address horses' agency? By asking horses and people to participate, are we "experimenting" in ways potentially detrimental to the horse? To what extent does the research entrench existing patterns of horse "ownership" and "use"? While I can ponder how to do research better, I also recognize that our efforts do not live up to our political ideals.

Whatever methods researchers deploy, they do not always adequately address existing systems of domination and human–animal separation. On the contrary the various methods in which academics have been trained, make assumptions about "animals" and about whose voice counts in academic inquiry. "Real" animals are, in that sense, often absent from scholarship, even in many animal studies (including Critical Animal Studies);[3] animal agency in the research process is very rarely apparent. And research that does include actual animals seems often to require that the animal participate in a structured investigation for us to observe – which can seem close to doing experiments, with all the negative connotations that may imply. Are we thus asking research questions of these voiceless others, while doing little or nothing to challenge their exploitation? How much can including animals in such ways really give them a voice?

So, in the latter part of this chapter, I turn to critique, and consider the limitations of projects such as our horse–human study for critical animal politics. Doing investigations with already-captive or domesticated animals not only raises questions of whether or not such research reinforces exploitation, but also questions of ethics of consent, of participation and accountability, and of authorship and responsibility. Is such research justified, or does it once again lock nonhuman animals into human control?

On methods

However much we may try to "listen to animals' voices", we face methodological difficulties, not least through the heritage of disciplinary divisions. Although CAS prioritizes "interdisciplinarity", in practice research usually begins from particular sets of methodological and epistemological assumptions, firmly rooted in existing disciplines. Animals have long been studied within science (notably ethology, the study of animal behaviour), but often without significant reference to either the human context or systems of exploitation (including, of course, the direct role of animals in laboratory scientific practices). Other disciplines have historically excluded animals – the social sciences, for example – and are only belatedly trying to acknowledge them as architects of our shared sociality (Irvine 2007).

What is becoming very clear, however, is that there are many ways in which humans "relate" to nonhuman animals, and concomitantly many ways of studying such engagements. Our lives have long been closely entwined with other species, through multiple layers and histories, as Haraway (2008) emphasized – indeed, it makes little sense to separate them out for different types of inquiry. Scholarship across all kinds of animal studies broadly reflects that multiplicity, drawing on a wide variety of backgrounds and methodologies. But, amid the calls to "bring animals in" to diverse disciplinary traditions, what does it mean to study interspecies relationships, or myriad "relatings" to use Haraway's word?

The notion of relatings covers human–animal interconnections at many levels, a wide range of practices, including agriculture, and the place of animals within the production of scientific knowledge. Researchers can thus map interconnections, tracing for example the networks that form shared human-nonhuman worlds. By tracing networks, moreover, researchers arguably achieve some kind of symmetry: nonhuman animals become seen as crucial parts of the social networks investigated. That is an important consideration for scholars in critical animal studies, as it offers means of moving away from the persistent separation of human and animal (see Taylor 2011, 2012).

Tracing networks and doing ethnographies are important to understanding how interspecies' lives interconnect (see, for example: Buller 2012; Hamilton and Taylor 2013; Nimmo 2010). We, too, use ethnography in some of our current research, around horses and humans.[4] Ethnography allows an examination of sociocultural worlds in which horses are enmeshed, and permits us to try to "follow" the horse through her/his various encounters with specific people. But these are, of course, horses living within human contexts; it is people who determine the course of their lives and the quality of their relationships with humans. In that sense, the networks traced do not overtly allow or focus upon equine agency.

Moreover, following networks, which include other species, does not necessarily include a focus on the animals themselves, their behaviour and what they are actively contributing to making relationships work (or not).[5] Rather, the animals as active agents sometimes seem to fade into the background. They may be actors in these networks, but what do they actually do? While I consider tracing networks to be an immensely valuable approach in allowing us to understand human–animal connections, I remain unconvinced that it necessarily permits us to see animal actions, to hear their voices.[6]

Yet animals can and do contribute to making meaning within specific relationships – and it's that which I find fascinating to study. There is without doubt someone returning my gaze when I look into the eyes of animals involved in our research. Yet, although our ethnographic work tries to include encounters with actual animals, I do not necessarily feel that I am meeting horses as the co-producer of these social interactions, except through human narratives relating to them. To be sure, the fieldwork entails meeting people alongside horses, and we can make observations. But it is not easy to "do" ethnography when you can't directly interview a key participant, and we necessarily resort to conversations with the people. These are, to me, fantastic approaches to mapping human–animal interactions at a broad level – but arguably less good at understanding how specific, individual, relationships work.

For many people in animal studies, however, it is those *specific* relationships that motivate their work in this field – usually, the one-to-one relationship, the (sometimes) close bond between us and another being. These relatings are a microcosm, as Haraway (2008) reminds us, of the various ways our lives entwine with those of other species on this planet. Here, the relationship is a living-alongside, an intimate sharing of our selves. These relationships are politically (to some in CAS) problematic, but they nonetheless form the starting point for many of us – the ground of both our scholarship and our politics. While I acknowledge that this entails a problematic relationship (e.g. "ownership" of "pets"), I admit to living with dogs and horses. I, at least, gain a great deal from this arrangement – not least my passion for understanding them. Whether they gain much, only they can tell. In my view, however, and despite undeniable structures of domination, there is some possibility in these relationships for mutual benefit, even love (as Cudworth 2010, argued).

Such close relationships interest me greatly, both personally and as a researcher. I do live closely with other animals, and they are crucially part of my life and thought. Given histories of disciplinary divisions, however, it is not straightforward to study interspecies relationships in detail. While there are many investigations of how humans affect animals, or vice versa, it seems to me that there are relatively few studies of in-depth *relationships*, and how they develop. Significant exceptions include those based on a symbolic interactionist framework (e.g. Sanders' (1999) and Irvine's (2004) observational studies of dog–human bonds). Such inquiries are concerned with how meaning is produced in social engagements, and they take seriously the role of the animal participant as the co-producer of such meaning. In that sense, they seem to me to come closer to permitting the voice of the animal to be heard, and to be understood as a co-agent in the performativity of "relationships" (Birke *et al.* 2004).

Scientific investigations, while less concerned with interspecies relationships, can however reveal something about the animals themselves – how they behave in specific situations, what their priorities are – especially when the work involves understanding animals' worlds from their points of view (e.g. in the work of cognitive ethologists, such as Marc Bekoff (2002)). In this sense, they tell us something about how different species might experience their lives. But scientific approaches undoubtedly have limitations, partly because of their separation of the researcher from engagement with whoever is being studied, and partly because of prevalent reductionism. Ethologists often expect to focus on one specific aspect of an animal's behaviour – aggressive responses, for example – and then examine what/how/why/when such behaviour occurs. This enables them to concentrate on details, but it is also in danger of taking that behaviour out of the ongoing stream of animal life (which sometimes includes engagements with humans), thus losing much of the context of how the animal engages with the world.[7]

What is thereby often lost is an understanding of the dynamic processes that make up a relationship, especially when the relationship involves another species. An alternative approach, however, is to assess the overall "quality" of an interspecies relationship, rather than breaking individuals' behaviour up into component parts. This may be one way to bring divergent disciplinary frameworks into conversation and – possibly – could make research a little more sensitive to the active contribution of the animals. A focus on behavioural "quality", on the demeanour of the whole animal, has been the central concern of Francoise Wemelsfelder's work on animal subjectivity, for example (see Wemelsfelder 2012).

Such concerns do not always fit well with fellow scientists, who sometimes see it as too subjective (see Wemelsfelder 2012). Yet it is not "affiliative episodes" or "approach responses" that build relationships, but whole animals working in concert with one another (or sometimes with us), and it is the quality of *the behaver*, who s/he is, and her/his relationships that matter. Because of this emphasis on the whole behaving animal, I am happier with such investigations rather than those that focus on some specific aspect of behaviour in isolation. Thinking about behavers seems to offer ways of including and respecting animal being – even within an "experimental" remit. It does not, of course, necessarily challenge the systems of exploitation in which animals are embedded: Wemelsfelder has mostly studied pigs, for instance, kept within agricultural units. But that is a perpetual dilemma for researchers in animal studies – for we are always studying what is already there. The more that research makes explicit that there is clearly "someone at home" whenever we see the battery chicken, the farrowing sow or the caged laboratory rodent, then the more it can potentially contribute to improving animals' lives.

Qualitative methodologies, however, are less commonly used in the natural sciences; at most, these would be descriptive accounts and would not usually include, say, interviews with (human) participants. Nevertheless, a few studies in animal behaviour are more qualitative, contributing a different slant on understanding how relationships work. Of particular interest are those based on observational studies grounded in dynamic systems theory, or DST. Barbara King, for example, in her book *The Dynamic Dance* (2004) uses DST to discuss research into communication between great apes. As she argues, there have been many important studies of how these species communicate, and what they understand. These tell us a great deal not only about the specificities of these animals' behaviour, but also give us hints about the evolutionary heritage of our own communications.

King points out the reductionism of many studies of animal communication, arguing that focussing on how participants use gestural signals is not enough. Rather, communicative signals are (for social species) usually used within specific contexts of ongoing relationships, which have histories. They entail expectancies and emotional connections, rendering it meaningless to focus on a specific gesture. In her own work with great apes, King emphasizes *patterns* and *co-regulation*; that is, both participants in a dyad are mutually influencing each other, and draw on histories of doing so: you simply cannot separate out who-did-what-to-whom. She notes that: "Long-term detailed analysis of social events, with a focus on the *co-regulated dance between social partners in an emotional system*, represents a genuine departure from primatology's conventional reliance on statistical analysis of signals produced and received" (ibid.: 231, my emphasis). This is why her book's title refers to a dynamic dance; it is as though the patterns of social behaviour are choreographed, in webs of emotional connection.

Emotional connection is what most of us who live closely with companion animals would report. I could be a sceptical scientist and suggest that the dogs and horses with whom I live, are merely conditioned to respond "emotionally", perhaps in expectation of food. Or I could laugh at that caricature, and recognize that these are beings with minds, whose feelings and abilities to manipulate my behaviour are quite incredible. It is, I hope, clear where my beliefs and expectations lie. But passionate beliefs and emotional connections aside, I am also a researcher. While this means that I am inevitably embroiled in ideologies and histories of research that deny animal subjecthood, I am also fascinated by how such interspecies engagements are made; how they work. So, I want to try to understand interspecies encounters, even despite the inbuilt limitations of research processes, as we know them.

A case study: horses and their humans

People who live with or around horses often talk about developing a close relationship, how important this bond is for both them and the horse, and how one can easily see whether or not someone else has this closeness (Birke *et al.* 2011; Birke 2008). This does not necessarily just happen; on the contrary, many people living in close association with companion species believe that good relationships must be worked at, so that the negotiations which form the basis of any inter-individual relationship are mostly positive ones and horse and person "work well" together (Birke 2007). This, then, is the starting point for my other research project, into details of ongoing human–horse relationships[8] – that the *quality* of interactions between different horses and different people might be discernible. Can we observe the "working well" together, to which so many horse people strive?

In this project, we focus on detailed mutual engagements of horses and people. We are interested in how both participants work together (or not) – and, in particular, we want to know if we can say something about the "quality" of the relationship.[9] What happens when each is engaged with the other, carrying out actions? What kind of mutuality, if any, might be recognized? Is there, in King's words (sometimes) a dynamic dance between human and horse/dog or other animal?

In focusing on what makes relationships "tick" between horses and their people, we try to consider both overall quality and details of interaction, to pay attention to how both horse and human construct the relationship. Now, there is an immediate methodological problem: for how can such interaction be tracked with some degree of symmetry? In many ways, our approach combines qualitative and quantitative, ethological and sociological, methodologies – but there will always be some asymmetry, simply because we cannot verbally ask the horse questions. This strategy is similar to the multilevel approach advocated by Franklin *et al.* (2007) for investigating dog–human interactions; these authors suggested detailed observation in naturalistic settings and interactions at home, combined with – from the human side – use of interviews and diaries. In this way, the dogs' actions are observed specifically within the relationship and become directly meaningful for the research, while documents such as diaries allow the human caretaker to give voice to their own feelings and responses based on engagement with the animal. Through such layered work over time, these authors suggest, we might understand the part played by both partners in making relationships grow, in producing biographies. Relationships, after all, are usually ongoing and performative – they must be co-produced (Birke *et al.* 2004).

Combining methods is one example of multi-strategy research. But it does present practical problems in trying to integrate methods from quite different traditions. The purpose of doing so is to use strategies that take the animal's behaviour seriously (drawing on ethological traditions) as well as employing tried-and-tested techniques, such as interviews, to assess human perspectives. While the difficulties of such integration are considerable, we nevertheless argue that bringing them together permits a deeper understanding of the many facets that make up a "relationship" between person and animal. Here, my concern is to think about methodological issues raised if/when we do research on *specific* human–animal relationships, and I am – for the moment – putting on hold the question of what kind of politics such studies (and methodologies) support.

We do, admittedly, focus on a specific moment in time – thus, on the outcome of possible previous interactions, rather than on the process (see Noske 2003). It is not so much a biography, in that sense, as a story about how the horse and the person already work together. Nevertheless, we aim to use the differing methods and map them onto one another. Video recordings, for example, are treated as though they were written transcripts of interviews – from such detailed observation and note-taking, key themes emerge, which can be further analysed, both qualitatively and quantitatively. What I want to emphasize is that this becomes open-ended and descriptive, rather than hypothesis-led, and includes multiple components, including ethological observation of horse-plus-human, monitoring of physiological changes and sociological approaches (interviews, fieldwork observations of horses and people in stable yards).

Like Franklin *et al.* (2007), we believe that several – and simultaneous – levels of analysis are methodologically essential for research into human–animal relationships. Undoubtedly, the pursuit of qualitative assessments is likely to meet criticism from many fellow scientists, while reviewers of articles may well look askance at studies that do not seem to "fit" prevailing disciplinary standards. Nevertheless, we would argue that the juxtaposition of such different approaches does offer the means to challenge and extend traditional methods of studying individual relationships. We are already seeing how Animal Studies challenges traditional disciplines, as the social sciences and humanities have begun to think about including nonhumans, while scientists used to working with animals have begun to consider how to include, say, the perspectives and behaviour of animals' caretakers. Furthermore, these divergent methodologies, incorporated within a single study, can be an important means of validating each other, offering a form of triangulation within a multi-strategy study (Bryman 2004).

Although the physical structure of our research setting constrains both human and horse (and the horse is further constrained by being led), I am fascinated by how much each pays attention to the other, in a kind of silent communication – especially when the person is familiar to the horse, and the relationship is relatively established. This mutuality often seems reminiscent of King's "dynamic dance" (King 2004; Fogel 2000), in which the behaviour of each partner cannot readily be broken down into separable acts. Rather, they happen together, they are co-regulated. It is not a case of when A does X, then B does Y, but rather one of: *while* A is beginning to do X, B is beginning to do Y. These highly integrated sequences are almost always with pairs familiar to each other; who have expectations of each other. They are a form of choreography, a mutual field of influence, and give the impression of coordination, of a well-practised routine – a dance. They lend themselves to narrative description, which links the actors; they do not lend themselves so readily to a numerical evaluation of separate components.

Here, the specific findings and my own fascination are less important than the implications. On the plus side, what we are trying to do is to bring divergent methodologies into dialogue, combining ethological observation with sociological approaches such as interviews and ethnography – a crucial combination for animal studies, I would argue. Trying to understand how "good" relationships between humans and animals work can perhaps potentially benefit (some kinds of) animals by helping us to understand what they contribute to, and get out of, close engagements with us. Human failures to understand what nonhumans are trying to tell us is a significant cause of animal suffering, so it seems to me to be important to know what happens when relationships *do* appear to work well. Taking animals seriously must, I believe, include finding ways to observe just how they themselves produce human–animal encounters, and so we to track what the horse is doing in these apparently coordinated movements. We can't ask them verbal questions, so we must pay very close attention to what they are trying to say through their behaviour – whether or not that is in a research context.

On the minus side, trying to combine methodologies is an extremely difficult task in practical terms. We can add them together, but they are very hard to integrate – and there is then the danger that the research cannot be published because it does not fit readily into pre-existing disciplinary categories. And if research is not published, then it was (1) a waste of time; and (2) ethically dubious, for why are we requiring efforts from participants of whatever species if the study never sees the light of day?

Of rather more significance here, however, is that our research does constitute a type of manipulation (asking horses and people to move around together), and in that sense is an "experiment". But even doing ethnography and tracing networks involves some manipulation (interviewing people, observing what they and the nonhumans are doing), so I cannot escape this dilemma simply by concentrating on our other strand of research on the grounds that there, I am not directly manipulating animals while doing ethnography. I do, after all, move animals around on a daily basis in our shared lives: dogs need to be put on leads at times; horses on lead-ropes; manipulation is part and parcel of how we deal with companion animals too. And in research, arguably my very presence as a researcher observing people and horses will have some effect on all of them. If research into human–animal engagements is justified at all, then some of it will involve degrees of manipulation.

Nor does our study overtly challenge the very situation in which horses are kept and controlled by humans. While we try in the research to pay attention to what the horses do, they are undoubtedly already domesticated, living their lives within human-dominated frameworks of ownership. They cannot readily determine how to spend the day, but are always at the behest of human requirements – and would almost certainly prefer to be out munching grass than plodding around being led or ridden. Still, there is a profound effect of these relationships on the well-being of horses who are being so used, and – for me – it matters that we try to understand these processes better.

So, what this example illustrates, I hope, is the dilemma researchers face in studying human–animal relationships. On the one hand, we want to use methodologies that permit (to some extent) animal agency, that acknowledge that the animal is her/himself an actor in producing interspecies relationships. This means combining methods that arguably do not fit well together. Still, we can try to cross those bridges, and include careful observation of the animal to ensure some parity. Yet on the other hand, we are well aware of the politics of working with domesticated animals. Any research investigating their interactions with humans can be seen as failing to challenge the very systems of exploitation in which such animals are enmeshed. Ours is no exception.

Exploiting research

In many ways, doing research inevitably constitutes exploitation, whoever the subjects and whatever the questions. If you do observations of specific animals, as we do, then you are requiring these animals to do whatever you say (with or without their human guardian). If you do studies at a macro level, of (say) structures and practices of modern agriculture, then it can be said that you help to perpetuate animal exploitation precisely because the research, by its nature, does not immediately challenge the very practices under scrutiny. Demonstrating exploitation may help to raise public awareness, but it does not, in itself, explicitly challenge structures of power.

How, then, do methods contribute to animals' exploitation? No doubt there are many answers, but one particularly springs to mind and that is the problem of *distancing*: we study others, but not often do those others participate in the research (of whatever species they are). This is most evident in scientific research, where researchers have been taught to put a premium on objectivity – but many other kinds of research position the researcher as standing apart. Those who are studied are thereby denied agency, they do not directly produce the research, even while they may be contributing to knowledge. And nor are they usually given credit in processes of research publication and career development. As feminists have often noted, it has too often been "women" who have been studied, in ways that did not allow them participation or acknowledged their oppression. So too with our studies of nonhuman animals – too often seen through an uncritical lens, and rarely as participants in, and producers of, the research.

In this section, I want to consider some of these questions posed by feminists in debating what constitutes ethical research. Among these are issues of participation, of consent and accountability, of authorship and responsibility. Feminist authors emphasized research explicitly informed by feminist values; they were (and are) critical of research that fails to address, and be responsible towards, the needs of the people and communities being studied. Research, they argued, should above all be accountable to its subjects (Skeggs 2001). Amongst other things, it should reject objectivist assumptions that researchers can stand apart from their subjects. It should, furthermore, be research that involved the subjects at different stages, as well as involving the researcher in the communities being investigated.

Ultimately, of course, accountability is unattainable, whoever the subjects are, research will always entail some degree of exploitation. With humans, at least, researchers can encourage participation and collaboration with those who are studied, asking them their responses to the research process. Yet, however participative, it is ultimately researchers who make decisions about the conclusions and dissemination of the research, so leaving participants out of the loop. Including subjects is, furthermore, sometimes difficult to negotiate, even if they are human, and negotiation is markedly more problematic if they are not. Even so, our studies should at least aim towards some form of accountability towards the animals that we study, some acknowledgement that research *practices* have implications (see Birke (2009) for further discussion of accountability).

To be accountable means, at the very least, that research must never construct those animals who are studied as objects, but as subjects. But in addition, it must mean considering possible consequences of the research. Could there be a direct improvement, for *those specific* animals? Has the research been intrusive, or presented those animals with difficulties or stressors? Might the research have implications for the lives of *other* animals? What are the broader political consequences of our research? Even if investigations do not obviously entail participation or challenge existing institutions, researchers can still think about what implications their studies might have for animals' lives – indeed, this should be required, in my view, of any research in animal studies.[10] While the micro-level analysis I describe here does not, overtly, seem to challenge systemic exploitation, it can, I believe, contribute to political debate *precisely because* it is only by studying actual animals and their responses that we can begin to understand their experiences.

Feminist debates about accountability of research with people position them as subjects who *can engage with* the research – feeding back comments, for example, or being involved with further developments. Now that is clearly not possible when research involves animal subjects, for whom direct resistance or refusal to cooperate may be impossible or lead to dire consequences. But perhaps it is something we can be thinking about. What role do our animal subjects play in producing knowledge when the research is ostensibly "about" them? Can we find ways in which they, too, can engage with the research, without detriment to themselves? How would we recognize, or even encourage, their participation? And what can we learn from the ways in which they do, at times, resist?[11]

Animals undoubtedly do contribute to the processes and outcomes of human–animal research: they are, after all, co-producers of the relationships we seek to study. But on the whole, they contribute without obviously agreeing to do so. That will probably always be true of our investigations. Writing about such issues in relation to animals, Linzey (2010) notes that "... we do often (rightly or wrongly) presume consent, but *presuming* consent is still a long way from *voluntary*, verbal consent as we know it between human beings" (page 1, emphasis in original). But, he notes, this is just as true of research involving some human beings, who may for various reasons be unable to give informed consent. But rather than shying away from doing anything at all, Linzey emphasizes that working with any subjects unable to give verbal agreement means being particularly responsible and ultra-sensitive to the ethical dilemmas.

In Animal Studies broadly, we produce knowledge, often jointly with those nonhuman others who are part of the subject matter of our inquiries. To me, however, merely generating information (or papers) is not enough: it matters that we try to foreground the part animals play in knowledge production, and think about how our research treats them, or how it might impact upon others of their kind. If we cannot get "informed consent", then at least we must ask ourselves whether they are likely to cooperate (or want to) in the situation we study. Otherwise, it is knowledge produced *by* us, *about* them, produced coercively – thus further entrenching differences of power. It is important to acknowledge that we humans may be privileged producers of knowledge, but we are not the only ones, and animal roles in knowledge-making may bring many surprises.

Trying to make your voice heard

> We've got quite a one-to-one relationship ... we are, like, together most of the day and she's, just, relaxed around me, and, like, she'll just follow me around the school and I leave her to it.
>
> > (from research interviews: L, commenting on his relationship with his horse, Kir)

Certainly, animals can contribute to, and alter, research progress in unforeseen ways, as Michael (2004) acknowledged in his account of sociological interviews interrupted by resident nonhumans (the cat overturning the tape recorder, for instance). Let me recount one example from our work with horses/people. One day, we had begun filming Kir, rather earlier than we usually did, before she was asked to walk around with my colleague. This early start, which meant that her usual person ("L") was still nearby, was serendipity: watching the video at very slow speeds I was fascinated to realise how closely Kir and L were – very subtly – attending to each other and seeking to maintain contact. He was standing behind her, but both he and she were constantly making small movements of head, eyes, ears (in her case!) and bodies to allow this fleeting contact at a distance. Kir was noticeably more relaxed when she was able to sustain contact, to know where L was. These were very slight movements of each (not easily seen at usual film speed), which were inseparable from those of the other – truly a dynamic dance, in King's terms (2004). It was this togetherness that L acknowledged, and which meant that Kir would often simply follow him around.

At this point, I was glad that we were trying to be qualitative and open-ended, rather than following our scientific training and doing the hypothesis-testing so important to scientists (in which case we would likely have missed the exchange or seen it as irrelevant). I would say that Kir altered our perception of the research, and its direction, by making us think about these minute, subtle ways of paying-attention-to-each-other in cases of established human–animal relationships – how Kir was as much an architect of the process as was her person. She was, incidentally, also adept at indicating that she was less happy working with my colleague: she made her views plain by resisting.

No doubt Kir was coerced into doing these tasks, in the sense that L came along and put a head collar on her, at our request, to lead her into an arena. In that sense, doing any research involving actual animals has an element of coercion. But watching the filmed sequences of Kir and L, I wondered about how important it apparently was to her to maintain proximity to him. This raises questions about how research could be developed, which remains sensitive to what *the animal* might want or prefer in specific situations. Kir's apparent anxiety when moving away from L made us aware that having a familiar person around matters to many domestic animals. Indeed, many companion animals do seem to build relationships of mutuality or trust with some humans, which could be seen as integral to research processes. Within that context, I find it hard to say that the horse was forced: she did, apparently, want to be with L. This is not to deny the wider context, of domestication and domination of horses; domestic horses have little choice but to accept their subordinate role, and to make the best of it. But, like Erika Cudworth (2010), I accept – with reservations – that some degree of mutuality is possible.

Earlier, I posed questions about the implications of research studies for animals. To continue using Kir's example, I would say that there was probably little of direct benefit to her in doing what we asked: on the contrary, she was not entirely happy to walk alongside my colleague. Her body tension was greater, and so was her heart rate. In that sense, she was, apparently, slightly stressed. Perhaps we can argue over whether this is ethically acceptable, although I suspect that most attempts to move animals around during research involve such levels of stress. If it is not acceptable, if we always perceive research as tarred with the brush of experimentation, then we must abandon *any* attempt to include animals in research. Given that I accept, with reservations, that observing animals with people – and sometimes manipulating the setting – is necessary if we are to include actual animals, then some degree of short-term stress seems likely. We cannot escape this risk simply by observing social settings, either, as even our presence on, say, a farm, might disturb the animals while we watch. But we do, I would argue, have to think of ways to mitigate that risk, as well as to include them with open minds, and listen to how they might change the direction of the study.

Kir's temporary anxieties aside, I would like to think that research into how animals relate closely to/with humans, has the potential to benefit the lives of other animals. I strongly believe that we need to understand the minutiae of day-to-day engagements with other animals, to understand what is being *said* when we gaze into their eyes. I am perfectly happy to use anthropomorphism, if it helps us to ask more nuanced questions. But I am not happy to rely on it solely, to infer from my own feelings that that is how another species feels (not least because my experiencing is freighted with histories of human/Western sociocultural life). For that understanding, I do think we need to involve animals, directly in research, to find out a bit more about how they produce their worlds. I doubt that researchers can avoid all forms of exploitation in doing so (short of not doing research at all), while I think we have a long way to go before we really do include the voices of animals – and I am sure that this includes my own work. But what that means is that we have to try very, very hard to think about minimizing exploitation and maximizing space for the agency of subject animals.

At present, we humans assume the responsibility of speaking for animals, of championing what we think are their causes. If the constituency being championed were a human one, there might follow debate about the voices of those whose interests were not served by the prevailing position – much as black women struggled to make their voices heard against the predominance, in 1970s feminism, of white, middle-class experiences. Humans are inclined to think that nonhuman animals are not a political constituency, and that they have no voice on such matters (though see Donaldson and Kymlicka 2010). Perhaps – although that may simply be that we have not yet managed to hear what they are trying to say. We have been, as Winnie-the-Pooh so sadly recognized, too busy talking to, or about, animals and rarely listening.

And Kir might well agree: whatever the politics of research, she was quite clearly trying to make her presence felt and her voice heard.

Acknowledgements

Most of my ongoing research is done with other people, and I am grateful to Mara Miele for her collaboration and discussions on the "behaviour consultancy" research project, and to Jo Hockenhull and Tami Young for their collaboration/ discussion in studying human–horse relationships. I also thank the horses, whose participation – willing or otherwise – is essential. But special thanks must go to Kir, whose tuning-in to her human made us pay close attention to how she was creating the relationship.

Notes

1 Academia, however, does not necessarily provide a milieu sympathetic to animal activism – nor, indeed, to activism of any sort. As a caveat, I should say that I am also

nervous that animal studies becomes too divorced from animal activist politics over time – much in the way that I perceived women's studies to become too separated from feminist activism (Birke 2009). These are, of course, my opinions, and others may disagree.

2 This is not only domestic animals, who may be moved by means of ropes, leashes etc. It is also true of studies of wild animals, whose movements can be "experimentally" manipulated by (say) provisioning. It is still an experiment.

3 By this I mean that most studies tend to focus on the place of "animals" in social structures or representations. This is important for our understanding of how such structures work to perpetuate oppression, but it often seems to exclude actual, individual, animals. The irony is obvious. In conferences and lecture halls, we talk about relationships with other animals, while those very beings are noticeable by their absence (perhaps excepting very small organisms, who escape our gaze). The exception is service dogs, who might accompany their person to the meeting.

4 This is an ethnographic investigation of the role of "behaviour consultants", done in collaboration with Mara Miele. Horse "owners" may seek external expertise if they feel that they are confronting a behaviour "problem" with their horse (the horse barges them, or refuses to load into a trailer, for example). To some extent, such consultations are about seeking a better relationship between horse and human, and making it work.

5 Similarly, sociologists have been slow to recognize the role of nonhuman animals within work, as Porcher and Schmitt (2012) have argued with respect to the work played by dairy cows within wider relationships of dairy production.

6 Part of the problem here, it seems to me, is that some studies focus on generic animals (how "animals" are represented, for example, or how "agricultural animals" are implicated in networks of human systems of food production), while others are more concerned with very specific individuals. The concept of human–animal relationship is quite different in these two approaches to research.

7 Some ethologists do, of course, take seriously how to meet animals as subjects (e.g. Bekoff 2002; Smuts 2001; Wemelsfelder 2012). My point here is to stress that scientific approaches do prioritise a pursuit of objectivity and distancing, which makes meeting animals much more difficult – as Wemelsfelder notes.

8 This work is being done jointly with Jo Hockenhull, with the collaboration of Tami Young and Sarah Redgate.

9 We ask the human and horse to carry out a simple task, to see how they work together (navigating on foot, around obstacles). Our aim is to examine closely how well a horse works with people who are familiar or not: does this affect the closeness of the engagement? Admittedly, this task must be at the human's behest, as horses are led by humans. While there are horses who will willingly follow their human about without a lead rope, these are relatively rare, and there are safety issues if we were to ask participants to do this – another practical problem facing researchers studying human–animal relationships!

10 For some years now, academic organizations, like the Association for the Study of Animal Behaviour, have required that papers submitted to their journal, *Animal Behaviour*, make clear that ethical considerations have been addressed, and they have a specific committee dedicated to ensuring and overseeing this. This does not overtly challenge the fact of using animals in experimental settings, but it does require that ethics are scrutinized. Perhaps journals in human–animal studies might do something similar?

11 By this I do not mean simply that we might learn that they don't like it. Rather, I want to ask whether the ways in which they might resist can tell us something about the research process, or about the relationship.

References

Bekoff, M. (2002) *Minding Animals: Awareness, Emotions and Heart*, Oxford: Oxford University Press.

Birke, L. (2009) "Naming names—or, what's in it for the animals?", *Humanimalia*, 1(1), available at: www.depauw.edu/humanimalia/issue01/birke.html (accessed on 14 December 2013).

Birke, L. (2007) "'Learning to speak horse': the culture of 'natural horsemanship'", *Society & Animals*, 15: 217–240.

Birke, L. (2008) "Talking about horses: control and freedom in the world of 'natural horsemanship'", *Society & Animals*, 16: 107–126.

Birke, L., Bryld, M. and Lykke, N. (2004) "Animal performances: an exploration of intersections between feminist science studies and studies of human/animal relationships", *Feminist Theory*, 5: 167–183.

Birke, L., Hockenhull, J. and Creighton, E. (2011) "The horse's tale: narratives of caring for/about horses", *Society & Animals*, 18: 331–347.

Birke, L. and Hockenhull, J. (2012) "On investigating human–animal bonds: realities, relatings, research", in L. Birke and J. Hockenhull (eds) *Crossing Boundaries: Investigating Human-Animal Relationships*, Leiden: Brill.

Bryman, A. (2004) *Social Research Methods*, Oxford: Oxford University Press.

Buller, H. (2012) "Nourishing communities: animal vitalities and food quality", in L. Birke and J. Hockenhull (eds) *Crossing Boundaries: Investigating Human-Animal Relationships*, Leiden: Brill.

Cudworth, E. (2010) *Social Lives with Other Animals: Tales of Sex, Death, and Love*, Basingstoke: Palgrave Macmillan.

Dawkins, M. S. (2006) "Through animal eyes: what behaviour tells us", *Applied Animal Behaviour Science*, 100: 4–10.

Donaldson, S. and Kymlicka, W. (2011) *Zoopolis: A Political Theory of Animal Rights*, Oxford: Oxford University Press.

Fogel, A. (2000) "Systems, attachment, and relationships", *Human Development*, 43: 314–320.

Franklin, A., Emmison, M. Haraway, D. and Travers, M. (2007) "Investigating the therapeutic benefits of companion animals: Problems and challenges", *Qualitative Sociology Review*, 3: 42–58.

Hamilton, L. and Taylor, N. (2013) "Ethnography in evolution: adapting to the animal 'other' in organizations", *Journal of Organizational Ethnography*, 1(1): 43–51.

Haraway, D. (2008) *When Species Meet*, Minneapolis: University of Minnesota Press.

Hoff, B. (1982) *The Tao of Pooh*, London: Methuen.

Irvine L. (2004) *If you Tame Me: Understanding our Connection with Animals*, Philadelphia: Temple University Press.

Irvine, L. (2007) "The question of animal selves: implications for sociological knowledge and practice", *Qualitative Sociology Review*, 3: 5–22.

King, B. J. (2004) *The Dynamic Dance. Non-vocal Communication In African Great Apes*, Cambridge, MA: Harvard University Press.

Linzey, A. (2010) "Why animal suffering matters", *Critical Society*, 4, Autumn: 1–4.

Michael, M. (2004) "On making data social: heterogeneity in sociological practice", *Qualitative Research*, 4: 5–23.

Nimmo, R. (2010) *Milk, Modernity and the Making of the Human: Purifying the Social*, London: Routledge.

Noske, B. (2003) "Horse images and the human self-image in equine research", in F. de Jonge and R. van den Bos (eds.) *The Human-Animal Relationship: Forever and a Day*, Assen: Royal van Gorcum.

Porcher, J. and Schmitt, T. (2012) "Dairy cows: workers in the shadows?", *Society & Animals*, 20: 39–60.

Sanders, C. (1999) *Understanding Dogs: Living And Working With Canine Companions*, Philadelphia: Temple University Press.

Skeggs, B. (2001) "Feminist ethnography", in P. Atkinson, A. Coffey, S. Delamont, J. Lofland and L. Lofland (eds) *A Handbook of Ethnography*, London: Sage.

Smuts, B. (2001) "Encounters with animal minds", *Journal of Consciousness Studies*, 8: 293–309.

Taylor, N. (2011) "Can sociology contribute to the emancipation of animals?", in N. Taylor and T. Signal (eds) *Theorizing Animals: Re-thinking Humanimal Relations*, Leiden, Brill.

Taylor, N. (2012) "Animals, mess, method: post-humanism, sociology and animal studies", in L. Birke and J. Hockenhull (eds.) *Crossing Boundaries: Investigating Human-Animal Relationships*, Leiden: Brill.

Wemelsfelder, F. (2012) "A science of friendly pigs … carving out a conceptual space for addressing animals as sentient beings", in L. Birke and J. Hockenhull (eds) *Crossing Boundaries: Investigating Human-Animal Relationships*, Leiden: Brill.

5 Studying perpetrators of socially-sanctioned violence against animals through the I/eye of the CAS scholar

Jessica Gröling

Critical Animal Studies (CAS) scholars seek to counter the morally and politically corrosive tendencies of conventional scholarship and its mask of objectivity and value-neutrality by explicitly advocating and aligning themselves with intersectional struggles for animal, Earth and human liberation. Their intersectional analysis of oppression requires on the one hand a systemic and historically nuanced perspective, and on the other a micro-sociological engagement with the ways in which oppressive discourses and practices fostered and reproduced in the everyday personal and professional lives of those who engage in them. In this chapter I will attend to the latter project.

In the study of human attitudes towards other-than-human animals, notions of character and personality are frequently assumed to render behaviour intelligible by providing certain behavioural predispositions, giving rise to studies that have sought to correlate character traits, personality defects or demographic factors with attitudes towards animals and their exploitation. However, the idea that particular traits are conducive to certain attitudes and that these attitudes result in cross-situational behavioural consistency is more often assumed than observed. Many studies have used or evaluated, personality-based approaches to study attitudes to animals, finding weak or conflicting correlations (Broida *et al.* 1993; Mathews and Herzog 1997; Furnham *et al.* 2003). Where the attitudes and actions of perpetrators are concerned, socially-sanctioned violence, such as animal experimentation or slaughterhouse work, is more readily conceived in terms of situational causes than the socially unacceptable abuse of animals, which, because of its pathology, has frequently been examined using an individualistic, psychopathological perspective. Even though the study of socially-sanctioned violence against animals does not have a legacy of focusing on the individual perpetrator as the locus of animal abuse, some members of the animal advocacy movement repeatedly call for research into the 'psychological makeup' of people engaged in animal experimentation or slaughterhouse work (e.g. in Calvert 2013: 75). While I reject this approach on epistemological grounds, I will argue that attending instead to situational factors carries its own epistemological, methodological and ethical constraints for the CAS scholar.

CAS scholars are committed to ideals and practices of research that reject domination and control. CAS further engenders a moral responsibility to promote reflexivity and awareness of our privileges as scholars with economic, political and representational advantages over some human and all other-than-human animals. It is therefore incumbent upon CAS scholars to avoid obscuring, sanitizing or neutralizing the reality of animal suffering in the quest for acceptance by the academic mainstream, which demands of the academic that they maintain the façade of the rational, unemotional observer. Most communities of critical researchers have struggled with the challenge of obtaining respect and scholarly acknowledgement from the mainstream, if for no other reason than to attract funding and ensure the continued viability of their fields, whilst avoiding the risk of becoming co-opted and colluding in the very ideas and activities they seek to oppose. One of the founders of CAS, in a recent proclamation of its demise, stated that the original intention was for CAS 'to thrive on the margins, not to luxuriate in the center' of academia (Best n.d.: 31). Whether and how CAS scholars are likely to become complicit in their own 'domestication' will depend largely on what they choose to study and how they negotiate the methodological and ethical constraints of scholarship in the academic–industrial complex (see Nocella *et al.* 2010). This chapter will illustrate some of the challenges that may be faced by scholars in the doing of CAS research by examining how CAS scholars might approach the study of perpetrators of socially-sanctioned speciesist violence – ontologically and epistemologically – drawing on examples from existing scholarship in human–animal studies, as well as feminist and race research.

Lessons from feminist scholarship

As a field of scholarship whose proponents are intersectional activist–academic hybrids, CAS is influenced by ideas from other struggles and critical research paradigms that have variously dealt with the theme of difference. Over the past several decades, feminists in particular have made many valuable contributions that define what critical scholars might seek to research and how they might deal with practical and ethical dilemmas in the field (e.g. DuBois in Bowles and Duelli Klien 1983; Stanley and Wise 1983; Harding 1987). For some, the language and experience of women is at the core of feminist research (DuBois in Bowles and Duelli Klien 1983). Kremer (1990), drawing on Maria Mies' (1981) guidelines for feminist research, argues that many of its components are not accessible to men, such as the view from below and the understanding of women's experiences as well as the ability to actively and fully participate in women's struggles. This notion that only women can do feminist research leads her to advocate separatist feminism. Early on, similar ideas gave rise to feminist standpoint theory (Smith 1974; Harding 2004), which departed from traditional research at the epistemological level but was soon criticized for presupposing a unitary category of 'woman' and ignoring the different standpoints of black feminists (Hill Collins 2000).

The call for sensitivity to the plurality of voices and heterogeneity of subjectivities met with a postmodern critique of the positivist obsession with dislocation, objectivity and value-neutrality and gave rise to a demand for reflexivity in research, which involves acknowledging the positionality of the researcher and how their experiences, values and intentions shape their research (Fraser and Nicholson 1990). For many feminist scholars the focus is on deconstructing and undermining oppressive knowledge structures and overcoming the androcentric bias that previously posed as a neutral account of human social life. Other feminist scholars have focused on initiating personal and emancipatory change through collaborative or participatory action research. Addressing the power hierarchies between researcher and researched and building relationships of trust and reciprocity in the research process, then, are defined as the distinctively feminist mode of inquiry (Maynard and Purvis 1994). Ann Oakley (1981), in her research on motherhood, critiques what she refers to as the masculine paradigm of interviewing, which demands that interviewers maintain a distance between themselves and their research subjects; that they keep their opinions to themselves and don't pass judgement. Her critique of this 'outsider doctrine' emphasizes that reciprocity in fieldwork relationships, in other words investing in the relationship by being willing to share one's own thoughts and experiences rather than treating the interview as an instrumental information-gathering exercise, is not only ethically desirable but also aids in the generation of rapport and a richness of research that would be otherwise unattainable. What unites these approaches is their emphasis on research that *works for* women.

However, a variety of epistemological trends have threatened and challenged feminist scholarship throughout its history. Whereas much feminist research replaced the positivist, ostensibly disinterested 'view from nowhere' (see Harding 2005 on 'weak objectivity') with a situated and politically committed viewpoint, the former is still widely regarded as necessary for the production of valid knowledge. Where a postmodernist rejection of theoretical abstractions, generalizations and claims to Truth in favour of cultural diversity has come to the rescue, it has simultaneously met with charges of relativism for paralysing already marginalized feminist perspectives.

Another challenge for feminist methodology was Nader's (1972) suggestion that feminists ought to focus more attention on the people and institutions they do not like, in other words studying 'up' as opposed to 'down'. The tendency within feminist research to study the oppressed and disempowered has led to a paucity of research on the powerful, with a few notable exceptions that I will refer to later (Luff 1999; Klatch 1988; Blee 1991, 1993, 1999, 2003; Neal 1995). By so closely associating 'critical' research with studies on and in support of advocacy groups, the potential benefits of an in-depth ethnographic focus on powerful or repellent groups have been largely ignored. The feminist commitment to empowerment and polyphony may be appropriate for giving voice to otherwise muted perspectives, but is usually less effective, I argue, for studies of powerful groups or individuals whose actions we fundamentally oppose, such as the deliberate taking of animal life for human gain. Before outlining the implications for methodology and ethics in relation to CAS research on perpetrators, all of whom have power over the animal body but are variously located on the spectrum of power within the human social context, I will discuss some of the ontological and epistemological dimensions of this challenge.

The ontology of the perpetrator

I have chosen to refer to animal experimenters, slaughterhouse workers and others engaged in violence towards other-than-human animals as perpetrators rather than offenders because the practices I am interested in are socially sanctioned (albeit often stigmatized). However, I acknowledge that the choice of this term obscures the fact that certain perpetrators may themselves be the victims of some form of oppression. Consider, for example, the oppression and suffering of workers in the animal slaughter industry (Eisnitz 2007). My main concern here is with the implications of a postmodern dissolution of the unified self for the ways in which scholars seek to explain what 'makes perpetrators tick'.

As alluded to above, the assumption that the animal movement would benefit from identifying particular personality traits or character defects that correlate with a greater likelihood to engage in or support exploitation and violence is, unfortunately, a sociological dead end. However, it prevails as an approach to research in much of contemporary human–animal studies. Situationists, on the other hand, refer to the power of situational variation to account for how inconsistent dispositions can seemingly coexist in a single person and become differentially instantiated, thus undermining the focus on individual psychopathology as the cause of individuals' direct involvement in behaviour injurious to animals. Social interactionist and postmodern scepticisms of the notion of a unified and integrated self emphasize that the self is socially constructed in interaction with others and that socialization processes and situational forces rather than 'mysterious internal states' instil 'which motives are acceptable for which action' (Cohen 2001: 59). Studies of perpetrators of racial violence and the Holocaust, particularly in the fields of sociology and social psychology, also demand greater appreciation of social structural and situational forces. Doris (2002), Arendt (1964) and Browning (1992) for instance have written about the normalization and routinization of violence and the denial of suffering that is fostered by particular constellations of situational forces. Zygmunt Bauman famously wrote about boundary construction in modernity, the social construction of 'moral distance' and the manufacture of 'moral sleeping pills' (1989: 26). He suggests that humans have a strong instinctive aversion to the suffering of others but that 'overcoming animal pity' (ibid.: 24) involves internalizing socially constructed stories and justifications for violence. Although Bauman does not specifically refer to other-than-human others, his meta-ethical theory would seem to lend itself to their inclusion. Some might venture to suggest that the oppression of other-than-human animals then represents a repression of a natural human tendency towards empathic engagement with the suffering of others.

Building on the work of well-known social psychologists Milgram (1974) and Zimbardo (2004), Albert Bandura's (1999) theory of moral disengagement sheds light on the significance of social-psychological factors for deactivating self-sanctioning mechanisms that would otherwise prevent the instantiation of violent behaviour. These include the use of euphemistic labelling and pejorative language to describe a victim, the displacement and diffusion of responsibility and attribution of blame to the victim, disregard for and distortion of the consequences of detrimental behaviour, moral justification and advantageous comparison. Bandura's work has been applied to the animal farming context by Mitchell (2011), who suggests that a number of these conditions are embedded in the practices and discourses surrounding the animal farming industry itself, ranging from the justifications and cost-benefit analysis provided by the industry, to the physical and psychological distance created between consumers and sites of animal slaughter, and the facilitated ignorance of the general population (see also Wicks 2011). I (Gröling 2011) have subsequently also applied it to the context of animal experimentation at British universities.

One might, however, counter that a theory of moral *disengagement* is committed to a particular ontological view of human nature. It assumes that humans are not, in their 'natural' state, essentially cruel or positively disposed towards violence, and that 'immoral' behaviour (so defined outside of any particular socio-historical context) requires a process of disengaging the mechanisms that would ordinarily guard against 'immoral' action. The naturalization of pity and empathy has critiqued by historical sociologist Norbert Elias (1969), who disputes the notion that a particular emotive response to the suffering of others is ingrained and argues that empathy and repugnance only become refined throughout the civilizing process. If the starting point is one of original apathy as opposed to original empathy, then the question may not be how moral self-sanctioning mechanisms become *dis*engaged but perhaps how one becomes morally *en*gaged. Nevertheless, even if one is to deny the existence of a natural human response to suffering, there is still the need to account for the 'complicated, frustrating, ambiguous, paradoxical' (Bekoff 2009: xxix) nature of our relationship with other-than-human animals. Whether it is labelled 'moral schizophrenia' (Francione 2000) or cognitive dissonance or whether one merely acknowledges the inconsistency in the ways in which some people tout their love for certain animals whilst harming other animals (often of the same species) (Arluke 1988; Birke 2003), there is a need to account for the apparent lack of compunction experienced by those involved. Scientists involved in harm-inducing experimental procedures on animals in laboratories, I argue, do not enter the profession with a pathological desire to deliberately inflict suffering on sentient animals, but become socialized into an institutional culture and form of rationality that necessitates and approves of such actions and allows individuals to return home to their companion animals without experiencing a sense of guilt or hypocrisy.

To examine how this compartmentalization is facilitated in the workplace demands a rejection of accounts of violence that focus too much on the perpetrator and not enough on situational and contextual forces. The stereotyping and demonizing of individual perpetrators of socially-sanctioned violence against other-than-human animals gives us little understanding and interventionist power to subvert or dismantle the structures of speciesist practice; in fact, it may serve to cover up these structures and their imperative towards violence. In research on people engaged in animal experimentation, a situationist focus would enable us to outline how the defensive devices and ideologies that offer justifications and rationalizations for experimentation on live animals are embedded in the collective consciousness, cultural tools, regulatory practices and infrastructure of the institution in question and give rise to interactional dynamics that can account for particular mediations of meaning and identity.

Ethnographic research in laboratories and slaughterhouses

Two scholars whose work has brought them into close contact with slaughter-house and animal laboratory workers and whose ethnographic accounts have attracted attention from animal advocates, critical scholars as well as the academic mainstream, are Timothy Pachirat and Arnold Arluke. However, it must be noted that neither refer to themselves as CAS scholars. My exploration of their work is an attempt to assess their compatibility with the CAS paradigm.

Arluke (1988, 1989, 1990, 1991a, 1991b, 1994, 2004; see also Arluke *et al.* 2006; Arluke and Sanders 1996) set out to study the occupational lives of laboratory scientists, showing a particular interest in the 'emotional labour' (Hochschild 1983) they engage in, including their efforts to suppress unpleasant emotions that arise from their encounters with animal suffering:

> [P]eople who work with them [animals] in laboratories must find ways of coming to terms with their actions. They must do so not only in terms of their own feelings about animals and animal suffering, but also in terms of their awareness of wider public antipathy.
>
> (Arluke *et al.* 2004: 4)

Through a combination of interviews, participant observation and analysis of official documents and public debates, Arluke *et al.* shed light on the process of enculturation which demands that scientists '[acquire] not only skills and knowledge but also beliefs and ethics' (ibid.: 13). Instead of engaging in the heavily polarized ethical and scientific debates on animal experimentation, theirs is an exploration of the social processes at work in the laboratory and its surroundings to identify how laboratory workers influence and reinforce each other's identities. The authors give a convincing account of socialization into the world of the animal experimenter, which involves distancing and desensitization processes, psychological defences, coping strategies and rationalizations of behaviour. New scientists learn to deal with their 'emotional and moral repugnance' (ibid.: 86) and familiarize themselves with the boundaries of acceptable behaviour in the lab. Through interaction with others in this social setting, they are taught that science is supposed to be devoid of emotion and that a career in science demands that they separate out their emotional from their professional selves. They 'learn that it is acceptable, even necessary, to suspend asking "tough" questions in order to get on with their "real" learning, which they do with a sense of excitement and awe rather than moral trepidation' (Arluke and Hafferty 1996: 223).

Socialization also implies acquiring 'motive talk', being able to express the beliefs and expectations of the scientific community in an effort to justify one's actions to oneself as much as to the outside world. Moral justification and psychological adjustment then are socially situated processes that are necessary for acquiring the identity of a biomedical scientist. The ability to evade responsibility and deny wrongdoing is further facilitated by the division of (emotional) labour in the laboratory, where tasks are broken down into manageable components and laboratory technicians and animal care staff can reasonably conceive of themselves as carers, as opposed to facilitators of animal exploitation. Staff may adopt 'discourses of ignorance' (Michael 1992), choosing not to know what happens elsewhere, and focusing solely on their particular job. Arluke et al.'s (2004) work provides valuable insights into the ways in which those engaged in animal experimentation create shared meanings and manage the contradictions and ambiguities that surround their work and their relationship to other-than-human animals, depending on their professional position in the social structure and facilitated by a variety of tools and situational factors. The methods they (Arluke in particular) used to gain such insight however could be heavily criticized from a CAS perspective because they involved direct infliction of harm:

> He found himself shaving the hair on mice, punching holes in their ears for identification, measuring the size of their tumors, injecting them with experimental drugs, 'sacrificing' them, dissecting them, and disposing of them. This degree of participation proved to be invaluable for both strengthening rapport and learning firsthand how it felt to work with animals as research tools.
>
> (Arluke and Sanders 1996: 25)

Pachirat (2011) conducted in-depth ethnographic work in an American slaughterhouse to which he gained covert access, adding to a number of other covert studies of slaughterhouse work (Thompson 1983, Fink 1998). Similar to Arluke's motivations for participatory ethnographic work, Pachirat sought to remove the distance that ordinarily separates academics from the people and social worlds that are the subject of their research. By deceiving the slaughterhouse managers to gain access not as an academic but as a *bona fide* employee, he not only intended to acquire an authentic experience of the work of industrialized killing but also wanted to avoid the added attention and potential suspicion that he may have attracted had he presented himself as an academic. He wanted to avoid becoming implicated in the power hierarchy and being seen by other line workers as someone with greater social and economic capital than them. During his time at the slaughterhouse, Pachirat gradually moved between jobs, some of which involved processing dismembered body parts and others that put him into direct contact with live animals immediately before their death. He also spent some time as a quality control worker, which gave him an insight into the internal hierarchies of the slaughterhouse.

Pachirat's focus was on the relationship between power and sight in the slaughterhouse and the ways in which the reality of animal slaughter is made tolerable for all involved. The slaughter of animals is typically hidden from public view, something that is thought to perpetuate the public's ignorance of the origins of their food and often leads activists to suggest that 'if slaughter-houses had glass walls, the world would be vegetarian/vegan'. Cohen (2001: 185) argues that the animal advocacy movement represents one of the 'living relics of Enlightenment faith in the power of knowledge: if only people knew, they would act'. Indeed, 'the impulse to link sight and political transformation is strong' (Pachirat 2011: 242). Pachirat reveals 'not only how distance and concealment segregate the slaughterhouse from society as a whole but also how surveillance and concealment sequester the participants from the work of killing within the walls of the slaughterhouse itself' (2011: 236). His ethnography draws attention to the internal divisions within the slaughterhouse that create physical, social and linguistic distance:

> The divisions of labour and space on the kill floor work to fragment sight, to fracture experience, and to neutralize the work of violence. But what I realize as I settle back into the hypnotic rhythms of wiping hooks and hanging livers by their posterior venae cavae is that this fragmentation, fracturing, and neutralization also creates pockets of refuge, places of safety and sanity even here in the heart of the slaughterhouse.
>
> (Pachirat 2011: 159)

A particularly important observation is that all slaughterhouse employees can reasonably deflect responsibility for their part in the killing as long as they are not the person who puts the bolt gun to the animal's head.

> 'Only the knocker'. It is simple moral math: the kill floor operates with 120 + 1 jobs. And as long as the 1 exists, as long as there is some plausible narrative that concentrates the heaviest weight of the dirtiest work on this 1, then the other 120 *kill floor workers* can say, and believe it, 'I'm not going to take part in this. I'm not going to stand and watch this.'
>
> (Pachirat 2011: 160, emphasis in original)

Pachirat draws on Foucault's (1974) work on visibility and power, where the removal of barriers to sight itself has the power to exert a new form of control, much as in Bentham's panopticon. The work of the kill floor is broken down into a long chain of isolated tasks but one would expect that the total visibility afforded to the quality controllers in charge of overseeing the work of the kill floor from the vantage point of the catwalk above would render the process more emotionally or psychologically challenging for them:

[Y]et it is simultaneously this work of surveillance that deflects the QC's attention from an experiential grasp of the underlying work of killing that is taking place, despite a view from above which renders each step of the killing process visually accessible. (...) [T]he QC becomes an exemplary instance of how experiential compartmentalization is produced even, and perhaps especially, under conditions of total visibility.

(Pachirat 2011: 232)

The quality controllers' job description demands that they focus their attention on monitoring hygiene practices and filling in audit forms. Similar to the animal laboratory example, animals here become easily objectified and viewed only as data inputs and statistics.

Hamilton and Taylor (2013), whose research involved work-shadowing meat inspectors in British abattoirs (Hamilton) and interviewing slaughterhouse workers (Taylor), are similarly interested in the boundary-drawing activities of those involved in the killing industry and believe that 'exposing the workings of the killing floor is the best way to foster a broader discussion of the politics involved' (ibid.: 66), for the detail that is often missed out in meta-theoretical analyses:

tell[s] us how it is that individual humans can do what they do to individual animals, how the 'peculiar relationship' to animals exists in a vague state somewhere 'between meat and mercy' and how that relationship is embedded in a whole host of daily practices, routines and repetitions.

(Ibid.: 66)

Arluke and Pachirat's research, however, is based on detailed first-hand accounts by academics who have chosen to *become* their subjects in order to gain 'intimate familiarity' (Blumer 1969) with their motivations, activities, struggles and relationships with others in their social setting. In light of the many ethical issues raised by this form of research, particularly from a CAS perspective, the epistemological value of participant observation in relation to other potential methods such as interviews deserves further exploration.

The ethnographic I/eye

Ideally, ethnography is about teasing out the hidden, symbolic, moral, or pragmatic logics that underlie certain types of social action; the way people's habits and actions make sense in ways in which they are not themselves completely aware.

(Graeber 2009: 111)

In contrast to the traditional scientific method, which requires separation and emotional distance, good ethnography involves establishing rapport, developing empathy and stepping into the shoes of those whom we seek to study to see the world from their eyes. Sanders argues (in Arluke 2004) that there is a consider-able difference between what people say they do (and why) and what they actu-ally do. Participating in the setting or using the self as an ethnographic resource then enables the researcher to evaluate people according to their actions, not merely their own verbal accounts of these actions:

> Ethnographic research like Arluke's avoids this source of error by allowing the investigator to, in a sense, see attitudes in action. The ethnographer can use events in the field to validate, refine, and question participants' verbal accounts of the way things are and how they would act in certain circum-stances (...).
>
> (Sanders in Arluke 2004: x)

Presence and participation in the events being described then are thought to lend ethnographic authority. Aside from their presence in the field, the ethnographer's perceptual ability and objectivity are also thought to be partly dependent on their prior dispassionate or neutral stance with respect to the actions of the group of people being studied. Objectivity is variously understood as an attribute of reality, of the knower, and of the knowledge they produce, where the objectivity of the latter is predicated on the prior neutrality of the epistemic agent with respect to the reality being observed (Lloyd 1995). The power of Haraway's 'modest witness', 'the authorized ventriloquist for the object world' (1997: 24), gives rise to the differentiation between expert knowledge and mere opinion or propaganda. For an openly critical scholar to engage in participant observation may raise suspicions by unsympathetic audiences as to the sincerity of the obser-vations being made, just as animal activists engaging in covert investigations may come under fire for supposedly cherry-picking their evidence of animal cruelty. Cassell (1988: 92–93) argues that the ability to empathize is essential to truly know the 'Other' and appreciate their perspective and that for empathy to be possible requires a degree of neutrality. In other words, to understand and empathize with the animal experimenter or slaughterhouse worker, and thereby gain ethnographic insight, would require a prior neutral attitude with regard to animal experimentation and slaughterhouse work. Criticisms of the possibility and desirability of a 'neutral, dispassionate, cool and rational, objective observer of the human condition' (Shepher-Hughes 1995: 410) are not new, whether it be from the feminist perspective previously outlined, or from within interpretivist (Blumer 1969) and critical theory traditions (Horkheimer 1937; Adorno 1966; Gouldner 1964). Despite the fact that many qualitative researchers now agree that value-neutrality is impossible, their desire to produce objective research still causes them to claim that they can 'bracket' their own assumptions and values (Strega in Brown and Strega 2005).

However, asserting their ostensible neutrality by not taking sides often merely has the effect of obscuring or re-embedding existing power relations. Becker argues that 'the question is not whether we should take sides, since we inevitably will, but rather whose side we are on' (1967: 239). The dilemma of whether to

take sides then becomes more a question of whether researchers should strategically safeguard their private convictions or be open about them. Whether it is academically acceptable for the researcher to side with the underdog depends in
large part on whether the research takes place in an already politicized situation.
However, where not taking a stand on contentious issues such as animal experimentation serves inadvertently to perpetuate the status quo, the masquerade of
neutrality is just as much a committed position as any other (see Marcuse 1969
on 'repressive tolerance').

Nevertheless, it is reasonable to expect that disapproval of practices involving violence towards other-than-human animals may indeed get in the way of
CAS scholars wanting to engage in participant observation in the first place.
The question of how distant researchers must be from the Other in terms of
their values and motivations then is related to the question of how close they
need to get to them and their actions in order to understand. The social desirability bias causes some social scientists to be suspicious of verbal accounts
given in interview situations. Goldhagen, author of the controversial *Hitler's
Willing Executioners* (1996), for instance, is very outspoken on the need to not
accept at face value some of the self-exonerating accounts of perpetrators and
has criticized Browning for basing his famous work *Ordinary Men* (1992)
largely on the testimony of Germans involved in the Holocaust. The suggestion that where stigma or serious social criticism is attached to particular
actions or professions, their representatives may produce dishonest and therefore ostensibly useless accounts is an important one to consider also in our
context (see also Hurn, forthcoming). The modernist notion of the rational,
self-reflexive research subject conceives of the interview as a simple snapshot
of that subject's perspective. However, the social constructionist and interactionist critique counters that the performative nature of the interview situation
is too frequently ignored and that accounts are just as much about the space
created for the respondent to occupy as they are about his or her ideas and
experiences. It is important then to bear in mind that interviewers and interviewees are jointly involved in the production of accounts, eliciting presentations of self that are likely to conform to the dominant discourses of
subjectivity in a particular context (Clifford and Marcus 1986: 107). Participants are likely to narrate themselves either in terms that they perceive to be
appropriate and acceptable to the researcher or 'yielding little more than organizational slogans repeated as personal beliefs' (Blee 2003: 201; see also
DeSoucey in Cherry *et al.* 2011). Interviewees may come to occupy particular
subject positions as a situational response to the presence of the particular
interviewer, re-emphasizing the notion of the dialogic self, which is narrated
as much for the individual themselves as it is for the interviewer (Denzin
1997).

It may well be the case that Pachirat and Arluke would have elicited very different presentations of self from their knowing and unknowing research participants had they framed themselves and their involvement differently. This further
underlines my previous argument about the susceptibility of the self to situational

and interactional variation. Reflexive interpretative analysis of interviews therefore requires taking into account the researcher's position as perceived by the
interviewee.

However, these accounts are manifestations of a negotiation of the self that
occurs not just in the interview but also in the interviewee's day-to-day activities. Contrary to survey- and attitude-based studies of perpetrators, and '[r]ather
than interpreting actions and language as external manifestations of subjective
and deeper-lying elements in individuals, the research task is the locating of particular types of action within typal frames of normative actions and socially situated clusters of motive' (Mills 1940: 913). The Freudian concern for
deeper-lying, 'real' motives is obviated when the focus of the research is the
situation and its typical vocabularies of motive. The latter function as cues and
justifications in constraining and inducing particular actions and explain not *why*
but *how* such actions came about. History can help explain why humans came to
experiment on other animals, but what we are presently interested in are the
factors that create continued involvement in the practice, thereby demonstrating
the intrinsically social character of motivation. Such an investigation helps to
explain how situations give rise to or facilitate the compartmentalization of
motives and the resulting attitudinal and behavioural ambiguity:

> In interviews, scientists tended to present themselves through a kind of
> rational emotionality – that is, they defended animal use, including their
> own, but acknowledged past and present emotional reactions to it in the
> form of admitting that there were some things they could not do (...). By
> expressing ambivalence, these scientists are indicating their engagement in
> quite different networks, allowing them to claim identities as both scientist
> and simultaneously as respecter of animals.
>
> (Arluke *et al.* 2006: 92)

C. Wright Mills dispels as a distraction the focus on whether the verbalized
motive is representative of the 'real' motivation as opposed to an *ex post facto*
rationalization or justification for an action:

> Motives are accepted justifications for present, future, or past programs or
> acts. To term them justification is *not* to deny their efficacy. Often anticipa-
> tions of acceptable justifications will control conduct. ('If I did this, what
> could I say? What would they say?') Decisions may be, wholly or in part,
> delimited by answers to such queries. (...) [A]ll we can meaningfully be
> asking for is the controlling speech form which was incipiently or overtly
> presented in the performed act or series of acts. There is no way to plumb
> behind verbalization into an individual and directly check our motive-
> mongering, but there is an empirical way in which we can guide and limit,
> in given historical situations, investigations of motives. That is, by construc-
> tion of typal vocabularies of motives that are extant in types of situations
> and actions. Imputations of motives may be controlled by reference to the

typical constellation of motives which are observed to be societally linked with classes of situated actions.

(Mills 1940: 907–910)

What this means is that the discursive vocabularies of motive presented in an interview and the interviewee's 'real' motivations are only *temporally* different. Although the suspect nature of retrospective accounts is something that does need to be considered, the crucial point to emphasize is that different audiences will require different accounts and that by prompting interviewees into different 'discourse registers' (Cameron and Frazer 1989), the researcher can ascertain the interviewee's familiarity with and reliance on particular cultural frames and forms of rationality in their everyday professional lives, generating important insights into the situational forces that influence behaviour and highlighting potential points for intervention. Interviews, then, are not to be discounted as a potentially useful method for research into violence against animals, where more participatory research is not considered appropriate.

Conventional and critical research ethics: dilemmas and possibilities

There are ethical and practical issues with studying perpetrators that will be felt by scholars coming from an ostensibly neutral perspective, but there are additional ethical dilemmas involved in such research that are perhaps perceived more acutely by critical scholars. Just as feminist scholars conducting research that can be understood as being for women but not just about women, especially those 'studying up', tend not to draw on the feminist ethics and methods of inquiry previously outlined, CAS scholars may struggle to conduct research in settings such as laboratories and slaughterhouses, not least because scholars engaged in empirical research have immediate obligations towards diverse and often opposed parties: other-than-human animals and other disadvantaged groups, activists, academics, institutions, funding bodies and human research participants.

Most scholars are bound by the ethical guidelines and codes of practice of professional associations, such as (in the UK context) the British Sociological Association's (BSA) *Statement of Ethical Practice* (2002) and the Association of Social Anthropologists of the UK and the Commonwealth's (ASA) *Ethical Guidelines for Good Research Practice* (1999), or guidelines set by funding bodies and ethics committees at research institutions. Research ethics are increasingly defined in terms of the rights of the research subject and the obligations of the researcher. According to the BSA, '[s]ociologists have a responsibility to ensure that the physical, social and psychological well-being of research participants is not adversely affected by the research. They should strive to protect the rights of those they study, their interests, sensitivities and privacy, while recognizing the difficulty of balancing potentially conflicting interests' (2002: 3). In other words, it is the researcher's role to ensure confidentiality and anonymity, minimise the impact of questions that may trigger the recollection or reporting of stressful events and avoid deceiving research participants about aims, methods or risks, in order to protect the social, psychological and economic interests of the research subjects (Denscombe 1998). Most institutional ethics committees or review boards assessing proposals for empirical research with human subjects therefore require thorough risk assessments and assurances from the researcher that the necessary safeguards have been put in place to protect participants. Committees usually request that participants are given written consent forms that disclose risks, give information about what the research is about, who is undertaking it and why it is being undertaken, who is funding it, how the research will be used and disseminated and how participants' personal data will be securely stored and protected. Consent, it is routinely emphasized, is an iterative process. In other words, researchers must inform participants of any changes to the above and allow participants to withdraw from the research at any time.

Researchers also have ethical obligations to the institution they are a member of, as well as to funding bodies and the research community in general. Unethical behaviour may have the implication of damaging existing relationships in the field or closing off access for future research. Ethics committees also recognize that research is not a unidirectional process and that researchers may themselves be put at risk.

Problems arise where there are conflicts of interest. The ASA states that 'when there is a conflict, the interests and rights of those studied should come first' (1999: 2), adding that with research on vulnerable groups 'it may be necessary to withhold data from publication or even to refrain from studying them at all' (1999: 2). However, the BSA suggests that 'where power is being abused, obligations of trust and protection may weigh less heavily' (2002: 3). This raises the obvious question of who is to decide where and when power is being abused. The power abuses that CAS scholars want to draw attention to are usually not considered thus by the rest of the academic community. How then, within the ethical remit set out by ethics committees and codes of practice, are CAS scholars to address their ethical responsibilities to those who are at the other end of speciesist violence? All too frequently research is assumed to be an innocent pursuit, part of a 'telic' enterprise, where the advancement of knowledge contributes to the 'common good' and inevitably generates social progress. Where 'ethical research' is that which safeguards the inviolable interests of those being studied irrespective of the political context and the values of the researcher *vis à vis* the research population, research becomes inherently conservative. The principle of beneficence coupled with an expectation as to the objectivity and value-neutrality of the researcher may give the impression that researchers ought not take a stand on ethical questions, such as whether animal experimentation is justified and whether there is a utilitarian argument for overriding some of the interests and rights of animal experimenters as a result.

The fallout from covert social scientific research that has been deemed unethical (e.g. Humphreys 1970) has raised questions as to the acceptability of violating the principle of informed consent in research in general, even where the research design necessitates that participants are deceived about the true purpose of the research so as to generate valid data (e.g. Milgram 1963). Many institutional ethics committees therefore will not grant approval for covert research of this kind. Douglas (1976), a key proponent of 'investigative social research', which advocates using deception to build affection and trust, has been heavily criticized for advancing adversary research that relies on morally dubious techniques and deliberate manipulation (Gold 1977). When considered together, all of these factors make it considerably less likely that powerful groups or institutions will be studied from an adversary angle, and that such studies would receive the necessary funding.

Ethics as defined from within a critical paradigm such as feminism or CAS, may come with additional challenges and recommendations for acceptable behaviour in the field:

> Whereas previously ethical considerations were believed to set boundaries to what researchers could do in pursuit of knowledge, now ethical considerations are treated by some as constituting the very rationale of research. For example, its task becomes the promotion of social justice.
>
> (Hammersley 1999: 18)

From a CAS perspective, a major practical obstacle to research with perpetrators relates to the difficulty of gaining physical and social access to the research setting. Access to a financially or politically powerful institution or a group that is the subject of stigma, secrecy, political or ethical debate is most likely to be granted if the research seems harmless.

> To penetrate the shields of the powerful the social scientist has to be lucky and/or devious.
>
> (Burawoy 1998: 22)

Where the subject is a powerful and state-sanctioned institution, members may have little reason to tolerate the investigative presence of an academic. Gaining access therefore usually requires ingratiating oneself or being deceitful as to one's motivations and intentions:

> If an antagonistic researcher wishes to use the traditional anthropological technique of participant observation to study the culture of power, the investigator's approach must, of necessity, involve a certain falseness. Motives must be concealed: the ethnographer cannot say, 'this research is designed to expose your misbehaviour'.
>
> (Cassell 1988: 90–91)

In the USA, the introduction of 'ag-gag' laws may impede access to animal-exploiting industries, signalling defensiveness on their part and a recognition of the importance of the flow of information for the future course of human–animal relations. Marginalized and disempowered groups may also respond with suspicion to requests by academics to study them and may be more welcoming if they perceive the research process to be a *quid pro quo* relationship where, in allowing themselves to be studied, the group becomes entitled to the social power of the researcher who, as a result, risks becoming a symbolic member of the group and lending it legitimacy.

To avoid having to make promises of reciprocity or neutrality, some advocate going undercover. Nigel Fielding (1990), in his research on the National Front in England, consciously presented himself as a potential convert, which enabled him to 'pass' as a sympathizer and afforded him an insider's view into the movement. However, covert research requires skill in the presentation of self and can be a dangerous undertaking:

> As in other situations where identities have to be created or established, much thought must be given by the ethnographer to 'impression management'. Impressions that pose an obstacle to access must be avoided or countered as far as possible, while those that facilitate it must be encouraged, *within the limits set by ethical considerations.*
>
> (Hammersley and Atkinson 1995: 83, my emphasis)

Another important consideration for covert or deceptive research is that false identities may be tested in the field, a problem that afflicts undercover activists and academics alike (Cherry *et al.* 2011). This is especially so in the age of the internet, where a simple internet search of a name may undermine attempts at masking subversion.

Where covert research is not an option – either for practical reasons, because an institutional ethics committee won't endorse it, or because it would necessitate first-hand involvement in harm-inducing practices that the researcher is not willing to engage in – a more distanced, interview-based approach may be necessary. This, as we have seen, need not undermine the research from an epistemological point of view; however, it does not remove the potential sense that one is colluding in or giving tacit approval to violent or abhorrent practices, as the following reflections from feminist and anti-racist scholars exemplify:

> How does a researcher in this position offer objections or challenges to racism without jeopardizing the interview and perhaps, if the respondent is powerful, access to the whole case-study. Even if the outcome is not so extreme, an objection could result in a significant alteration in the nature of what the respondent is prepared to tell the researcher. Yet to remain silent implicitly implies consent. [...] When I did remain silent when racism was encountered in the interview process my political dilemma was acute. That I had 'sold out', colluded and compromised were feelings that I continually dragged along with me through the fieldwork.
>
> (Neal 1995: 528)

> Listening to views, nodding or saying simple 'ums' or 'I see', to views that you strongly disagree with or, ordinarily, would strive to challenge, may be true to a methodology that aims to listen seriously to the views and experiences of others, but can feel personally very difficult and lead to questioning of the whole research agenda.
>
> (Luff 1999: 698)

Herman (1994), Klatch (1988) and Blee (1999) have similarly expressed a concern that friendly gestures, smiles, nods and a sympathetic tone of voice wrongly convey concurrence with others' beliefs. Throughout her research, Blee (1999) notes the conflict she experienced in balancing her anti-racist values with her desire to gain a deeper insight into female membership of the Ku Klux Klan.

> Many interviews include polite sparring, with informants making laudatory comments about the Klan and seeking a positive response in return. Once informants decided that I was unlikely to denounce their Klan membership during the interview, though, they were forthcoming with opinions, prejudices, and memories.
>
> (Blee 1991: 5)

The oft-noted concern that research on and with perpetrators or advocates of violence risks collapsing into collusion also manifests itself as a fear of uncomfortable and unwelcome resemblances between researcher and researched. Klatch (1988) writes about building rapport with her informants, female members of the American New Right, on the basis of the common ground they shared as women. However, both Klatch (1988) and Blee (1991) express their fear of recognition:

> I was prepared to hate and fear my informants. My own commitment to progressive politics prepared me to find these people [female members of the Klan] strange, even repellent. I expected no rapport, no shared assumptions, no commonality of thought or experience. What I found was more disturbing. Many of the people I interviewed were interesting, intelligent, and well informed. Despite my prediction that we would experience each other as completely foreign, in fact I shared the assumptions and opinions of my informants on a number of topics (...). These former Klan members were not the 'other', with strange, incomprehensible ways of understanding the world, as I had earlier assumed.
>
> (Blee 1991: 6)

However, Back (2002), whose research brought him into uncomfortable contact with proponents of racism and nationalist politics, advocates for the possibility of a 'mutual destabilization (...) on both sides of the ethnographic divide (...) that disrupts the ethnographer's preconceptions, while at the same time critically evaluating the taken-for-granted predispositions of the communities under study' (ibid.: 43). Defenders of this view may indeed argue that the perceived need to denounce the views of perpetrators and advocates of violence at every opportunity stems from an indulgent desire to maintain one's personal feeling of moral purity. It can be argued that to repeatedly challenge views that one finds objectionable may be either futile or counterproductive in a research setting and that for reasons of access and understanding, it is necessary to refrain from or limit confrontation. However, as I have already suggested, prompting one's research participants into different 'discourse registers', be it through questioning or challenging another's claims or views, can itself be a useful exercise and can contribute to an understanding of the situational forces that person is subjected to.

Conclusion

The questions posed here are many and complex and this chapter intends only to be the start of a conversation. How can scholars maintain their dual positionality as animal liberationists and *bona fide* scholars if their interest is in empirical research involving people whose actions they fundamentally oppose? How do scholars speak to, and on behalf of, diverse audiences made up of research participants, CAS scholars and other academics, while also maintaining their primary commitment to other-than-human animals? The temptation to cultivate an image of the dispassionate observer and masquerade as the 'liberal' scholar (Root 1993) becomes greater the more scholars are determined to produce research that will be acceptable to academics, activists and lay people with different and often competing commitments. This general acceptability, however, remains a mainstream academic requirement for epistemic credibility. The possibility and desirability of participant observation is further undermined by ethical issues and although a case can be made for the value of interview-based research in place of first-hand experience, this does not negate the numerous practical and ethical challenges that CAS scholars are likely to face. Non-neutrality can improve epistemic validity in circumstances where sympathetic subjectivities aid in the building of trust and rapport and lend depth and credibility. However, the silence or deception strategically necessitated by antagonistic research to gain and maintain access to a research setting could result in a sense of collusion or betrayal in the practices whose desired abolition was a motivating factor for the research. The fusion of the role of activist and scholar is furthermore difficult to reconcile with the academic demands for disclosure and informed consent. Nader (1972) instead recommends using document analysis and methods that do not necessitate direct engagement through deception, which is an option I have also considered elsewhere (Gröling 2013).

It is clear that the epistemological criteria for what is considered 'good' empirical research were constructed according to a particular canon of research that assumes for scholarship an untenable amoral relativism with respect to all but its procedural norms. In reality, the pursuit of knowledge is never sheltered from the exercise of power and this is rightly foremost in the minds of those pursuing empirical research from a CAS perspective. It is becoming common practice for contemporary ethnographers to acknowledge where their own 'biases' might have influenced their research, but this is often done in a somewhat apologetic fashion. The current situation with regards to reflexivity is one of imbalance, where some 'biases' (e.g. feminist, pro-animal) are reflected upon more frequently than others that have enjoyed some universal privilege (e.g. male, humanist). This has the effect of giving the impression that some biases are more distorting than others. The invisibility of mainstream biases, which is often the unintended consequence of reflexivity, is problematic. CAS is building a case for moving beyond reflexivity, and for advocacy in the social sciences, with a view to shifting the markers of epistemological acceptability so that validity is no longer determined by the purported absence, or indeed the apologetically acknowledged presence, of the researcher, but rather by their committed presence and their informed and consistent attitude towards injustices. In conclusion, the challenges faced by critical scholars are as much political as they are technical. Research that seeks to challenge relations of dominance with respect to human and other-than-human animals must simultaneously challenge the hegemony of existing research paradigms, their ontological foundations and markers of validity and reliability.

References

Arendt, H. (1964) *Eichmann in Jerusalem: A Report on the Banality of Evil*, New York: Penguin Books.

Arluke, A. (1988) 'Sacrificial symbolism in animal experimentation: Object or pet?', *Anthrozoös*, 2: 98–117.

Arluke, A. (1989) 'Living with contradictions', *Anthrozoös*, 3: 90–99.

Arluke, A. (1990) 'Uneasiness among laboratory technicians', *Lab Animal*, 19: 21–39.

Arluke, A. (1991a) 'The ethical thinking of animal researchers: problems and prospects', *The New Biologist*, 3(1): 1–2.

Arluke, A. (1991b) 'Going into the closet with science: information control among animal experimenters', *Journal of Contemporary Ethnography*, 20: 306–330.

Arluke, A. (1994) 'The ethical socialization of animal researchers', *Lab Animal*, 23(6): 30–35.

Arluke, A. (2004) *Brute Force: Animal Police and the Challenge of Cruelty*, West Lafayette, Indiana: Purdue University Press.

Arluke, A. and Hafferty, F. (1996) 'From apprehension to fascination with "dog lab": The use of absolutions by medical students', *Journal of Contemporary Ethnography*, 25: 201–225.

Arluke, A. and Sanders, C. R. (1996) *Regarding Animals*, Philadelphia: Temple University Press.

Arluke, A., Birke, L. and Michael, M. (2006) *The Sacrifice: How Scientific Experiments Transform Animals and People*, West Lafayette, Indiana: Purdue University Press.

Adorno, T. (1966/1973) *Negative Dialectics*, New York: Seabury Press.

Association of Social Anthropologists of the UK and the Commonwealth (1999) *Ethical Guidelines for Good Research Practice*, available at: www.theasa.org/ethics/Ethical_guidelines.pdf (accessed 20 May 2013).

Back, L. (2002) 'Guess Who's coming to dinner? The political morality of investigating whiteness in the gray zone', in V. Ware and L. Back (eds) *Out of Whiteness: Color, Politics and Culture*, London: University of Chicago Press.

Bandura, A. (1999) 'Moral disengagement in the perpetration of inhumanities', *Personality and Social Psychology Review*, 3: 193–209.

Bauman, Z. (1989) *Modernity and the Holocaust*, Ithaca: Cornell University Press.

Becker, H. (1967) 'Whose side are we on?', *Social Problems*, 14(3): 239–247.

Beckoff, M. (2009) 'Increasing our compassion footprint: it's simple to make changes to accrue compassion credits', *Human Ecology Review*, 16(1): 49–50.

Best, S. (n.d.). *The Rise (and Fall) of Critical Animal Studies*, available at: www.liberazioni.org/articoli/BestS-TheRise(and%20Fall)ofCriticalAnimalStudies.pdf (accessed 20 May 2013).

Birke, L. (2003) 'Who–or what–are the rats (and mice) in the laboratory?', *Society & Animals*, 11: 207–224.

Blee, K. (1991) *Women of the Klan: Racism and Gender in the 1920s*, Berkeley: University of California Press.

Blee, K. (1993) 'Evidence, empathy, and ethics: lessons from oral histories of the Klan', *Journal of American History*, 80: 593–606.

Blee, K. (1999) 'From the field to the courthouse: the perils of privilege', *Law and Social Inquiry*, 24: 401–405.

Blee, K. (2003) *Inside Organized Racism: Women in the Hate Movement*, California: University of California Press.

Blumer, H. (1969) *Symbolic interactionism: Perspective and method*, Englewood Cliff, NJ: Prentice Hall.

British Sociological Association (2002) *Statement of Ethical Practice for the British Sociological Association*, available at: www.britsoc.co.uk/media/27107/StatementofEthicalPractice.pdf (accessed 20 May 2013).

Broida, J., Tingley, L., Kimball, R. and Miele, J. (1993) 'Personality differences between pro-and anti-vivisectionists', *Society & Animals*, 1: 129–144.

Browning, C. R. (1992) *Ordinary Men: Reserve Police Battalion 101 and the Final Solution in Poland*, New York: HarperCollins.

Burawoy, M. (1998) 'The extended case method', *Sociological Theory*, 16(1): 4–33.

Calvert, C. (2013) 'Academics and activists: responses and reflections', in C. D. Calvert and J. S. Gröling (eds) Proceedings of the Critical Perspectives on Animals in Society conference 2012, available at: http://hdl.handle.net/10871/8582 (accessed 20 May 2013).

Cameron, D. and Frazer, E. (1989) 'Knowing what to say: the construction of gender in linguistic practice', in R. Grillo (ed.) *Social Anthropology and the Politics of Language. Sociological Review Monograph 36*, London: Routledge.

Cassell, J. (1988) 'The relationship of observer to observed when studying up', in R. G. Burgess (ed.) *Studies in Qualitative Methodology*, London: JAI Press.

Cherry, E., Ellis, C. and DeSoucey, M. (2011) 'Food for thought, thought for food: consumption, identity, and ethnography', *Journal of Contemporary Ethnography*, 40(2): 231–258.

Clifford, J. and Marcus, G. E. (1986) *Writing Culture: The Poetics and Politics of Ethnography*, California: University of California Press.

Cohen, S. (2001) *States of Denial: Knowing About Atrocities and Suffering*, Cambridge: Polity Press.

Denscombe, M. (1998) *The Good Research Guide*, Buckingham: Open University Press.

Denzin, N. K. (1997) *Interpretive Ethnography: Ethnographic Practices for the 21st Century*, Thousand Oaks, CA:Sage.

Doris, J. M. (2002) *Lack of Character: Personality and Moral Behavior*, Cambridge: Cambridge University Press.

Douglas, J. D. (1976) *Investigative Social Research: Individual and Team Field Research*, Beverly Hills: Sage Publications.

DuBois, B. (1983) 'Passionate scholarship: notes on values, knowing and method in feminist social sciences', in G. Bowles and R. Duelli Klien (ed.) *Theories of Women's Studies*, London: Routledge and Kegan Paul.

Eisnitz, G. (2007) *Slaughterhouse: The Shocking Story of Greed, Neglect and Inhumane Treatment Inside the U.S. Meat Industry*, New York: Prometheus Books.

Elias, N. (1969) *The Civilizing Process, Vol. I. The History of Manners*, Oxford: Blackwell.

Fielding, N. (1990) 'Mediating the message: affinity and hostility in research on sensitive topics', *American Behavioral Scientist*, 33: 608–620.

Fink, D. (1998) *Cutting Into The Meatpacking Line: Workers And Change In The Rural Midwest*, London: Chapel Hill.

Foucault, M. (1974) 'The eye of power', in C. Gordon (ed.) *Power/Knowledge: Selected Interviews and Other Writings*, New York: Pantheon.

Francione, G. (2000) *Introduction to Animal Rights: Your Child or the Dog?* Philadelphia: Temple University Press.

Fraser, N. and Nicholson, L. J. (1990) 'Social criticism without philosophy: an encounter between feminism and postmodernism', in L. J Nicholson (ed.) *Feminism/Postmodernism*, London: Routledge.

Furnham, A., McManus, C. and Scott, D. (2003) 'Personality, empathy and attitudes to animal welfare', *Anthrozoös*, 16(2): 135–146.

Gold, R. L. (1977) 'Review: investigative social research: individual and team field research by Jack D. Douglas; Doing Social Life: The Qualitative Study of Human Interaction in Natural Settings by John Lofland', *Contemporary Sociology*, 6(6): 654–656.

Goldhagen, D. J. (1996) *Hitler's Willing Executioners: Ordinary Germans and the Holocaust*, New York: Knopf.

Gouldner, A. (1964) 'Anti-minotaur: the myth of a value-free sociology', *Social Problems*, 9(3): 199–213.

Graeber, D. (2009) 'Anarchism, academia, and the avant-garde', in R. Amster, A. DeLeon, L. Fernandez, A. Nocella and D. Shannon (eds) *Contemporary Anarchist Studies: An Introductory Anthology of Anarchy in the Academy*, Oxon: Routledge.

Gröling, J. S. (2011) ' "When science speaks, let no dog bark": moral disengagement in animal experimentation at British universities', paper presented at the Reconfiguring the 'Human'/'Animal' Binary – Resisting Violence conference, Prague, October.

Gröling, J. S. (2013) 'University ethical review committees and the Animals (Scientific Procedures) Act: using the Freedom of Information Act as a research tool', in C. D. Calvert, and J. S. Gröling (eds) Proceedings of the Critical Perspectives on Animals in Society conference 2012, available at: http://hdl.handle.net/10871/8582 (accessed 20 May 2013).

Hamilton, L. and Taylor, N. (2013) *Animals at Work: Identity, Politics and Culture in Work With Animals*, Leiden: Brill.

Hammersley, M. (1999) 'Some reflections on the current state of qualitative research', *Research Intelligence*, 70: 16–18.

Hammersley, M. and Atkinson, P. (1995) *Ethnography: Principles in Practice*, London: Routledge.

Haraway, D. (1997) *Modest_Witness@Second_Millennium.FemaleMan(c)_Meets_Onco-Mouse: Feminist and Technoscience*, London: Taylor and Francis.

Harding, S. (1987) 'The method question', *Hypatia*, 2(3): 19–35.

Harding, S. (2004) 'Introduction: standpoint theory as a site of political, philosophic and scientific debate', in S. Harding, (ed.) *The Feminist Standpoint Theory Reader: Intellectual and Political Controversies*, London: Routledge.

Harding, S. (2005) 'Rethinking standpoint epistemology: what is "strong objectivity?"', in A. Cudd and R. Andreasen (eds) *Feminist Theory: A Philosophical Anthology*, Oxford: Blackwell Publishing.

Herman, D. (1994) *Rights of Passage: Struggles for Lesbian and Gay Legal Equality*, Toronto: University of Toronto Press.

Hill Collins, P. (2000) *Black Feminist Thought: Knowledge, Consciousness, and the Politics of Empowerment*, London: Routledge.

Hochschild, A. (1983) *The Managed Heart*, Berkeley: University of California Press.

Horkheimer, M. (1937/1976) 'Traditional and critical theory', in P. Connerton (ed.) *Critical Sociology: Selected Readings*, Harmondsworth: Penguin.

Humphreys, L. (1970) *Tearoom Trade: Impersonal Sex in Public Places*, London: Duckworth.

Hurn, S. (forthcoming) *Human-Animal Farm*, Guildford: Ashgate.

Klatch, R. E. (1988) 'The methodological problems of studying a politically resistant community', in R. Burgess (ed.) *Studies in Qualitative Methodology*, London: JAI Press.

Kremer, B. (1990) 'Learning to say no: keeping feminist research for ourselves', *Women's Studies International Forum*, 13(5): 463–467.

Lloyd, E. A. (1995) 'Objectivity and the double standard for feminist epistemologies', *Synthese*, 104(3): 351–381.

Luff, D. (1999) 'Dialogue across the divides: "Moments of rapport" and power in feminist research with anti-feminist women', *Sociology*, 33(4): 687–703.

Marcuse, H. (1969) 'Repressive tolerance', in R. Wolff, B. Moore Jr and H. Marcuse (eds) *A Critique of Pure Tolerance*, Boston: Beacon Press.

Mathews, S. and Herzog, H. A. (1997) 'Personality and attitudes toward the treatment of animals', *Society & Animals*, 5(2): 169–175.

Maynard, M. and Purvis, J. (1994) *Researching Women's Lives From A Feminist Perspective*, London: Taylor and Francis.

Michael, M. (1992) 'Lay discourses of science: science-in-general, science-in-particular and self', *Science, Technology and Human Values*, 17(3): 13–333.

Mies, M. (1981) 'Towards a methodology for feminist research', in G. Bowles and R. Duelli Klein (eds) *Theories of Women's Studies II*, Berkeley: University of California.

Milgram, S. (1963) 'Behavioural study of obedience', *Journal of Abnormal and Social Psychology*, 67: 371–378.

Milgram, S. (1974) *Obedience to Authority: An Experimental View*, London: Tavistock.

Mills, C. W. (1940) 'Situated actions and vocabularies of motive', *American Sociological Review*, 5: 904–913.

Mitchell, L. (2011) 'Moral disengagement and support for nonhuman animal farming', *Society & Animals*, 19: 38–58.

Nader, L. (1972) 'Up the anthropologist – perspectives gained from studying up', in D. H. Hymes (ed.) *Reinventing Anthropology*, New York: Pantheon Books.

Neal, S. (1995) 'Researching powerful people from a feminist and anti-racist perspective: a note on gender, collusion and marginality', *British Educational Research Journal*, 21(4): 517–531.

Nocella, A. J., Best, S. and McLaren, P. (eds) (2010) *Academic Repression: Reflections from the Academic Industrial Complex*, Edinburgh: A.K. Press.

Oakley, A. (1981) 'Interviewing women: a contradiction in terms', in H. Roberts (ed.) *Doing Feminist Research*, London: Routledge and Kegan Paul.

Pachirat, T. (2011) *Every Twelve Seconds: Industrialized Slaughter and the Politics of Sight*, London: Yale University Press.

Root, M. (1993) *Philosophy of Social Science: The Methods, Ideals, and Politics of Social Inquiry*, Oxford: Blackwell.

Shepher-Hughes, N. (1995) 'The primacy of the ethical: propositions for a militant anthropology', *Current Anthropology*, 36(3): 409–420.

Smith, D. E. (1974) 'Women's perspective as a radical critique of sociology', *Sociological Inquiry*, 44: 7–13.

Stanley, L. and Wise, S. (1983) *Breaking Out: Feminist Consciousness and Feminist Research*, London: Routledge.

Strega, S. (2005) 'The view from the poststructural margins: epistemology and methodology reconsidered', in L. Brown and S. Strega (eds) *Research As Resistance: Critical, Indigenous, and Anti-Oppressive Approaches*, Toronto: Canadian Scholars' Press.

Thompson, W. E. (1983) 'Hanging tongues: a sociological encounter with the assembly line', *Qualitative Sociology*, 6: 215–237.

Wicks, D. (2011) 'Silence and denial in everyday life: the case of animal suffering', *Animals*, 1: 196–199.

Zimbardo, P. (2004) 'A situationist perspective on the psychology of evil: understanding how good people are transformed into perpetrators', in A. G. Miller (ed.) *The Social Psychology of Good and Evil*, New York: Guildford Press.

6 Doing critical animal studies differently

Reflexivity and intersectionality in practice

Nathan Stephens Griffin

"The master's tools will never dismantle the master's house"
(Audre Lorde 1979)

"It's words and pictures, you can do anything you like with words and pictures"
(Harvey Pekar 1986)

Introduction

Over the past decade Critical Animal Studies (CAS) has established itself as a legitimate and pressing strand of contemporary critical theory. Academics and activists, across nations and disciplines, critically examine the oppression and domination of human and non-human animals under the banner of CAS. Books, journal articles, blogs and opinion pieces continue to be published, examining the many intersecting issues surrounding the exploitation of non-human animals, and 2013 marked the *Journal for Critical Animal Studies'* benchmark tenth anniversary. CAS has been successful in adopting appropriate and often distinctive research methods in an effort to achieve the goal of social transformation. Harper's (2010) study of Black female vegans provides a pertinent example of this. The author utilized a participant-centred methodology, which focused on the lives, voices and experiences of participants above all else, in order to make visible the experiences of Black vegan women. The work was the first of its kind, challenging commonly held assumptions and pushing forward critical understandings of race, gender and veganism, using distinctive methods.

This chapter examines a number of reflexive, biographical and visual methods and methodological frameworks that have emerged in qualitative social research, and explores their current and potential value to CAS research. It also considers the use of visual research methodologies, such as "ethno-mimesis", as a means of producing reflexive CAS research, which combine the principles of biographical and visual research. In presenting this broad overview of biographical and visual research, I hope to illustrate the value and potential of reflexive research methodologies, and the compatibility between biographical, visual, and CAS research.

When the Institute for Critical Animal Studies (ICAS) set out its core principles (Best *et al.* 2007), they included a rejection of "pseudo-objective academic analysis by explicitly clarifying its normative values and political commitments" and the importance of "eschewing narrow academic viewpoints and the debilitating theory-for-theory's sake position in order to link theory to practice, analysis to politics, and the academy to the community" (ibid.: 4–5). These are all values that could be fulfilled and strengthened through the use of methods focused on specificity, subjectivity and accessibility, such as the reflexive, biographical and visual methods discussed in this chapter. Chief among the core principles of ICAS was a commitment to "interdisciplinary collaborative writing and research" (ibid.: 4). Interdisciplinarity means crossing boundaries between distinct academic disciplines, with the aim of accessing new ideas and achieving unified knowledge (Klein 1990). Leitch (2005) argues that through its promotion of difference and heterogeneity, interdisciplinarity has laid the foundations for the emergence of various critical 'interdisciplines', such as black studies, queer studies, post-colonial studies and women's studies (Leitch 2005, cited in Chettiparamb 2007). Through exploring how reflexive, biographical and visual methods and modes of representation have been used in other disciplines, CAS can further contribute to the interdisciplinarity project.

This overview is not intended to be exhaustive or definitive. Instead, I offer a snapshot of the present and future trajectory of biographical and visual methods within CAS research. The chapter begins by discussing the current academic research context, in particular the tensions felt by CAS researchers regarding audit culture, the academic–industrial complex and the lingering influence of positivism and empiricism within in social sciences. 'Reflexivity' is offered as an important counter-point to these pressures, and one that is compatible with the foundations of CAS. Following this, the potential practical application of various 'reflexive' research methodologies is explored. Various research strategies are discussed, such as biographical research, autoethnography, performative social science, visual methods (including the use of 'comics' in academia), participatory action research and ethno-mimesis. In each case, I consider some examples of the use of the method, within and beyond CAS. These methods are offered in an attempt to illustrate the current and potential value of reflexive, visual and biographical research methodologies to the continuing growth and development of CAS. In addition, 'Tips' for scholars hoping to get doctoral research funded are offered in the form of several short comics interspersed throughout the chapter.

These comics provide an illustrative example of how visual and reflexive modes of representation might be used, to enhance or supplement conventional academic writing.

Context

Academic culture

Academic life is increasingly permeated by an audit culture of university league tables, journal publication ratings (e.g. journal star ratings in the UK, and 'ERA

Figure 6.1 Tips 1 and 2.

ratings' in Australia) and various systems for assessing the research of higher education institutions (e.g. Research Excellence Framework (REF) in the UK) (Sparkes 2007). It is easy to feel lost and alienated by the bureaucracy of the system, the jumping through hoops, ticking boxes and numerous other distractions from the substantive tasks at hand. We find ourselves wondering whether or not we are cut out for it, we routinely ask ourselves "how can I survive in it?"

(ibid.: 542). For scholars working within a CAS paradigm, these feelings can be even more pronounced. Critical analyses are often invisible, overlooked or ignored in mainstream academic discourse and, as with any critical theory, the life of a CAS academic can be difficult and conflicted. The progress achieved in establishing the interdisciplinary field of CAS remains mediated by the "academic–industrial complex" (Bowley 2010) in which we exist day-to-day. Del Gandio (2010) defines the 'academic–industrial complex' as the growing trend among senior university staff to allow a conflict of interests between their educational commitments and private interests; for example, through sitting on the boards of corporations, or to reflect industrial priorities in other more tacit ways. As CAS scholars, we find ourselves working within institutions that support and even fund oppression and exploitation (for example through vivisection or agricultural science). Of course, for many activists outside of academia, those who work inside it are complicit in animal suffering, sexism, racism, heteronormativity, elitism etc. through our association with less-than-perfect, market-oriented academic institutions. This may be compounded by the perception of academics as neutral and removed from the issues they study. These tensions underpin our existence as individuals attempting to navigate academia, working towards an end to animal exploitation, within the increasingly marketized, hypercompetitive world of the academy, where financial security becomes ever more uncertain. Yet, emergent academic interest in the value of reflexivity in social science pursuits represents a potential source of optimism.

Reflexivity

> *For me sociology ought to be meta but always* vis-à-vis *itself. It must use its own instruments to find out what it is and what it is doing, to better know where it stands.*
>
> (Bourdieu and Wacquant 1992: 191)

Put simply, "reflexivity" is a process of self-reflection in scholarly work. Typically, one may describe research as reflexive if the author has attempted to acknowledge the research context, their own subjectivity and how both have impacted upon the work in hand. The influence of empiricism on the social sciences tends to rely upon various rigid dualisms such as fieldwork/writing, researcher/researched, expert/layperson etc. The so-called 'Crisis of Representation' in social sciences sought to break down these divides; it has seen (amongst other developments) increased focus on the importance of self-reflection, and a more deliberate blurring of the boundary between fieldwork and writing (Denzin and Lincoln 2005). For many years, dominant ideas surrounding social research held the natural sciences as a suitable model for social study (Bryman 2008). However, the previously unshakeable "positivistic" values of reliability, validity, replicability and generalizability, have increasingly been contested and problematized (ibid.). Indeed, it is worthwhile to continue to reflect on CAS research methodologies in response to the "linguistic", "narrative", "performative",

Figure 6.2 Tip 3.

"visual", "digital" and other so-called "turns" that are purported to have occurred in qualitative sociology over recent decades (Roberts and Kyllonen 2006: 4). This changing theoretical landscape created potential conceptual space for alternative ways of thinking about ontology, epistemology and methodology.

CAS and reflexivity

CAS is a compassionate, interdisciplinary approach to working against oppression, and reflexive research practice allows a level of self-awareness and critical reflection that can be invaluable in this pursuit. Many CAS scholars have already highlighted the value of reflexivity (Grubbs 2008; Nocella 2011; Twine 2010), and this process should continue. CAS is ideally suited to embracing reflexivity for several reasons. Epistemologically and ontologically, CAS has valued compassion and subjectivity over "neutrality" or "hard data". As such, the principles of reflexive research (including acknowledging the research context and the impact of the author) are compatible principles. Reflexive research is well placed to pursue goals of social transformation through research.

The following sections offer some examples of the application of reflexive methodologies within and beyond CAS.

Reflexive methodologies

Here, I consider a number of different, interconnected research strategies, each of which are well suited to CAS and reflexive academic practice. There are several recurring characteristics that make these strategies compatible with CAS research, for instance their compatibility with the tenets of reflexivity, their interdisciplinarity and their focus on marginal voices and emancipatory aims. I begin by discussing biographical research broadly, before focusing on autoethnography as a developing field, and finally considering the overlap between CAS and biographical research.

Biographical research

Biographical research has been defined as:

> an exciting, stimulating and fast-moving field [seeking] to understand the changing experiences and outlooks of individuals in their daily lives, what they see as important, and how to provide interpretations of the accounts they give of their past, present and future.

> (Roberts 2002)

Generally it entails the collection and interpretation of "personal documents" (including interview transcripts, diaries, autobiographies) to attempt to understand individuals' life narratives, and see events, actions, value systems etc. from the perspective of the participant (ibid.: 1, 3).

Biographical research can take many forms. Roberts (2012) offers a non-exhaustive list that includes life history, oral history, narrative, life/course cohort, discourse-biography, biography, autobiography, auto/biography, auto-ethnography and 'testimonio'. Each approach is distinctive, but they all share a common concern for the specificity and subjectivity of lived experience. Biographical methods now represent an established interdisciplinary set of research practices (ibid.). This expands beyond the traditional boundaries of academia into professional practice, for example the increasing prevalence of biographically-influenced practice within health and welfare professions (ibid.). This trend only looks set to continue with increasing focus on the biographical dimensions of life experience, that is, the sensual, emotional, performative and visual, coming to the fore. This has been concomitant with a problematization of traditional researcher/author-participant/audience relations, and more awareness and acknowledgement of reflexivity.

Biographical research tends to be interdisciplinary in nature, and implies an approach in which "theory and empirical investigation are interwoven ... during or at the end of field work rather than as a precursor to it" (Bryman 1988: 81). It attempts to regard social life in "processual" rather than static terms (ibid.). Chamberlayne *et al.* (2000) argued that the "cultural/linguistic turn" in social science has been followed by various turns, including the "narrative/biographical turn".[1] The emphasis on language and representation during the cultural/linguistic turn signified a problematization of the nature of scientific knowledge. The "biographical turn", focusing on the study of lives, has raised various epistemological and political issues for social researchers and for social inquiry generally. Biographical research appeals because it explores individual accounts of life experience, within a given social and cultural context; it charts and records social and cultural developments, not just at a broad general level, but at a personal and inter-subjective level (Roberts 2002: 5). Hence, Denzin and Lincoln (1994) argue that theories must be understood according to narrative and hierarchies between researcher and researched must be broken down and replaced. These principles are also features of autoethnographic approaches to research, discussed shortly.

The seminal methodological text of biographical research is probably Thomas and Znaniecki's (1918) *The Polish Peasant in Europe and America*. This was notable in using biographical materials as sociological data, capable of allowing insight into both the lives of the individuals, as well as the broader social, political and economic structures in which these lives were lived (Apitzsch and Siouti 2007). This and the work that followed (particularly of their fellow Chicago School thinkers) had a lasting impact on how social research is conducted and what is deemed valuable data. Feminism's huge contribution to biographical research has included the use of oral histories to highlight marginalized voices in society (Gluck and Patai 1991). Underpinned by many of the same foundational values, autoethnography has emerged from biographical research as a reflexive approach to academic study.

Autoethnography

Autoethnography strives to challenge the conventions of social research, placing the individual researcher at its centre (Muncey 2010; Sparkes 2002). Thus, it provides an accessible way of representing research, which is "grounded in everyday life" (Plummer 2003: 522). Autoethnography entails a focus on specificity and subjectivity, and represents a bourgeoning and exciting methodological apparatus in social research (Jones 2005; Sparkes 2002). Autoethnography has been defined as "a self-narrative that critiques the situatedness of self with others" (Spry 2001: 710). For Jones and Adams (2010) autoethnography is a "queer" method, highlighting the theoretical overlap between autoethnography and queer theory; each entails a deliberate focus on fluidity, inter-subjectivity and "the particular", and each questions and challenges normative discourses. In practice, autoethnography frequently entails highly personalized accounts of the research process, in which the author has a central narrative presence (Sparkes 2002). Much like biographical methods in general, autoethnography tends to offer opportunities for accessing

Figure 6.3 Tip 4.

silenced and marginal voices, particularly where authors belong to groups who have been subject to oppression and domination (Muncey 2010). Richards (2008) argues that one significant way that so-called "abnormal" lives are controlled and normalized is through being written about. People who do not belong to a dominant ideology tend be presented as homogenous, or the heterogeneity of their experience is under-represented. This produces accounts that are not just inaccurate, but also limiting to the people whose identities are being "re-inscribed, often in subtle, but damaging ways" (ibid.: 1720). Autoethnography can address this problem of representation, and can also help to open up academic accounts of particular groups and experiences to laypersons (ibid.: 1726). Muncey's (2010) study of the emergence and progress of autoethnography is grounded in her own context and gendered experience, and professional experience as a nurse. She points to various trends, like the increasing academic focus on the individual and giving a voice to health service users and carers, as being indicative of the growing relevance of narrative inquiry.

In placing the author at the centre of the research project, autoethnography offers novel prospects for new modes of representation and expression within academic settings. This biographical focus should continue to be embraced within CAS research, particularly due to the various commonalities between biographical and CAS methodologies.

Biographical research and CAS

Biographical research's interdisciplinary scope, increasing emphasis on reflexive research practice and its deliberate focus on accessing silenced/marginalized voices, makes it highly compatible with the tenets of CAS. The following examples illustrate successful combinations of biographical research techniques, within animal advocacy contexts.

The meat and dairy industry deliberately distances consumers from the realities of animal exploitation and slaughter. The subjectivity of non-human animals is systematically erased from significance. Thus, biographical research provides an obvious point of resistance. In focusing on the silenced and the marginalized, biographical research defies processes of normalization. Therefore, the idea of (non-human) animal biographies warrants consideration. Whilst unquestionably invaluable to animal advocacy movements, Tsovel (2005) argues that conventional reporting strategies tend to rely on short-term, methodologically uniform and narrow approaches, which can preclude in-depth biographical focus on individual animals, or worse, disregard the experience of non-human animals altogether. Tsovel (2005) praises journalist Lovenheim's (2002) groundbreaking book *Portrait of a Burger as a Young Calf.* In tracing the lives of calves in minute detail, the reader is not distanced from an emotional attachment to the animal. This promotes a level of empathy and compassion, which quantitative approaches would (1) be unable to achieve and (2) would not seek to achieve, due to the underlying positivistic rationale.

Autoethnography has also been utilized in an explicitly CAS context. Grubbs' (2008) autoethnography of her six-week internship with Farm Sanctuary examined social movement theories through her focus on the animal rights movement, producing a thoughtful, narrative account of the inner workings of an activist organization. Gingrich-Philbrook (2005: 311) argues that such autoethnographic accounts can allow access to the "lost arts, and hidden experiences" of the marginalized – stories of struggle, oppression and humiliation. CAS should access these stories to promote empathy, compassion and solidarity among oppressed groups, towards the aim of total liberation. For Jones and Adams (2010: 197), autoethnography is a "thoroughly political" methodological approach, entailing a marked commitment to questioning, challenging and undermining normative and oppressive discourse. Nocella (2011) also utilized autoethnography within mixed methods research into the North American animal advocacy movement (discussed later).

These projects represent novel means of resisting animal exploitation and promoting CAS principles within academia, as does performative social science, which is explored next.

Performative social science

A growing convergence between the social sciences and arts is highlighted by Jones (2006), who notes that diversities of possibilities for representing research data are expanding. More research projects challenge traditional distinctions between research and representation, observing and reporting. These may be united under the banner of performative social science. Denzin's (2003) notion of "Performative Ethnography" connects critical pedagogy with performative research methods as an emancipatory research technique. Alexander (2005) defines Performance Ethnography as "the staged re-enactment of ethnographically derived notes ... work[ing] towards lessening the gap between a perceived and actualized sense of self and the other" (ibid.: 411). For Denzin, performative methods can be "an act of intervention, a method of resistance, a form of criticism, a way of revealing agency" and can facilitate democratic social change (Denzin 2003: 9 cited in Kincheloe and McLaren 2005). Performative-arts projects also challenge the typical separation of fieldwork and writing:

> The distinction between gathering and reporting data begins to dissolve, producing something new and quite interesting, adding to our capabilities.
>
> (Jones 2006: 69)

Conventional academic practice is to present data in the form of a written research paper, or perhaps an oral conference presentation. But these accepted practices could benefit from being shaken up. Alternative research methods and modes of representation, such as poetry, film-making and performance, can provoke new ways of thinking about issues (Bagley and Cancienne 2001). Atkinson and Coffey (1995) charted the ways that academics have increasingly challenged accepted conventions and started to experiment with alternative

Figure 6.4 Tip 5.

research and representation practices. Richardson (1992) argues that this enables both researcher and reader to engage with the participant's life in deeper, more sensitive, open and meaningful ways. The distinctive, performative methods espoused by a growing body of authors, like Denzin and Jones, offer exciting possibilities for the presentation of data.

Performative social science and CAS

Perhaps understandably, given the narrative, dramatic and fictive principles of performative social science, we may usefully look beyond academia for examples of performativity and CAS effectively coalescing. Documentary filmmaking has been a valuable tool within the animal advocacy movement, almost always informed by academic thought in some way, for example, through the use of evidence-based argument, expert testimony or the influence of animal advocacy ethics and philosophical perspectives. Recent documentaries such as *The Cove* (2009) and *Project Nim* (2011) may not be regarded as CAS projects, due to their narrow, non-radical focus; however, they each proved effective in promoting issues at the heart of animal advocacy (Shoard 2009). *Bold Native* (2010) offers an interesting case study as a work of fiction, with relatively high production values and a highly professional aesthetic; it also promotes a radical animal advocacy agenda. The film, which follows one activist's journey through a thought-provoking and entertaining narrative, was critically well received and earned plaudits within and beyond the academic community (Weitzenfeld 2012). *Bold Native* opens the door for performative, fictive filmmaking, informed by academic research and practice, thus enabling CAS to reach beyond its typical academic/activist audience and promote animal advocacy effectively and on a broader scale.

Visual methods

The development and increased use of so-called "visual methodologies" in the social sciences is, in many ways, tied to the development of critical sociology in Germany during the 1930s and 1940s (cf. the "Frankfurt School"), and more recently "Cultural Studies" in the UK, during the 1970s and 1980s (cf. the "Birmingham School") (Rose 2012). The latter gave way to what has become widely known as the "cultural turn" in social science, which entailed an increased focus on "the visual" and heightened efforts to incorporate and explore the visual within and through social research. For some authors, "the visual" is the most significant human sense (ibid.). According to Berger (1972: 7), "seeing comes before words. The child looks and recognizes before it can speak". Jay (1993) has used the term "Occularcentrism" to describe the centrality of the visual to contemporary Western life. Berger uses the expression "ways of seeing" to signify the complex process by which we engage with the visual; "we never look just at one thing, we are always looking at the relation between things and ourselves" (Berger 1972: 9, in Rose 2012: 13):

Visual imagery is never innocent, it is constructed through various practices, technologies and knowledges. A critical approach to the visual is therefore needed: one that takes into account the agency of the image, considers the social practices and effects of its viewing, and reflects on the specificity of that viewing by various audiences, including the academic critic.

(Rose 2012: 17)

It is important, however, to approach the visual with an awareness of critical analyses, to avoid replicating oppressive practices in visual research. The dominance of the visual has also been challenged as problematic, particularly through feminist critiques of the "male gaze", wherein, heterosexual male experience is assumed to be universal and women are objectified and disempowered (Mulvey 1975). Visual methods are, arguably, attuned with biographical research. Denzin argues that within biographical research "it is necessary to get as close to actual experience as possible" in order to "capture, probe, and render understandable problematic experience" (Denzin 1989: 69). This focus on specificity and in-depth analysis also lends itself well to visual methods, where a clearer image of the subject in question frequently relies on proximity and specific focus (Rose 2012).

Visual methods are at somewhat of a crossroads, according to Grady (2008), in the sense that they currently represent a niche in social research methods, but have the potential to be accepted as part of mainstream research practice. Within this context, those working within the "visual research" paradigm find themselves reflecting upon the ways in which their research impacts on this trend

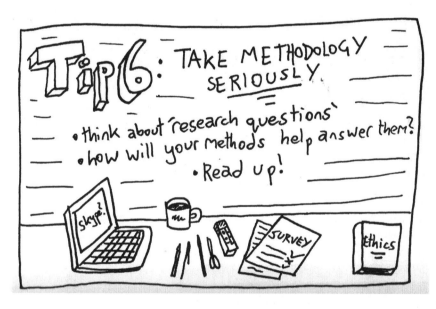

Figure 6.5 Tip 6.

towards acceptance. In their study of exhumations of mass graves in contemporary Spain, Ferrándiz and Baer (2008) highlight the value of visual recording methods. They consider a variety of visual and audio-visual interventions, such as photography, videos and video-testimony, produced by both academic and non-academic actors working in the region, and explore the significance of these visual methods in recording and as a triggering device in shared social memory and understandings of Franco's legacy. Friend (2010) utilized visual methods in her research with incarcerated asylum seekers. Her four-year research project sought to document the experience of "invisible" immigration detainees (that is, appellant or failed asylum seekers incarcerated in UK "removal centres"). Friend utilized sound/image frameworks to produce an exhibition entitled "Border Country", where visual methods are effective in promoting empathy and countering processes of dehumanization and marginalization.

Visual methods and CAS

The animal advocacy movement has long since relied on visual modes of communication in promoting its causes. Static and moving images, illustrating the horrors of animal exploitation, have been used to shock and promote empathy, and to make visible the experience of non-human animals (for example, through footage obtained covertly from slaughterhouses and vivisection labs). Graphic design and aesthetics also represent important aspects of engaging the public

Figure 6.6 "Arriving at the Slaughterhouse" (2010) (source: © Jo-Anne McArthur/We Animals – weanimals.org).

with animal rights issues (Jasper and Poulson 1995). Artists such as Sue Coe (1995) have used their work effectively as a means of illuminating the suffering and exploitation of animals (see www.graphicwitness.org/coe/enter.htm).

Photographers, including Jo-Anne McArthur (2010), have used photography to explore the nature of animal exploitation through images that "bear witness to animals at their moment of death". Visual methods, arguably, advance the aims of CAS in promoting empathy and compassion through striking visual stimuli. However, it is important to consider the complexity of the visual and to acknowledge that images may be interpreted in multiple ways; not always those intended by the creator (Berger 1972). Nevertheless, CAS research could benefit further from continuing to engage with visual methods and modes of representation and through a commitment to exploring beyond the boundaries of prose, especially through applying visual principles within academic contexts.

Comics and academia

Here, I discuss the potential use of comics (particularly biographical comics) in an academic and CAS context. Magnussen and Christiansen (2000) define comics as "a narrative medium utilizing visual and/or textual modes of expression". Eisner (1985) defined the medium more broadly, simply as "sequential art". Acknowledging these definitions, the term comic is used here to refer to a literary medium using words and images to tell a story or communicate ideas. Comics are now increasingly accepted as a legitimate and dynamic artistic medium, having been dismissed for years as childish and trivial. Critically acclaimed authors, including Art Spiegelman, Alison Bechdel and Marjan Satrapi, have helped to raise the profile of comics and have them increasingly accepted in literary and journalistic circles (Duncan and Smith 2009). One might argue that comics and biographical research are particularly well suited to one another, given the importance of narrative, subjectivity and experience in each medium. Works like Alison Bechdels' *Fun Home*, Marjan Satrapi's *Persepolis* and Art Spiegelman's *Maus* all represent creative, critically acclaimed works that combine biographical narratives and visual storytelling (Bechdel 2006; Satrapi 2008; Spiegelman 1991).

Whilst comics remain fairly marginal in an academic and professional context, there are some notable exceptions. Comics have been used in a therapeutic context as a means of enabling individuals living with illness to express themselves. Art therapy is an established mode of psychotherapy that uses visual art media as its primary mode of communication (BAAT 2011). Its concern is not the aesthetic quality of images produced by clients. The aim is to facilitate change and personal improvement through the use of art materials. The "better, drawn" project is a website that accepts submissions in the form of comics or illustrations, from people who have experienced chronic mental or physical illness. "The site is a way for people to write and draw about their experiences that might otherwise be difficult to talk about openly. In fact, we think that sometimes things can be said **better** when they're **drawn**" (Better Drawn 2011).

Comics are also increasingly regarded as a valuable tool elsewhere, particularly in education.

In the USA, the National Association of Comics Art Educators (NACAE) promotes the use of comics in an educational context, both as a teaching tool and also as a subject in and of itself. Comics have long since been used in the classroom as a means of helping young people to better engage with teaching materials, however, there are calls for comics to be taught as a stand-alone subject. The nuances of using words and images together in conjunction, and the creative potential of their juxtaposition, mean that the comic medium deserves wider recognition as a creative process:

> While many schools still hold antiquated notions of what comics are, a growing number of schools have started offering programs and classes in comics (or "sequential art").
>
> (Centre for Cartoon Studies 2013)

Similarly, the Centre for Cartoon Studies in Vermont offers postgraduate courses that centre on the "creation and dissemination of comics, graphic novels and other manifestations of the visual narrative" (CCS 2011).

Comics have been used in academia in a more functional context. Plowman and Stephen (2008) used comics as a means of representing video data in a static print journal. This, in effect, is an example of comics being used for practical necessity and as a substitute for another mode of communication, rather than for their own merits. However, this highlights some advantages of the medium in simultaneously presenting visual, verbal and descriptive data. It is evidence of the strengths of comics and of their potential academic legitimacy.

More recently, comics have been used to literally and symbolically emphasise the "messy" parts of a research process that are usually deliberately hidden (Jones and Evans 2011). This autoethnographic example illustrates how the use of comics in research has the potential to allow for greater researcher reflexivity, and can highlight the subjectivity and "situatedness" of the researcher (Spry 2001: 716). Likewise, comics may help to emphasise the contribution and creativity of participants in the research process. This is again tied to biographical research, in particular the idea of "performance narratives" in research, which mute the voice of the narrator and broadcast the participant's voice (Chase 2005: 665). It is potentially a way to dispel the "myth of the invisible, omniscient author" (Tierney 2002), and the inclusion of visual representations of the authors, within the research, works towards that same end.

Comics and CAS

There are high-profile examples of visual narratives within animal advocacy movements. One might look to the work of comic creators such as J. T. Yost to see effective examples of comics and CAS coming together, albeit outside of an academic context.

Figure 6.7 "Roadtrip" (2009) (source: J. T. Yost).

My own research attempts to combine autoethnography, comics and CAS by including reflexive, autoethnographic comics detailing the research process alongside my (ongoing) doctoral research into the biographies of animal rights activists and vegans (Stephens Griffin 2011, 2012).

Recent academic focus on issues of "research impact" and "public engagement" means that the use of non-traditional, eye-catching modes of representation, such as comics, have a heightened potential in promoting the ideas and issues that they espouse. The use of comics, therefore, represents a particular point of opportunity for CAS. In embracing and exploring the value of comics, CAS research could contribute to emerging and distinctive modes of representation and communication within academia, as well as expanding the reach and profile of CAS issues in the process. It may also represent an opportunity within "audit culture" to promote the aims of CAS through visual methods.

Participatory action research

Participatory Action Research (or PAR) is distinct from typical research in that it deliberately aims to intervene and positively impact upon the environment/context in which it is conducted; in short, PAR "seeks to understand and improve the world by changing it" (Baum *et al*. 2006: 854). The principles underpinning PAR are inclusion, participation, valuing all local voices and community driven and sustainable outcomes (O'Neill and Mansaray 2012). PAR is founded upon the necessity for reflexivity within research and on an awareness of the impact on, and interplay between, researcher and research topic. Many research projects are conducted with the aim of "improving" a particular social phenomena, but claims made about how this is practically achieved tend to be abstract and removed from the sort of strategic, reflective, collaborative interventions evidenced in PAR. Much like biographical research, PAR problematizes the epistemological assumptions of mainstream research, and sets out to produce knowledge and practice that is transformative in nature (Baum *et al*. 2006). PAR entails working with members of the community at whom the research is aimed, and engaging in a process of constant reflection and adaptation. PAR aims to empower research participants, and successful PAR projects require constant reflection on the power dynamics within the research, and a process of adaptation in response to these dynamics; in this regard, PAR is a truly reflexive research methodology. The specific methods used in a PAR project are selected in conjunction with the research strategy, in order to best fulfil the desired outcomes of the project.

Arts based participatory action research was conducted by O'Neill and Mansaray (2012), in participation with Regional Refugee Forum North East and Purple Rose, Stockton. This project sought to make visible the lived experiences of women in Stockton, Middlesbrough and Hartlepool, and highlighted their conceptions of community safety. The project utilized "story walks", in which women were asked to draw a map from their home to a special place, marking along the way places and spaces that were important to them. The researchers

Figure 6.8 Tip 7.

and community co-researchers then undertook the walks together, discussing the places chosen, taking photos and recording the voices of the women talking about their places. They then met at a workshop, to discuss photographs and choose images to form part of an exhibition. The story walks "helped to create individual and collective narratives about what it is like to be a new arrival, an asylum seeker or refugee, and for some women a refused asylum seeker" (O'Neill and Masaray 2012). PAR has also been utilized within a CAS context.

PAR and CAS

PAR is rooted in ethical and political praxis and so it is suited to the principles of CAS research. Nocella's (2011) partly auto-ethnographic research, on the animal advocacy movement, utilizes a Critical Pedagogical methodology informed by PAR practice. Nocella explored why and how activists respond to the process of "terrorization", that is, the stigmatization of being labelled as or associated with terrorists (Nocella 2011: 130). To do this, the author actively engaged and collaborated with animal advocacy activist movements in the USA, a methodology fundamentally rooted in respect and inclusion. Merskin (2011) has argued for the extension of PAR beyond research with human beings into a trans-species approach to research and communication. It therefore, presents a potential way for us to "hear" the marginalized voices of non-human animals and "fully envision those who are silenced", including non-human animals (ibid.: 144). Naturally, this poses some very important questions, for example, how one might attain something close to informed consent from a non-human research participant, but these are questions that warrant further exploration and discovery.

PAR can be a daunting methodological framework to dip into, and tends to be particularly effective when conducted in conjunction with practitioners or those with connections to the field. The Centre for Social Justice and Community Action (CSJCA – based at Durham University and headed by Sarah Banks) has produced a useful set of guidelines, particularly relating to the ethical implications of conducting PAR research (CSJCA, 2012). CAS scholars wishing to utilize PAR might benefit from engaging with these guidelines, or indeed working to adapt them to suit CAS research contexts (e.g. participatory work with non-human animals).

A research technique called "ethno-mimesis" has the potential to combine elements of biographical, visual and participatory social research, and this in turn, has the potential to benefit CAS research.

Ethno-mimesis

Here, I define and explain "ethno-mimesis" as a research strategy that combines aspects of biographical, visual and critical methodologies. I consider some effective examples of "ethno-mimetic" research, and conclude by exploring the potential for ethno-mimesis in a CAS setting.

The value of narrative and biographical methods in creating space for the voices of the politically marginal, in challenging negative stereotypes, raising public awareness around social issues and in producing political "texts' with real potential for social transformation, is highlighted by O'Neill (2010: 22). O'Neill and Hubbard (2010) have observed emotion and sensuousness in tension with reason, rationality and objectivity within social research, highlighting why it is important to embrace methods that help to access the emotionality of a research project. To better understand our social world and political contexts, it is important to consider the lives and experiences of those on the margins. The daily struggles of those whose ideas are incompatible with mainstream values, therefore, represent a valuable source. Biographical research allows us to produce potentially new meanings from the lives of participants, and to challenge existing discourses and the "dominant knowledge/power axis", which maintains and reproduces the marginal status of certain ideas and individuals (O'Neill 2010: 108). Further to this, the combination of biography and art has transformative potential in social research that could help to "create spaces for the marginalized to speak for themselves" (ibid.: 145).

The combination of art and biographical/ethnographic methods is described as 'ethno-mimesis' (O'Neill 2010; O'Neill and Hubbard 2010: O'Neill 2012). Etymologically, the term emerges from the combination of "ethnography", and "mimesis", which refers to 'sensuous knowledge', knowledge produced through the visual/artistic or poetic representation of research (ibid.: 158). In practical terms, ethno-mimesis involves a combination of ethnographic methods, the data from which goes on to form the basis for artistic and or visual outputs, which represent the lived experiences of those lives encountered within the ethnography. Typically, this involves working in conjunction with artists, poets, designers, writers, photographers and other creators, in order to produce works of art; "Artistic, visual, and sensuous representations of life experiences can be transformative, providing recognition, voice, a means of sharing identities through interdisciplinarity and hybridity" (O'Neill and Hubbard 2010: 48). Biographical narratives, as represented through artistic modes of representation, are useful in creating "multi-vocal, dialogic texts, which make visible emotional structures and inner experiences such as sensuous knowledge" (Kuzmics 1994: 9 quoted in O'Neill and Hubbard 2010: 47). Several research projects have used ethno-mimesis, and projects have worked with vulnerable, othered, or otherwise socially and politically marginal groups, such as sex workers, refugees, asylum seekers and the homeless (O'Neill 2012). Further to the internal methodological justifications for using ethno-mimesis, O'Neill (2012) argues that artistic methods are useful in ensuring that research is accountable to wider publics.

Ethno-mimesis and CAS

My own research utilizes a form of ethno-mimesis in asking animal rights activists and vegans to produce visual documents relating to their lives and experiences taken from a participant's submission to the project. It supplemented a large amount of narrative data, produced through biographical interviews. The

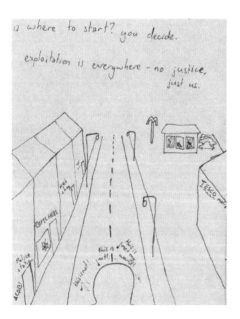

Figure 6.9 Comic created by research participant (source: Stephens Griffin 2012).

visual documents work in tandem with the interview data to provide a fuller and more nuanced, rich account of the lives of animal rights activists. Allowing participants to produce their own visual documents allows them to make a more active contribution than they otherwise would have. Furthermore, the use of images allows the project to appeal and have relevance to a wider audience, and aims to break down traditional boundaries between academic research and lay audiences.

My research aims to utilize the reflexive, visual, sensual and biographical tenets of ethno-mimesis, to present specific, dynamic, multi-modal accounts of the experiences of animal advocates. This stands in contrast to the kinds of large scale, quantitative, attitudinal surveys that have tended to dominate mainstream understandings of the movement (Herzog and Golden 2009; Plous 2006; Jerolmack 2003). These quantitative studies often decontextualize and simplify the aims and motivations of those involved. It is my hope that my project can provide one example of the sort of integrated reflexive, biographical, visual research that CAS stands to benefit from embracing.

Conclusion

This chapter has examined various research strategies in order to highlight and support the value of reflexive, biographical and visual methods and methodologies within the context of CAS research. This meant that I focused on biographical, visual and participatory research methods, as being particularly well attuned to both reflexive and CAS research practice. In presenting this broad overview of biographical and visual research, I aimed to illustrate the importance of reflexive research methodologies, and the compatibility between biographical, visual, and CAS research.

Biography, and the study of lives, has a vital role in the future of CAS, in particular through breaking the traditional moulds of biographical research and reaching beyond human biography, as a means of advocating for non-human animals. This can allow a level of insight and empathy that other methods may scarcely be able to achieve. Similarly, visual modes of representation (including comics) have been used frequently within animal rights activism as a means of promoting empathy and engaging people with the values and aims of the movement. However, owing to their effectiveness in communicating ideas, prose and text still dominate as the most acceptable means of academic representation. Obviously, there are good, practical reasons for this. Still, visuals do have a place in academic contexts and it is interesting to explore the use of visuals within academia as a means of overcoming exploitation and domination, in the same way that the animal rights movement has more broadly. Integrated reflexive, biographical, visual approaches to CAS research, such as those outlined in this chapter, represent a fascinating point of further exploration and development, and one full of potential for positive social transformation within and beyond the boundaries of academia.

Figure 6.10 Tip 8.

Acknowledgement

I would like to thank J. T. Yost and Jo-Anne McArthur for very kindly granting me permission to reproduce their artwork in this chapter.

Note

1 We may also consider the "practice turn" here; with its focus on overcoming "individualist/societal" dualisms, and affording both the social and the individual appropriate status in social theory (Whittington 2006).

References

Alexander, B. K. (2005) "Performance ethnography. The re-enacting and inciting of culture", in N. Denzin and Y. S. Lincoln (eds) *Handbook of Qualitative Inquiry, Third Edition*, Thousand Oaks, CA: Sage.

Atkinson, P. and Coffey, A. (1995) "Realism and its discontents: on the crisis of cultural representation in ethnographic texts", in B. Adam and S. Allan (eds) *Theorizing Culture: An Interdisciplinary Critique After Postmodernism*, London: UCL Press.

Bagley, C. and Cancienne, M. (2001) "Educational research and intertextual forms of (re) presentation: the case for dancing the data", *Qualitative Inquiry*, 7: 221.

Baum, F., MacDougal, C. and Smith, D. (2006) "Participatory action research", *Journal of Epidemiol Community Health*, 60(10): 854–857.

Bechdel, A. (2006) *Fun Home: A Family Tragicomic*, New York: Houghton Mifflin.

Berger, J. (1972) *Ways of Seeing*, London: BBC Press and Harmondsworth: Penguin.

Best, S., Nocella, A. J., Kahn, R., Gigliotti, C. and Kemmerer, L. (2007) "Introducing Critical Animal Studies", *Journal for Critical Animal Studies*, 5(2): 4–5.

Better, Drawn, (2011) *About Better, Drawn*, Tumblr, available online at: http://better-drawn.tumblr.com/about (accessed 20 September 2011).

Bold Native (2010) Motion Picture, Gather Films, United States.

Bourdieu, P. and Wacquant, L. J. D. (1992) *An Invitation to Reflexive Sociology*, Chicago: Chicago University Press.

Bowley, G. (2010) "The academic-industrial complex," *New York Times* (31 July 2010).

British Association of Art Therapists (BAAT) (2011) "What is art therapy?", BAAT, available online at: www.baat.org/art_therapy.html (accessed 20 September 2011).

Bryman, A. (1988) *Quantity and Quality in Social Research*, New York: Routledge.

Bryman, A. (2008) *Social Research Methods: Third Edition*, Oxford: Oxford University Press.

Chamblerlayne, P., Bornat, J. and Wengraf, T. (eds) (2000) *The Turn to Biographical Methods in Social Science*, London: Routledge.

Chettiparamb, A. (2007) 'Interdisciplinarity: a literature review', *Higher Education Academy*, available at: www.heacademy.ac.uk/assets/documents/sustainability/interdis-ciplinarity_literature_review.pdf (accessed 1 May 2013).

Chase, S. E. (2005) "Narrative inquiry: multiple lenses, approaches, voices", in N. Denzin and Y. Lincoln (eds) *The Sage Handbook of Qualitative Research: Third Edition*, London: Sage.

Coe, S. (1995) *Dead Meat*, New York: Four Walls, Eight Windows.

CSJCA (2012) *Community-Based Participatory Research: A Guide To Ethical Principles And Practice*, Connected Communities, AHRC.

Del Gandio, J. (2010) "Neoliberalism and the academic-industrial complex", Truthout. org, available at: http://archive.truthout.org/neoliberalism-and-academic-industrial-complex62189 (accessed 19 November 2012).

Denzin, N. (1989) *Interpretive Biography*, Vol. 17, Newbury Park, California: Sage.

Denzin, N. K. (2001) "The reflexive interview and a performative social science", *Qualitative Research*, (1): 23–46.

Denzin, N. K. (2003) *Performance Ehnography: Critical Pedagogy and the Politics of Culture*, Thousand Oaks, CA: Sage.

Denzin, N. K. and Lincoln, Y. S. (eds) (1994) *Handbook of Qualitative Research*, Thousand Oaks, CA: Sage.

Denzin, N. K. and Lincoln, Y. S. (2005) "Introduction: the discipline and practice of qualitative research", in N. Denzin and Y. Lincoln (eds) *Sage Handbook of Qualitative Research*, 3rd edition, Thousand Oaks, CA: Sage.

Duncan, R. and Smith, R. J. (2009) *The Power of Comics: History, Form and Culture*, New York: Continuum.

Eisner, W. (1985) *Comics and Sequential Art*, Florida: Poorhouse Press.

Ferrandiz, F. and Baer, A. (2008) "Digital memory: the visual recording of mass grave exhumations in contemporary Spain", *Forum Qualitative Sozialforschung/Forum: Qualitative Social Research*, 9(3).

Friend, M. (2010) "Representing immigration detainees: the juxtaposition of image and sound in 'Border Country'", *Forum Qualitative Sozialforschung/Forum: Qualitative Social Research*, 11(2).

Gingrich-Philbrook, C. (2005) "Autoethnography's family values: easy access to compulsory experiences", *Text and Performance Quarterly*, 25: 297–314.

Gluck, S. and Patai, D. (1991) (eds) *Women's Words: The Feminist Practice of Oral History*, New York: Routledge Press.

Grady, J. (2008) "Visual research at the crossroads", *Forum Qualitative Sozialforschung/ Forum: Qualitative Social Research*, 9(3).

Grubbs J. D. (2008) "Farm sanctuary: creating a space where theory meets practice", *Masters Thesis*, Cincinnati, USA: University of Cincinnati.

Harper, B. (2010) *Sistah Vegan: Food, Identity, Health, and Society: Black Female Vegans Speak*, Brooklyn: Lantern Books.

Herzog, H. and Golden, L. (2009) "Moral emotions and social activism: the case of animal rights", *Journal of Social Issues*, 65(3): 485–498.

Jasper, J. M. and Poulson, J. D. (1995) "Recruiting strangers and friends: moral shocks and social networks in animal rights and anti-nuclear protests", *Social Problems*, 42(4): 493–512.

Jay, M. (1993) *Downcast Eyes: The Denigration of Vision in Twentieth-Century French Thought*, Berkeley, CA: California University Press.

Jerolmack, C. (2003) "Tracing the profile of animal rights supporters: a preliminary investigation", *Society & Animals*, 11(3): 245–263.

Jones, K. (2006) "A biographic researcher in pursuit of an aesthetic: the use of arts-based (re)presentations in 'performative' dissemination of life stories", *Qualitative Sociology Review*, 2(1): 66–85.

Jones, P. and Evans, J. (2011) "Creativity and project management: a comic", *ACME: An International E-Journal for Critical Geographies*, 10(3): 585–632.

Jones, S. H. (2005) "Autoethnography: making the personal political", in Denzin and Lincoln (2005) *Sage Handbook of Qualitative Research*, London: Sage, pp. 763–792.

Jones, S. H. and Adams, T. E. (2010) "Autoethnography is a queer method", in K. Browne and C. J. Nash (eds) *Queer Methods and Methodologies: Intersecting Queer Theories and Social Science*, Cornwall, England: Ashgate.

Kincheloe, J. and McLaren, P. (2005) "Rethinking theory and research", in N. Denzin and Y. Lincoln (eds) *The Sage Handbook of Qualitative Research: Third Edition*, London: Sage.

Klein, J. T. (1990) *Interdisciplinarity: History, Theory, and Practice*, Detroit: Wayne State University.

Lorde, A. (1979) "The master's tools will never dismantle the master's house", in A. Lorde (ed.) *Sister Outsider: Essays and Speeches*, Berkeley, CA: Crossing Press.

Lovenheim, P. (2002) *Portrait of a Burger as a Young Calf: The Story of One Man, Two Cows, and the Feeding of a Nation*, USA: Three Rivers Press.

Magnussen, C. and H. Christiansen (eds) (2000) *Comics and Culture: Analytical and Theoretical Approaches to Comics*, Copenhagen: Museum Tusculanum Press.

McArthur, J. (2010) *The Slaughter* (Photo Collection), available at: www.weanimals.org/gallery.php?id=92 (accessed 21 November).

Merskin, D. (2011) "Hearing voices: the promise of participatory action research for animals", *Action Research*, 9(2): 144–161.

Morris, J. (2010) "Animal biographies" (Presentation), *International Council of Archaeozoology (ICAZ) Conference*, Paris: Pierre et Marie Curie University, August 25th.

Mulvey, L. (1975) "Visual pleasure and narrative cinema", *Screen*, 16(3): 6–18.

Muncey, T. (2010) *Creating Autoethnographies*, London: Sage.

Nocella, A. (2011) "A dis-ability perspective on the stigmatization of dissent: critical pedagogy, critical criminology, and critical animal studies", *Social Science–Dissertations*, Paper 178.

Nocella, A. (2012) "A dis-ability perspective on the stigmatization of dissent: critical pedagogy, critical criminology, and critical animal studies", Doctoral Thesis, New York: Maxwell School of Citizenship and Public Affairs, Syracuse University.

O'Neill, M. (2010) *Asylum, Migration and Community*, Bristol: Policy Press.

O'Neill, M. (2012) "Ethno-mimesis and participatory arts", in S. Pink (ed.) *Advances in Visual Methodology*, London: Sage.

O'Neill, M. and Hubbard, P. (2010) "Walking, sensing, belonging: ethno-mimesis as performative praxis", *Visual Studies*, 25(1): 46–58.

O'Neill, M. and Mansaray, S. (2012) "Women's lives, well-being and community", Durham: *Durham University, Regional Refugee Forum and Purple Rose Stockton.*

Plous, S. (2006) "An attitude survey of animal rights activists", *Psychological Science*, 2(3): 194–196.

Plowman, L. and Stephen, C. (2008) "The big picture? Video and the representation of interaction", *British Educational Research Journal*, 34(4): 541–565.

Plummer, K. (2003) *Intimate Citizenship: Private Decisions and Public Dialogues*, Washington, DC: University of Washington Press.

Project Nim (2011) Motion Picture, BBC Films, Icon Film Distribution, United Kingdom.

Richards, R. (2008) "Writing and the othered self: autoethnography and the problem of objectification in writing about illness and disability", *Qualitative Health Research*, 18(12): 1717–1728.

Richardson, L. (1992) "The consequences of poetic representation: writing the other, writing the self", in C. Ellis, and M. G. Flaherty (eds) *Investigating Subjectivity: Research on Lived Experience*, Los Angeles: Sage.

Roberts, B. (2002) *Biographical Research*, Buckingham: Open University Press.

Roberts, B. (2012) "Biographical research: history and innovation (Presentation), Advances in Biographical Methods Research Symposium: Creativity, Innovation and Application, Durham University, United Kingdom, 17 May.

Roberts, B. and Kyllonen, R. (2006) "Biographical sociology", *Qualitative Sociological Review*, 2(1).

Rose, G. (2012) *Visual Methodologies: An Introduction to Researching with Visual Materials*, London: Sage.

Satrapi, M. (2008) *Persepolis I and II: The Story of a Childhood and the Story of a Return*, London: Jonathan Cape.

Shoard, C. (2009) "The Cove: review", *Guardian*, 22nd October, available at: www.guardian.co.uk/film/2009/oct/22/the-cove-review (accessed 30 November 2012).

Sparkes, A. C. (2000) "Autoethnography and narratives of self: reflections on criteria in action", *Sociology of Sport Journal*, 17: 21–43.

Sparkes A. C. (2001) "Myth 94: qualitative health researchers will agree about validity", *Qualitative Health Research*, 11(4): 538–552.

Sparkes, A. C. (2002) "Autoethnography: self indulgence or something more?", in A. Bochner and C. Ellis (eds) *Ethnographically Speaking: Autoethnography, Literature and Aesthetics*, Oxford: Rowman and Littlefield.

Sparkes, A. C. (2007) "Embodiment, academics, and the audit culture: a story seeking consideration", *Qualitative Research*, 7: 521.

Spiegelman, A. (1991) *The Complete Maus: A Survivors Tale*, New York: Pantheon Books.

Spry, T. (2001) "Performing autoethnography: An embodied methodological praxis", *Qualitative Inquiry*, 7: 706–732.

Stephens Griffin, N. (2011) "A queer approach to speciesism" (Comic), *Journal for Critical Animal Studies*, 10(3): 119–127.

Stephens Griffin, N. (2012) "Queering veganism: adventures in reflexivity" (Poster), *Journal for Critical Animal Studies*, 10(1): 14.1.

The Centre for Cartoon Studies (2013) "About CCS", available online at: http://teachingcomics.org/ (accessed 3 January 2014).

The Cove (2009) Motion Picture, Oceanic Preservation Society, Lionsgate, United States.

Thomas, W. I. and Znanieki, F. (1918) *The Polish Peasant in Europe and America: Monograph of an Immigrant Group*, Boston: R. G. Badger.

Tierney, W. (2002) "Getting real: representing reality", *International Journal of Qualitative Studies in Education*, 15(4): 385–394.

Tsovel, A. (2005) "What can a farm animal biography accomplish? The case of portrait of a burger as a young calf", *Society & Animals*, 13(3): 245–262.

Twine, R. (2010) *Animals as Biotechnology: Ethics, Sustainability and Critical Animal Studies*, London: Routledge.

Weitzenfeld, A. (2012) "Bold Native: review", *Journal for Critical Animal Studies*, 10(1): 213.

Whittington, R. (2006) "Completing the practice turn in strategy research", *Organization Studies*, 27: 613.

Witek, J. (1989) *Comic Books as History: The Narrative Art of Jack Jackson, Art Spiegelman and Harvey Pekar*, Mississippi: University of Mississippi Press.

Yost, J. T. (2009) "Roadtrip", in *Old Man Winter Anthology*, New York: Birdcage Bottom Books.

Other links

British Association of Art Therapists (BAAT), available at: www.baat.org (accessed 14 December 2013).

Center for Cartoon Studies, Vermont (CCS), available at: www.cartoonstudies.org (accessed 14 December 2013).

National Association for Comics Art Educators (NACAE), available at: www.teachingcomics.org (accessed 14 December 2013).

The "We Animals" Project, available at: www.weanimals.org (accessed 14 December 2013).

Part III
Critical animal studies and anti-capitalism

7 Labourers or lab tools?

Rethinking the role of lab animals in clinical trials

Jonathan L. Clark

In *The Professional Guinea Pig*, an account of an anarchist community in Phila-delphia whose members support themselves by participating in clinical trials, Roberto Abadie (2010) argues that human subjects enrolled in phase one trials should be classified as workers. A phase one trial is designed to test the safety, as opposed to the efficacy, of an experimental drug; its main objectives are to discover any side effects and to better understand how the body metabolizes the drug. Unlike participants in later phases, who typically stand to benefit from the drug, participants in phase one trials do it mainly for the money, often because they have few other options, their involvement constituting not so much an instance of altruistic volunteering as an "economic draft" (Weinstein 2010: 122; see also Elliott 2010; 2008; Elliott and Abadie 2008). Drawing on Marx's cri-tique of the legal fiction of the labour contract, which presumes that workers with nothing left to sell but their labour-power choose freely to part with it, Kaushik Sunder Rajan (2007: 83) reminds us that the very purpose of the informed consent form that human subjects typically sign is to establish, as a legal matter, that they are not "coerced guinea-pigs", even if, as Melinda Cooper (2008: 88) puts it, they have "nothing left to sell but exposure itself". Increas-ingly, the pharmaceutical industry is outsourcing clinical trials to such places as China, India, and Eastern Europe, where there is a surplus of economically des-perate people whose bodies are said to have the added advantage, from the per-spective of the industry, of being relatively free of other drugs that might interact with the experimental drug, an advantage that presumably stems, in many cases at least, from their inability to afford such drugs (Cooper 2008; Petryna 2009; Prasad 2009; Rajan 2007).[1] In the United States and countries with similar laws, participation in clinical trials is not classified as work, at least not insofar as labour laws are concerned. In response to this lack of labour rights, several scholars have argued that participating in phase one trials should be regarded as a form of "experimental or clinical labour" (Cooper 2008: 73; see also Abadie 2010: 165–166; Carney 2011: 180; Elliott 2008; Rajan 2007; Waldby and Cooper 2010, 2008).

"Guinea-pigging", as it is often called, is a peculiar kind of work; instead of doing things it mainly entails having things done to you (Elliott 2008: 41; see also Abadie 2010: 2; Carney 2011: 180). With a few notable exceptions, such as the long history of physicians experimenting on themselves (but also on family members), human experimentation has been a more-or-less coercive relationship involving slaves, prisoners (both in penal institutions and, most infamously, in the Nazi concentration camps), soldiers, the poor, the mentally ill, racial minorities, children, women, and other vulnerable groups (see, e.g. Guerrini 2003; Lederer 1995; Washington 2006). And as the term "human guinea pig" suggests, this is a vulnerability that human subjects share, at least in some respects, with their nonhuman counterparts. "[W]hat generates our moral response to animals and their treatment", Cary Wolfe (2008: 11) writes, "is our sense of the mortality and vulnerability that we share with them, of which the brute subjection of the body – in the treatment of animals as mere research tools, say – is perhaps the most poignant testament" (see also Acampora 2006: 130). Whether this subjection is achieved through economic coercion, as in the case of many human subjects in phase one trials, or, as in the case of the nonhuman animals on whom drugs are tested, through physical confinement and restraint, it is the reduction of the body to a mere object of labour that strikes such a powerful ethical chord (Acampora 2006: 97–103; Greenhough and Roe 2011). Indeed, it was the recognition of this shared vulnerability that helped give rise to the movement against human experimentation in the United States (Lederer 1995). In the early twentieth century, Susan Lederer explains, it was the antivivisection movement, a movement concerned primarily with the plight of nonhuman animals, that led the charge against what it called "[h]uman vivisection", or non-therapeutic experimentation on vulnerable human groups (Lederer 1995: xiv, italics omitted). "That investigators would treat human subjects like laboratory animals was precisely what antivivisectionists feared", Lederer (1995: 123) writes.

In the US, the UK and countries with similar laws, before a drug may legally be tested on humans in phase one trials, it must first be tested for toxicity on nonhuman animals in what is called the pre-clinical phase. The purpose of pre-clinical toxicity testing is to reduce the risk to which human subjects in phase one trials are subjected. The connection between human and nonhuman subjects raises an intriguing question: if human guinea pigs are engaged in clinical labour, what about actual guinea pigs? To date, scholars working on the concept of clinical labour have had little to say about nonhuman lab animals, and most of the scholars who have written about these animals have classified them as part of the means of production, or "living laboratory equipment", not as labourers (Birke 2003: 213; see also Cetina 1999: 138–158; Kohler 1994: 6–8). Yet a close reading of some of these accounts reveals another way of thinking about lab animals, one that affords them greater agency. For example, in his history of the fruit fly as a model organism, Robert Kohler describes the flies most often as lab tools; occasionally, however, they figure as the scientists' "co-worker[s]" in the lab, though Kohler does not develop this theme in any depth (Kohler 1994: 1, see also 23; for other examples, see Pemberton 2004; Russell 2004). Picking up on this theme, this chapter examines the question of whether it makes sense, both analytically and politically, to regard the participation of nonhuman animals in pre-clinical toxicity testing as a form of clinical labour.

One of the hallmarks of Critical Animal Studies (CAS) is its engagement with Marxist categories such as labour (Best 2009a; Best and Gigliotti 2007; for early examples of such work, see Benton 1993; Noske 1997; Tapper 1988). Yet much of Marx's thought is firmly anchored in a paradigm of human exceptionalism that is inconsistent with what we are now learning about the lives of other animals (Bekoff 2011; Best 2009b). Given that one of the main goals of CAS is to critique human exceptionalism in all its various forms (Best 2009b; Gruen 2011: 1–43), what is needed is not simply an extension of Marx's categories to include nonhuman animals; what is needed is a reconstruction of those categories from the ground up in posthumanist terms (Haraway 2008: 46, 67, 73; Potts and Haraway 2010). We need a "more-than-human" Marxism (Whatmore 2006), a "posthumanist political economy" (Raulerson 2010: 152). Exploring the possibility that humans may not be the only lab animals engaged in clinical labour is as good a place to start as any.

Indeed, Donna Haraway (2008) has already begun to explore this possibility in her work. "The posthumanist whispering in my ear reminds me that animals work in labs", she writes, "but not under conditions of their own design, and that Marxist humanism is no more help for thinking about this for either people or other animals than other kinds of humanist formulae" (Haraway 2008: 73). Using the concept of labour to describe what human subjects do – and, more importantly, have done to them – already stretches Marx's categories; arguing that nonhuman lab animals labour, too, bends them beyond the breaking point. That's because using the concept to classify what nonhuman animals do – let alone what they have done to them – runs up against the "human exceptionalism" that undergirds Marx's analysis of the labour process (Haraway 2008: 46). My goal in this chapter is to interrogate Marx's human exceptionalism through an exploration of the concept of clinical labour. The chapter is a first step towards the development of a posthumanist understanding of the labour process.

Human exceptionalism

For Haraway, human exceptionalism is "the premise that humanity alone is not a spatial and temporal web of interspecies dependencies" (Haraway 2008: 11; see also Pickering 2008; Tsing 2012). This definition has much in common with what Riley Dunlap and William Catton called "[h]uman [e]xemptionalism" (Dunlap and Catton 1979: 250; see also Catton and Dunlap 1978). In two pioneering articles that helped launch the field of environmental sociology, Dunlap and Catton challenged what they described as sociology's then reigning "Human Exemptionalism Paradigm", which they had originally called the "'Human Exceptionalism Paradigm'" but quickly renamed in order to clarify that they were not denying the exceptional nature of human beings (Dunlap and Catton 1979: 250, italics and footnote omitted). "[W]hat environmental sociologists deny," they wrote, "is not that *Homo sapiens* is an 'exceptional' species but that the exceptional characteristics of our species (culture, technology, language, elaborate social organization) somehow *exempt* humans from ecological principles and from environmental principles and constraints" (Dunlap and Catton 1979: 250). Dunlap and Catton called for a "'New Ecological Paradigm'" that would acknowledge our dependence on the more-than-human world (Dunlap and Catton 1979: 250, italics omitted). The influence of their critique of human exemptionalism in sociology cannot be overstated. Yet looking back now, some three decades later, from the perspective of CAS, one is struck by their defense of human exceptionalism (for a similar critique, see Tovey 2003). If environmental sociology emerged as a challenge to human exemptionalism, the field of CAS has emerged as a challenge to the paradigm of human exceptionalism that pervades not just sociology but much of the academy.

Humans are, of course, unique, if by this we mean "having certain capabilities that all other [species] lack" (Ingold 1988b: 97). "The same goes for every species," Tim Ingold (ibid.) reminds us, "each of which is unique in its own particular way". "Uniqueness", in other words, "is not unique" (Garrard 2012: 149, italics omitted). But being unique is not the same as being exceptional, which means better than, not just different from, all the rest (Gruen 2011: 4–25). As Lori Gruen (ibid.: 2) explains, the ideology of human exceptionalism suggests not just that "[w]e engage in uniquely human activities" – which, of course, we do, just as pigs engage in activities that are uniquely porcine and cows engage in uniquely bovine activities – but that our uniqueness "elevate[s] us above animals". Gruen unpacks the logic of human exceptionalism, revealing its two implicit claims:

> The first is that humans are unique, humans are the only beings that do or have X (where X is some activity or capacity); and the second is that humans, by virtue of doing or having X, are superior to those that don't do or have X. The first claim raises largely empirical questions – what is this X that only we do or have, and are we really the only beings that do or have it? The second claim raises an evaluative or normative question – if we do discover the capacity that all and only humans share, does that make humans better, or more deserving of care and concern, than others from an ethical point of view? Why does doing or having X entitle humans to exclusive moral attention?
>
> (Ibid.: 4–5)

With respect to the empirical issue, mistaken claims that some trait is unique to humans often take the form of denials of evolutionary continuity (Bekoff 2011). "The anti-Darwinian view that humans are different in kind rather than in degree from other animals is a powerful conceit", James Gould and Carol Grant Gould (2007: 272) write, "but it does not stand up to scrutiny". Yet even if one manages to avoid this empirical error, one would still be engaging in human exceptionalism if one used differences in degree to elevate human beings above all other animals.

With human exceptionalism's two implicit claims in mind, we can now evaluate Marx's understanding of the labour process. Scholars have criticized

Marx on both grounds. They have criticized him for positing an exclusively human form of labour that is not, in fact, exclusive to humans, and for elevating human labour above that of all other animals. Before examining these critiques, however, we must first turn to Marx's analysis.

Marx's "exclusively human" form of labour

Before one can argue, as Jason Hribal does, that nonhuman animals "have laboured, and continue to labour, under the same capitalist system as humans", one must first make the case that what these animals are doing and having done to them in the capitalist labour process is in fact a form of labour (Hribal 2003: 436, footnote omitted). That a nonhuman animal is subsumed within the capitalist labour process does not necessarily mean that it is the animal's labour that is being exploited. The same goes for humans, who, besides being exploited as labourers, can also be exploited as instruments or objects of labour, to use Marx's terms (Guthman 2011). We need to start, then, with a definition of labour.

Marx (1990) defines labour in a key section of the first volume of *Capital*. Before analysing the *capitalist* labour process, he first analyses the labour process itself, regardless of how it is organized socially (Harvey 2010: 111–119; Marx 1990: 283–291). Understood in this universal sense, the labour process is, for Marx, part of our species' inescapable metabolic relation with the rest of nature. It is, as he puts it, "purposeful activity aimed at the production of use-values. It is an appropriation of what exists in nature for the requirements of man. It is the universal condition for the metabolic interaction [*Stoffwechsel*] between man and nature, the everlasting nature-imposed condition of human existence, and it is therefore independent of every form of that existence, or rather is common to all forms of society in which human beings live" (Marx 1990: 290, see also p. 133).

Just to go on living, we must labour, transforming nature to meet our needs. Although this passage suggests that humans interact *with* nature, understood as a separate realm, it is clear that Marx regarded us as parts of nature, a point he made in another key description of the labour process:

> Labour is, first of all, a process between man and nature, a process by which man, through his own actions, mediates, regulates and controls the metabolism between himself and nature. He confronts the materials of nature as a force of nature. He sets in motion the natural forces which belong to his own body, his arms, legs, head and hands, in order to appropriate the materials of nature in a form adapted to his own needs. Through this movement he acts upon external nature and changes it, and in this way he simultaneously changes his own nature. He develops the potentialities slumbering within nature, and subjects the play of its forces to his own sovereign power.
>
> (Ibid.: 283)

As this passage makes clear, Marx is not describing a metabolic relation between human beings and nature; he's describing a relation within nature, between different parts of it. The goal of the labour process, as Marx describes it here, is to transform external nature, understood as some part of nature that lies outside the labourer's own body. In the process of transforming this external nature, the labourer's "internal nature", including his or her own body, is also transformed (Dickens 2004: 258), but the labourer's body is not the object of his or her labour. This point will become important below, when we turn to the concept of clinical labour.

Marx divides the labour process into three main elements: "(1) purposeful activity, that is work itself, (2) the object on which that work is performed, and (3) the instruments of that work" (ibid.: 284). "In the labour process", he explains,

> man's activity, *via* the instruments of labour, effects an alteration in the object of labour which was intended from the outset.... The product of the process is a use-value, a piece of natural material adapted to human needs by means of a change in its form.
>
> (Marx 1990: 287)

Taken together, the instruments and objects of labour constitute the means of production (ibid.). Marx does not define an object of labour, but he does define an instrument of labour as "a thing, or a complex of things, which the worker interposes between himself and the object of his labour and which serves as a conductor, directing his activity onto that object" (ibid.: 285). The instruments of labour also include "all the objective conditions necessary for carrying on the labour process" (ibid.: 286). "These do not enter directly into the process", Marx explains, "but without them it is either impossible for it to take place, or possible only to a partial extent" (ibid.). Roads and canals are two examples he cites.

For Marx, whether something is classified as an instrument, an object, or both is determined not by its nature but by "its specific function in the labour process" (ibid.: 289). And this function is determined by the labourer's purpose, which forms the "intentional structure" that guides the process (Benton 1989: 66, italics omitted). Some of the things that (or who) become instruments or objects of labour are themselves products of past labour, whereas others are "provided directly by nature and do not represent any combination of natural substances with human labour" (Marx 1990: 290). For example, Marx (ibid.: 284) classifies uncaught fish as "objects of labour spontaneously provided by nature" because, unlike in the case of fish who are farmed in aquaculture, human labour plays no part in their creation – or at least not intentionally, as opposed to unintentionally, as in human-induced changes to aquatic ecosystems that affect fish populations. Marx does not mean to suggest that wild fish are, by their very nature, objects of labour, that nature created them to be eaten by human beings (cf. Helmreich 2008: 474). It is only when they are being pursued by anglers wielding nets, lines and other instruments of labour that they become objects of labour.

Marx uses nonhuman animals as examples to illustrate the exclusively human form of labour. As already mentioned, he classifies uncaught fish as objects of labour (Marx 1990: 284, 287 n.7), and he classifies other nonhuman animals, most notably domesticated animals, as instruments of labour. "In the earliest period of human history", he writes,

> domesticated animals, i.e. animals that have undergone modification by means of labour, that have been bred specially, play the chief part as instruments of labour along with stones, wood, bones and shells, which have also had work done on them.
>
> (Ibid.: 285–286, footnote omitted)

And in some labour processes, Marx explains, the very same animal is used as both an instrument and an object of labour. Take cattle feeding, in which, according to Marx, cattle are both objects of labour in the production of meat and instruments of labour in the production of manure (ibid.: 288). Why they are not instruments for producing both of these products of digestion is not entirely clear. In any event, what is most notable about Marx's examples is that nonhuman animals are always classified as part of the means of production; they are never labourers who work alongside human beings.

This brings us to the first element: purposeful activity. We are not the only organisms who transform external nature and are who themselves transformed in the process. "[A]nts do it, beavers do it, all kinds of organisms do it", David Harvey (2010: 112) reminds us (see also Lewontin and Levins 2007: 31–34). What Marx is primarily interested in, however, is what he describes as the "exclusively human" form of labour (Marx 1990: 284). What makes this type of labour exclusive to humans, and what elevates it above what Marx describes as "those first instinctive forms of labour which remain on the animal level", is that, unlike nonhuman labourers, human labourers are able to transform nature according to a mental blueprint or plan (ibid.: 283). Marx makes this crucial distinction by contrasting the architect with the bee:

> A spider conducts operations which resemble those of the weaver, and a bee would put many a human architect to shame by the construction of its honeycomb cells. But what distinguishes the worst architect from the best of bees is that the architect builds the cell in his mind before he constructs it in wax. At the end of every labour process, a result emerges which had already been conceived by the worker at the beginning, hence already existed ideally. Man not only effects a change of form in the materials of nature; he also realizes [*verwirklicht*] his own purpose in those materials...
>
> (Ibid.: 284)

It is here, in his discussion of this form of labour that, he claims, distinguishes humans from all other organisms, that Marx exhibits the human exceptionalism that has drawn so much criticism from scholars in the field of CAS (Nimmo 2011).

Critiques of Marx

Recall that, for Marx, it is not labour that sets humans apart. He acknowledges that at least some other animals labour, too. But they do so instinctively, he argues, and only humans are able to transform nature according to a plan. When engaged in this exclusively human form of labour, the labourer first creates a mental blueprint of the use-value that the labourer wants to create, and then the labourer sets out, wielding one or more instruments, to attempt to transform some external object into that use-value. It's the planning that allegedly makes this form of labour unique to humans. Richard Levins and Richard Lewontin (1985: 255) offer a useful summary of Marx's understanding of the uniquely human form of labour:

> it transforms the world of nature into a world of artifacts that serve human beings; this transformation is carried out socially rather than individually; and it is done by the producer first conceiving mentally the end to be achieved and the varied means of its achievement, thus action is teleological.

Given that some nonhuman animals transform nature into artefacts (e.g. birds' nests),[2] and others (e.g. ants) transform nature socially, they explain, "[w]hat seems to be unique to humans", at least in Marx's view, "is the conscious planning, the imagining of the result before it is brought into existence by deliberate teleological action" (ibid.). "This last element is what marks off human labour from the activities of *mere* animals", they write, implying with the word *mere* that human superiority, not just difference, lies at the heart of Marx's description of the labour process (ibid., italics added).

Anthropologist Tim Ingold has offered the most sustained critique of how nonhuman animals figure in Marx's analysis of the labour process (see, e.g. Ingold 2011; 2000; 1988b; 1987; 1986; 1983; 1980). Yet despite this cogent critique, Ingold seems ultimately to agree with Marx that only humans can transform nature according to a mental blueprint. As we will see, this puts Ingold's views, like those of Marx, at odds with the views of at least some students of animal cognition and consciousness.

In his critique of Marx, Ingold takes issue with Marx's use of domesticated animals as examples of instruments of labour (Ingold 2000: 307–308; 1980: 88). Ingold writes:

> [I]n *Capital*, domestic animals are classified alongside primitive tools as *instruments* of labour.... This, however, is to relegate animals to the status of mindless machines. In truth, the domestic animal is no more the physical conductor of its master's activity than is the slave: both constitute labour itself rather than its instruments, and are therefore bound by social relations of production...

(Ingold 1980: 88)

For Marx, you'll recall, an instrument of labour conducts the labourer's activity.[3] According to Ingold, classifying animals as conductors of human activity denies their agency. "[T]o regard the animal as a mere tool", Ingold writes, "is to deny its capacity for autonomous movement...; tools cannot 'act back' or literally *interact* with their users, they only conduct the users' action on the environment ..." (Ingold 2000: 307 citations omitted; see also Murray 2011). Yet later in *Capital*, as Ingold (2000: 308) acknowledges, Marx notes a fundamental difference between animals and inanimate tools. One of the problems with horsepower, Marx (1990: 497) writes, is that "a horse has a head of its own". So by suggesting that animals can be used as instruments of labour, Marx is not necessarily denying that they are capable of resisting their use as such.

Nor, for that matter, does Ingold deny that humans and other animals can become part of the means of production. Through the use of various "instruments of coercion", he explains, they can in fact be "virtually reduced to a machine existence through the systematic repression of their powers of autonomous action" (Ingold 2000: 307–308; see also Jones 2003). But, he goes on to explain,

> the essential difference between the human mastery over animals and over machines is that although both ... 'can be compelled to do work', the machine is compelled by the very nature of its construction whereas the animal is compelled by the external imposition of coercive force.
>
> (Ingold 2000: 308, citation omitted)[4]

That animals can be either labourers or means of production (or, perhaps, both simultaneously) reminds us that the role an animal plays in a particular labour process depends upon the intentional structure of that process. So when Ingold (2000: 307) says that "animals constitute labour itself rather than its instruments", or when Haraway (2008: 80) says that "animals are working subjects, not just worked objects" or the instruments with which that work is done, they are not saying that animals are inherently labourers, or even that it is only in their capacity as labourers that they are ever enrolled in the labour process. Rather, they are reminding us that, just like humans, other animals can be enrolled in the labour process not just as part of the means of production but also as labourers. In the end, then, Ingold's critique of Marx is that, in illustrating the exclusively human form of labour, Marx classifies other animals only as instruments or objects of labour (or both), ignoring the fact that they can also be fellow labourers, albeit labourers who are not engaged in the exclusively human form of labour.

Ingold also takes issue with Marx's description of nonhuman labour as instinctive. By characterizing it this way, Ingold (1983: 12) argues, Marx denies the capacity of at least some nonhuman animals to engage in "purposive labour". By purposive Ingold does not mean purposeful, at least not in Marx's sense of following a mental blueprint. He means that, in the process of transforming nature, the labourer is conscious of what he or she is doing. When it comes to planning, however, Ingold agrees with Marx. "Man is not uniquely purposive", Ingold writes, "but he is unique to the extent that he carries a conscious, symbolic representation of the *procedures* by which his purpose is to be executed" (ibid., italics in original). For Ingold, this unique capacity is grounded in another: the capacity for language. He writes:

> While accepting the Cartesian premise that thinking in the sense of the construction of prior intentions, being dependent upon language, is a uniquely human capacity, [I] rejec[t] the view that such planning is a condition for the intentionality of action. Thus, "the question of animal *consciousness* ... must be separated from that of animal *thinking*." The animal that does not premeditate and plan is not therefore an automaton, but a conscious agent and patient who acts, feels and suffers, just as we do. Like us, it is responsible for its actions, having caused them to happen, even though it lacks our human ability to render an account of its performance, whether beforehand as a plan or retrospectively as a report.
>
> (Ingold 1988a: 8–9, italics in original)

Lacking language, Ingold argues, animals cannot think, in the sense of "[a]ttending to concepts" (Ingold 1988b: 94). And without the ability to think, they cannot plan the transformation of nature. For all his disagreement with Marx, then, Ingold ultimately seems to agree that Marx's architect is engaged in an exclusively human form of labour. By distinguishing the architect's finished product from that of the bee, both Ingold and Marx imply that only the architect's qualifies as an artefact (Ingold 1988b). Ingold defines an artifact as "any object that results from the imposition of prior conceptual form upon material substance" (ibid.: 85). Given this definition, only Marx's exclusively human form of labour is capable of creating artefacts. Take, for example, the beaver lodge:

> Unlike the human house, the beaver lodge cannot be regarded as an artifact or work, since it is not the realization of a prior conception in the mind of the builder, any more than is the shell of a snail. But we have no reason whatsoever to deny that the beaver is acting intentionally, for ... the existence of a plan is not a necessary condition for the intentionality of action.
>
> (Ingold 1986: 315)

Beavers labour purposively, as Ingold defines it, because they experience building their lodges as something they do (Ingold 1988b: 95–96; 1986: 313). To distinguish the exclusively human form of labour from the purposive labour in which at least some other animals engage, Ingold borrows a distinction between "prior intentions" and "intentions in action" (Ingold 1988b: 96, citation and italics omitted; 1986: 312). "A prior intention is an imaginative representation of a future state that it is desired to bring about...[,]" whereas "intention in action ... corresponds to the experience of actually doing..." (Ingold 1986: 312).

"Conduct that is spontaneous, carried out without previous thought or reflection, but which we nevertheless experience as issuing from ourselves as agents, rather than being purely involuntary, carries intention, but is not motivated by prior intention", he explains (Ingold 1988b: 96). To be defined as labour, the transformation of nature must be accompanied by intention in action, but it need not involve prior intention (Ingold 1986: 321). What this means is that only organisms that are capable of intention in action can be said to labour. And among these purposive labourers, only humans are capable of prior intention (Ingold 1988b: 97). Ingold seems therefore to agree with Marx that human labour is different in kind, not just in degree, from that of all other animals (ibid.).

Yet for Ingold, the exclusively human form of labour is not the only (or even the main) type of labour in which humans engage. That we're the only ones who do it doesn't mean that it's all we do. Indeed, humans rarely plan things out, he argues, "except intermittently, on those occasions when a novel situation demands a response that cannot be met from the existing stock-in-trade of habitual behaviour patterns" (ibid.). Most of the time we are engaged in a form of labour that "does not differ all that substantially from the conduct of non-human animals" (ibid.: 85).

In his most recent work on the labour process, Ingold (2011) elevates the importance of intention in action and downplays the importance of prior intention:

> Perhaps ... the essence of production lies as much or more in the attentional quality of the action – that is, in its attunement and responsiveness to the task as it unfolds – and in its developmental effects on the producer, as in any images or representations of ends to be achieved that may be held up before it.... Conceived as the attentive movement of a conscious being, bent upon the tasks of life, the productive process is not confined within the finalities of any particular project. It does not start with an image and finish with an object but carries on through, without beginning or end, punctuated – rather than initiated or terminated – by the forms, whether mental or [material], that it sequentially brings into being.
>
> (Ibid.: 6; see also 1986: 321–324)

Yet even in this recent work, Ingold (2011) still seems to believe that it is only humans whose ongoing actions are ever punctuated by a mental blueprint of what is to be produced. Nevertheless, by elevating the importance of intention in action and downplaying the importance of prior intention, Ingold rejects the notion that what he apparently still takes to be our uniqueness elevates us above all other animals. As he puts it,

> ... once we dispense with the prior representation of an end to be achieved as a necessary condition for production, and focus instead on the purposive will or intentionality that inheres in the action itself – in its capacity literally to *pro-duce*, to draw out or bring forth potentials in the person of the

producer and in the surrounding world—then there are no longer any grounds to restrict the ranks of producers to human beings alone. Producers, both human and non-human, do not so much transform the world, impressing their preconceived designs upon the material substrate of nature, as play their part from within in the world's transformation of itself...

(Ingold 2011: 6, italics in original)

To summarize Ingold's view, then, at least as I understand it, Marx was not wrong to suggest that only human beings are able to plan the transformation of nature, but he was wrong to ignore the purposive labour of at least some other animals. Ingold also seems to want us to focus not on what makes our labour unique, but on what it has in common with that of other purposive agents.

But is it even true that Marx's exclusively human form of labour is exclusive to humans? "Until comparatively recently", Gould and Gould (2007: 4–5) explain in *Animal Architects*, "only the most unreserved romantics conceived of a role for planning and thought in any brain but ours". But thanks to recent research in cognitive ethology and related fields, our views of other animals are starting to change. Often described as an ecosystem engineer, the beaver is a prime candidate for challenging Marx's views (Cheng 2006). One of Marx's contemporaries, Lewis H. Morgan, was convinced that beavers were akin to human architects. "When a beaver stands for a moment and looks upon his work", Morgan (1986: 256) wrote, "evidently to see whether it is right, and whether anything else is needed, he shows himself capable of holding his thoughts before his beaver mind; in other words, he is conscious of his own mental processes". Ingold (1988b) rejected Morgan's notion that beavers – or any other nonhuman animals, for that matter – are capable of planning, but others are not so sure (see, e.g. Gould 2007; Gould and Gould 2007: 251–269, 278–279; Griffin 2001: 99–112). Donald Griffin (ibid.: 103) was puzzled that Ingold could recognize that at least some other animals are conscious of their activities, yet deny "any sort of foresight, even for a short time into the future – denying that animals have any thoughts about the likely results of their own activities". Describing how beavers excavate their burrows, Gould and Gould (2007: 257) observe that "[i]magination, an ability to plan, and a ready willingness to learn from experience seem the most realistic combination of cognitive faculties to generate this aspect of the beaver's life". "And this is just the burrow", they add. Offering a story about a group of beavers that repaired a dam that human vandals had damaged, Griffin (2001: 111) suggests that beavers are able to develop novel solutions to unexpected and unprecedented problems (see also Gould and Gould 2007: 266–268). "[W]hen an animal can repair unlikely damage to something it has built", Gould and Gould (2007: 278–279) explain, "the simplest interpretation is that it has some kind of picture of the goal or the structure of the finished product". Indeed:

[a]n ability to skip unnecessary steps, to take advantage of or compensate for unusual contingencies, to find alternative solutions to a problem, and to

use novel materials may suggest more than a picture; in these situations [beavers] may have some understanding of the goal, the needs to be met.

(Gould and Gould 2007: 279)

"In some sense", they continue, "the ability to skip steps in a process, or to repair damage in a flexible way, are acts of extrapolation, of seeing the consequences of actions before performing them … [This] is the ability … to formulate a plan" (ibid.: 279). Gould and Gould (ibid.: 268) therefore conclude that "[a]lmost everything about the actions of [beavers] suggests that they employ concepts and reasoning to power their behavior, with insight emerging when they encounter especially difficult challenges".

This and other recent research in cognitive ethology and related fields suggests that what Marx classified as an exclusively human form of labour may not be exclusive to humans after all. Jon Elster may well have been correct, then, when he observed nearly three decades ago that "Marx erred in *Capital I* when he denied to animals the capacity to work according to a mental plan" (Elster 1985: 65). As Sherryl Vint (2009: 124) puts it, Marx's "distinction between the imaginative labour of humans and the instinctual, responsive behaviour of animals has not held up..". (see also Levins and Lewontin 1985: 255; Smith 2001: 86). Of course, Marx did not have the benefit of this research when he was developing his understanding of the labour process. But the same cannot be said for the many contemporary Marxists who remain oblivious to this work, and its profoundly unsettling implications for their taken-for-granted human exceptionalism (Best 2009b). Yet given how deeply rooted this ideology is in Western thought, it is by no means clear that empirical evidence will ever settle the issue. As Gruen (2011: 12) observes, the perennial debate about human uniqueness tends to follow a "bar-raising dialectic". When it is no longer plausible to deny that at least one other species has a trait that was once said to be unique to humans, defenders of human exceptionalism can be expected to put forward another candidate to take its place as the uniquely human trait of the day. So even if it should someday become widely accepted that certain other animals are able to plan the transformation of nature, this new truth may simply be greeted with a claim that human labour is unique (and exceptional) in some other way. To be clear, I am not denying that human labour is unique, though I leave it for others to say what exactly makes it so. What I am suggesting is that the uniqueness of human labour is likely to be a matter of degree rather than kind.

Interestingly, once one accepts that at least certain other animals labour, it becomes possible to expand the concept of social relations of production to include "human-animal relations of production" (Tapper 1988: 52; see also Ingold 1980: 88). Another crucial implication of taking the labour of nonhuman animals seriously is that it unsettles the assumption that they are only ever enrolled in the labour process by humans (Smith 2001: 90–93). Yet perhaps at least some other animals are capable of initiating the labour process and enrolling human beings into it, either as fellow labourers or as instruments or objects of labour. In such cases, it would be the intentions of the nonhumans that would form the intentional structure of the labour process. In other cases, the intentional structure might be better understood as a combination of the intentions of all the purposive agents, whether human or nonhuman, who participate in the process. One of the great strengths of research in cognitive ethology and related fields is that it makes it possible to raise these kinds of possibilities and have them taken seriously (cf. Crist 2002).

Besides being criticized for his claims about the uniqueness of human labour, Marx has also been criticized for elevating human labour above that of all other organisms. Lawrence Wilde (2000) has defended Marx against this charge. Although Marx did regard human labour as unique, Wilde (ibid.) argues, he did not think that it was superior. Wilde questions the translation I quoted above, which has Marx placing the exclusively human form of labour on a higher level than "those first instinctive forms of labour which remain on the animal level" (Marx 1990: 283). Wilde (2000: 47) offers this translation instead: "We are not dealing here with the first forms of labour bounded by instincts as animals are." In claiming that the exclusively human form of labour is not bounded by instincts, Marx is not elevating it above the instinctive labour of all other animals, Wilde argues; he is just saying that it is different. Wilde also suggests that the contrast Marx draws between the architect and the bee conveys respect for the labour of other animals. "Marx selects the most intricate of animal productions to make his point, and in so doing reveals a respect for their endeavours and their nature" (ibid.). There is, however, a more plausible interpretation of this passage: that, according to Marx, even the worst human labourer has something that the best nonhuman labourer lacks. David Harvey, who has been teaching *Capital* for more than forty years, acknowledges the human exceptionalism that lies at the heart of Marx's description of the labour process. "The more we know about bees", Harvey (2000: 202) writes, "the more the comparison with even the best of human labour (let alone the worst of architects) appears less and less complimentary to our supposedly superior powers". Like Harvey, Barbara Noske (1997: 73) suggests that, for Marx, the exclusively human form of labour is superior to, not just different from, what he regards as the instinctive labour of all other animals. I think it is reasonably clear that Marx's analysis of the labour process is based on a foundation of human exceptionalism.

Clinical labour

Recall that in Marx's description of the labour process, the object of labour is part of the labourer's external environment, outside his or her own body. This, however, is not always the case. In body-building, for example, the body-builder's own body is the object (and an instrument) of his or her labour (Wacquant 1995). Using parts of his or her body as instruments, often in conjunction with such instruments as weights, the body-builder sculpts his or her own body, and like Marx's architect often does so in order to achieve a form that he or she first conceives of mentally (Wacquant 1995). As in body-building, in clinical trials it is the human subject's own body that is the main object of labour. It is also sometimes an instrument, as in the simple act of popping a pill. But unlike in body-building, neither the human subject nor anyone else, for that matter, has a particular bodily form in mind. Nor could they. For even after pre-clinical testing on nonhuman animals, it's still unclear what the drug will do. Indeed, the whole point of a phase one trial is to figure out what a drug is capable of doing to a human body.

Hearing guinea-pigging described as labour can be somewhat jarring. What makes it so difficult to think about participation in clinical trials as a form of labour is its passivity, at least as compared to the more active "bodily labour" of, say, the body-builder (Wacquant 1995: 65). "Guinea pigs do not do things[,]" Carl Elliott (2008: 41) writes, "so much as they allow things to be done to them". One professional guinea pig described the work as follows:

> [Y]ou are not asked to produce or to do something anymore, you are being asked to endure something. So, if you are a guinea pig you are enduring something, people are doing things to you and you are just enduring it, you are not actually producing something. I feel that I am a worker but it is not work…
>
> (quoted in Abadie 2010: 2)

In addition to enduring any procedures that are performed on their bodies, human guinea pigs in phase one trials must also endure being exposed to the risk of an adverse reaction to the experimental drug. For Melinda Cooper (2008), exposure to risk is what defines this peculiar form of labour:

> An essential component of clinical trial participation is what is referred to as "risk" in technical terms but which might be better rendered by the more suggestive term "exposure". Human subject experimentation in drug testing can be described as a form of transformative exposure, where the patient is called upon to both experience the sometimes unpredictable metabolic effects of pharmaceutical compounds and perform a number of second-order tasks such as adhering to a strict regime of diet and drug administration, self-monitoring and recording of information. This is a depiction of labour that places it somewhere between passive and active participation, experi- ence and self-experiment. Labour would then be defined as the experience of self-transformation—commodified. If we were to redefine labour in this way, the contribution of the clinical trial participant or tissue donor to the production of bioeconomic value would become more readily comprehens- ible. The clinical labourer is the person who "consents" to their own self- transformation for a certain return (although this return can be direct or indirect, monetary or in kind)…
>
> (Cooper 2008: 76; see also Waldby 2012)

For Cooper and her colleagues, guinea-pigging is one type of clinical labour, defined as "processes in which subjects give clinics and commercial biomedical institutions access to their in vivo and in vitro biology, the biological productivity of living tissues within and outside their bodies" (Mitchell and Waldby 2010: 339; see also Waldby and Cooper 2010, 2008). A particularly "onerous for[m] of clinical labour", guinea-pigging entails "lending [one's] bodily metabolism and everyday experience of health and illness to often risky pharmaceutical research..." (Mitchell and Waldby 2010: 339). Some of the second-order tasks that guinea pigs often perform could, in theory, be performed by someone (or something) else. The most essential service, the one that defines this labour process, is the guinea pig's willingness – if it makes sense to call it that, given the social constraints within which such choices are often made – to endure "risk exposure itself" (Cooper 2008: 88; see also Waldby 2012). For Cooper (2008: 90), "participation in clinical trials is doubtless one of the most extreme forms of contemporary experimental labour, simply because it invalidates any distinction between labour power and the body of the labourer". To put this in Marx's terms, the guinea pig's labour – if that is what we want to call it – consists of participating in the transformation of him- or herself into an object of labour, and then enduring the consequences of being utilized as such, including the risk of an adverse reaction and the consequences of any reaction that does occur.

Some scholars prefer to describe human guinea pigs as objects of labour rather than as labourers (Weinstein 2010). Building on the work of Joe Dumit, Matthew Weinstein (ibid.: 119) suggests that what is extracted from human guinea pigs in the capitalist labour process is not surplus labour but "surplus health". "Health beyond that needed for basic functionality...can be extracted by the pharmaceutical companies and converted into clinical knowledge", he writes (ibid.). But although "their work as human scientific objects" may not be work "in a classical Marxist sense of the term" (Weinstein ibid.: 124, 119), it is arguably labour in Ingold's sense of the term. After all, enduring the process of being worked on involves purposive action. Even so, using the term labour to describe enduring what one professional guinea pig described as "'the mild torture economy'" is unsettling (quoted in Abadie 2010: 2). Although the capitalist labour process often inflicts pain on the bodies of workers (Fracchia 2008), this is not typically the goal. In clinical trials, by contrast, the whole purpose is to expose subjects to the risk of harm (Cooper 2011). As I explain below, our understandable uneasiness about using the term labour to describe the process of enduring "torture" raises questions about whether making this analytical move is likely to be politically productive.

It's unclear whether all aspects of guinea-pigging would constitute labour in Waldby and Cooper's sense of the term. Ingold's (1983) useful distinction between eating and digesting can help highlight the relevant issue here. Eating is labour because we do it intentionally, Ingold (1983) argues, but the involuntary physiological processes involved in digesting the food are not. While eating, it is I who am labouring, Ingold argues, whereas it is my body, not I, that does the work of digestion (Ingold 1983). By this logic, ingesting a drug is labour, and so is enduring the process by which one's body metabolizes the drug, including any adverse reactions to it, but the involuntary biochemistry of metabolism itself is not a form of labour. This is because Ingold's definition of labour is tied to purposiveness, and "purely involuntary" activities do not count (Ingold 1988b: 96). Although Waldby and Cooper refer to the "biological labour" of the body, and although they seek to overcome the "mind/body split", their views on the issue of whether labour can be involuntary are not entirely clear (Waldby and Cooper 2008: 59; 2010: 9). One could of course argue that labour need not be purposive, that organisms are constantly labouring simply by being alive, and that anything a living organism does, whether intentionally or unintentionally, constitutes a form of labour. Labour would thus be defined as the dialectical metabolic relation between an organism and its environment. One could even extend the concept of labour to what cells and other biological materials do (see, e.g. Thacker 2005). But if we accept Ingold's view of labour as an ongoing process, then perhaps we need not break that process down into its constituent pieces and ask which of them are labour and which are not. As Ingold (2011) explains, labour is an ongoing process that is punctuated by different kinds of moments; some may be purposive, others may reflect a degree of planning, and still others may be entirely involuntary. Ingold (2011) does, however, seem to imply that the labour process must involve at least some degree of purposiveness – or, at the very least, that only those beings who are capable of purposive action can be said to ever labour.

Clearly, the human beings, dogs, primates and rodents who are involved in clinical trials typically endure the process as purposive agents. Nonhuman animals experience toxicity testing as something they do – and have done to them. Of course, unlike human subjects, they do not worry about the risks to which they are exposed, though they apparently can anticipate certain experimental procedures (Nuffield Council on Bioethics 2005: 166). But when it comes to the most essential aspect of the labour process – namely, purposively enduring exposure to an experimental drug and its effects – the participation of nonhuman subjects in pre-clinical trials would seem to be enough like the participation of human subjects in phase one trials that if we are going to call the latter labour we should probably call the former labour, too. Or should we?

What's in it for the (lab) animals?

Cooper (2008) wants the concept of clinical labour to do political, not just analytical, work. "To formulate human subject experimentation as a form of labour is at once an observation and a provocation", she writes (Cooper 2008: 76). "Perhaps 'labour' is always a critical concept", she argues, "and one that emerges from a point of view of non-compliance or contestation" (Cooper 2008: 77). "By reformulating human subject experimentation as labour", she is "hoping to open up the scope of political critique to include a consideration of this liminal, but essential moment in the production of biomedical value" (Cooper 2008: 77). One reason for characterizing guinea-pigging as a form of labour is that doing so might benefit human subjects in some way. Abadie (2010: 165–166) takes this position, urging us "to recognize that volunteers' participation is labour, even if it is what they call a 'weird type of work,' and provide better working conditions and proper compensation". Another reason for describing guinea-pigging as labour is that doing so highlights the crucial contributions that these human subjects make to drug innovation. "There is no medical efficacy, no patentable biomedical innovation, and thus no innovation value without the participation of living bodies in clinical trials", Cooper (2008: 78) reminds us. And this includes the bodies of the nonhuman animals on whom drugs are tested (Birke 2012; Waldby 2012). Indeed, given current scientific practices and legal structures, the profitability of the pharmaceutical industry depends upon the bodily contributions of both types of subjects.

Whether it is politically productive to apply the concept of clinical labour to human guinea pigs is a question I leave for others to address. For scholars in the field of CAS, the question is whether it's a politically productive move in the case of actual guinea pigs. If one's main goal is to demonstrate that nonhuman animals are a source of value in the capitalist labour process, then characterizing their contribution as labour makes sense. But it's hard for me to imagine how developing a posthumanist labour theory of value might help improve the plight of nonhuman lab animals. This is not an argument against developing such a theory, but we are still left with Lynda Birke's (2009) question, "What's in it for the animals?" As Birke explains,

> to ask "what's in it for animals?" is partly to plead for greater accountability to the animals we bring into our studies. To think about that question is not only to ponder what they might think about it, but also to consider whether our investigations can help to bring about change – in the ways we think about them and their abilities, in the ways we treat them, in the ways we respect – or not – the places they live. That is, perhaps, not the remit of many academic inquiries. But thinking about politics has always been within the remit of fields like women's studies, which sought to challenge – and change – the oppressions besetting women throughout the world. In my view, it should remain within the remit of animal studies. Animals may indeed be supremely indifferent to the names we give them: but they are not indifferent to the naming of oppression.
>
> (Ibid.: 7)

So I am left with the question of what, if anything, might be in it for "lab animals" when we reformulate what they do – and have done to them – as a form of labour. Donna Haraway suspects that there may, in fact, be something in it for them:

> My suspicion is that we might nurture responsibility with and for other animals better by plumbing the category of labour more than the category of rights, with its inevitable preoccupation with similarity, analogy, calculation, and honorary membership in the expanded abstraction of the Human. Regarding animals as systems of production and as technologies is nothing

new. Taking animals seriously as workers without the comforts of humanist frameworks for people or animals is perhaps new and might help stem the killing machines.

(Haraway 2008: 73, footnote omitted)

Refashioning the category of labour to include nonhuman animals helps challenge the paradigm of human exceptionalism that justifies so much violence against animals. Describing the animals who are subsumed within the capitalist labour process as labourers, instead of as part of the means of production, also helps remind us that they are subjects rather than objects (ibid.: 80). That said, the same point could just as easily be made by suggesting that they are not labourers but rather subjects who are treated as objects of labour.

Zipporah Weisberg (2009) is less optimistic than Haraway about refashioning the category of labour to include lab animals. "In reality", she writes, "animals in labs are not workers – not even alienated workers – but worked-*on* objects, *slaves* by any other name" (ibid.: 36). "To call them anything else is to gloss over the brutal reality of the total denial of their ability to act in any meaningful way—namely, as self-determining *subjects*", Weisberg (2009: 36) writes. This is not the place to examine the use of the slavery analogy, except to note that not all slaves have been reduced to mere objects that are incapable of resistance – a point I return to below, in my discussion of Foucault.[5] What I take from Weisberg is a deeper appreciation of the limitations of thinking about the labour process in abstraction from the "human-animal relations of production" within which it is organized (Tapper 1988: 52). Although focusing my analysis on the labour process itself enabled me to examine the human exceptionalism that lies at the heart of Marx's concept of labour, doing so also risked obscuring what Weisberg (2009) describes as the relations of domination to which many lab animals are subjected. To understand this domination, we need to turn now to Foucault.

Foucault distinguishes power relations from both "relations of violence" and "states of domination" (Foucault 2000b: 340; 1997: 283). As he explains,

... what defines a relationship of power is that it is a mode of action that does not act directly and immediately on others. Instead, it acts upon their actions: an action upon an action, on possible or actual future or present actions. A relationship of violence acts upon a body or upon things; it forces, it bends, it breaks, it destroys, or it closes off all possibilities. Its opposite pole can only be passivity, and if it comes up against any resistance it has no other option but to try to break it down. A power relationship, on the other hand, can only be articulated on the basis of two elements that are indispensable if it is really to be a power relationship: that "the other" (the one over whom power is exercised) is recognized and maintained to the very end as a subject who acts; and that, faced with a relationship of power, a whole field of responses, reactions, results, and possible inventions may open up.

(Foucault 2000b: 340)

Foucault makes two crucial points here. First, using force against someone's body is not an exercise of power, for power operates on conduct not the body (see also Lukes 2005: 86–87, 157 n.18). As he puts it, "[a] man who is chained up and beaten is subject to force being exerted over him, not power" (Foucault 2000a: 324). His second point is that "[p]ower is exercised only over free subjects, and only insofar as they are 'free'" (Foucault 2000b: 342). In other words, "in power relations there is necessarily the possibility of resistance…" (Foucault 1997: 292). Whereas power relations are fluid and reversible, states of domination are frozen and fixed; someone who has no room for manoeuvre, no capacity to resist, is locked in a state of domination. For example, "slavery is not a power relationship when [the slave] is in chains", Foucault (2000b: 342) writes; it is a power relationship "only when [the slave] has some possible mobility, even a chance of escape". Similarly, if someone "were completely at [another's] disposal and became [that person's] thing, an object on which [that person] could wreak boundless and limitless violence", Foucault (1997: 292) explains, "there wouldn't be any relations of power".

Clare Palmer (2001) has used Foucault's ideas about power and domination to think about human–animal relations. In the most extreme forms of domination, where there is no realistic chance of resistance, the animal who is being dominated is reduced to a mere thing, Palmer (2001) explains. As she puts it,

> … although animals can be thought of as individuals who react in a Foucauldian sense, when they are placed by humans in situations or environments where no reaction or response from them is possible, they are being treated *as things* – even though they, like the shackled slave, *could have* been treated as beings who react…. [W]hether a being falls into the category of thing/person on any particular occasion depends not on its "nature," but rather whether, on that occasion, it behaves as a being which reacts. Where reaction is not permitted, the being is treated in this context as a thing—an object to which things are done – however much one might want to maintain that, in other contexts, the being is not just a "thing".
>
> (Ibid.: 354)

When the extreme form of domination that Palmer describes occurs in the labour process, it transforms the labourer from someone who does things into something to which, or with which, things are done. To put this in Marx's terms, the dominated labourer becomes part of the means of production, an object or an instrument of labour.

This kind of domination is common in human–animal relations of production, including those that occur in many labs. Of course, "lab animal" is a nebulous category, the precise contours of which depend upon how one defines "lab" and "animal". And however one defines these terms, actual lab animals experience a wide range of living conditions and deaths. Thus, to speak about the condition of *the* lab animal, or even of lab animals in general, is too imprecise to be of much use in analysing power relations. Following Palmer (2001), we need to examine these relations on a case-by-case basis. This means analysing power relations on a lab-by-lab, or even an animal-by-animal, basis; one should not assume that every lab animal is subjected to the most extreme forms of domination that one can imagine. That said, one should also avoid fixating solely on the particular procedures to which a particular lab animal is subjected, or on the animal's living conditions, for even if these allow for the possibility of resistance, the animal may, at the very same time, be utterly expendable. For example, we learn from a recent UK report that "[a]ll animals used in toxicity testing are routinely killed immediately at the end of experiments for examination" (Nuffield Council on Bioethics 2005: 165). And as is well known, even perfectly healthy lab animals, including those who have been subjected to no experimental procedures whatsoever, are often "euthanized" when they are no longer needed, tossed out like any other piece of useless laboratory equipment.[6] So although certain lab animals may over the course of their lifetimes have various opportunities to resist, most lab animals never make it out of the lab alive. That lab animals' lives are ultimately in the hands of human laboratory workers reveals the overarching structure of domination within which any resistance is ultimately framed (Weisberg 2009; cf. Palmer 2001). Moreover, no matter how much a lab animal may be able to resist, it is also worth remembering that, *Rise of the Planet of the Apes* notwithstanding, reversing the power relation so that the nonhuman has the upper hand is quite simply impossible (on relations of domination as irreversible, see Palmer 2001). For real guinea pigs, then, "[g]uinea [p]ig [r]esistance", as Weinstein (2010: 113, italics omitted) calls it, is typically quite limited indeed (Birke 2012; Greenhough and Roe 2011).[7]

Concluding thoughts

In this chapter I examined whether it makes sense, both analytically and politically, to use the concept of clinical labour to understand the participation of nonhuman animals in pre-clinical toxicity testing. I started by using recent advances in our understanding of animal cognition and consciousness to challenge the human exceptionalism that undergirds Marx's analysis of the labour process. Next I made the case that, if human guinea pigs are engaged in a form of clinical labour, it makes a certain amount of sense to say that real guinea pigs are, too. But in thinking about this labour, I argued, drawing on the work of Zipporah Weisberg (2009), we must remain cognizant of the human–animal relations of production within which it is organized. Building on Clare Palmer's (2001) reading of Foucault, I suggested that many lab animals are subjected to relations of domination that render resistance futile and reduce these animals to part of the means of production, mere instruments or objects of labour. The main lesson that emerges from this analysis is that, in our efforts to develop a posthumanist perspective on the labour process, scholars in the field of CAS must focus not just on the labour process itself, but also on the human–animal relations of production. Otherwise, we risk "occlud[ing] the … state of unfreedom to which [many nonhuman animals] are subjected" (Weisberg 2009: 34).

I'd like to conclude by pointing to a significant limitation of this chapter, one that highlights the conceptual work that remains to be done. Challenging Marx's human exceptionalism is necessary but not sufficient to develop a posthumanist perspective on the labour process. Given that concepts like labour and rights tend to be based upon a particular, Western conception of what it means to be human, they are typically extended only to those nonhumans who are most like who these humans take themselves to be (Potts and Haraway 2010; Smith 2001: 240 n.33; Wolfe 2003: 21–43). Instead of extending Marx's humanist concept of labour to (certain) other animals, what we need to do, Haraway (2008: 67) argues, is rethink the concept from the ground up, refashioning it "in non-humanist terms...". Although I cannot take up this task here, one approach, which I've pursued elsewhere (Clark, under review), is to stop thinking of labour as something that is done by individuals, whether human or nonhuman, and start thinking of it as a manifestation of what Jane Bennett (2010: 20–38) describes as the distributive agency of heterogeneous assemblages of human and nonhuman actants. What labours, from this posthumanist perspective on agency, is not any particular actant in the assemblage, but rather the assemblage itself. In the end, then, rethinking the labour process should be understood as part of the broader posthumanist project of rethinking agency, a project that promises to deliver a far humbler understanding of what it means to live in a more-than-human world.

Notes

1 Weinstein (2010: 119) questions the presumption of pharmaceutical naivety, as it is called, noting "the history of India's and Eastern Europe's home-grown, patent-evading pharmaceutical industries," along with "the pharmaceutical ambiguity of local/herbal medication".
2 As I explain below, others argue that planning is a prerequisite for the production of artefacts.
3 Marx's other definition of an instrument of labour is not relevant here.
4 In some dystopian visions of biotechnology, animals have been genetically engineered to be completely docile, reducing them to living machines that are compelled by their nature to do work (Warkentin 2006). I should note that the use of words like "reduce" in this context is potentially problematic for at least two reasons: first, it assumes that all inanimate instruments are manageable; and second, it elevates those who labour above those who serve merely as instruments or objects of labour. This is perhaps another form of exceptionalism, one that privileges the "active" over the "passive" elements of the labour process.
5 For a recent discussion of the deployment of the analogy in animal rights discourse, see Kim (2011). For an historical perspective on the analogy, see Davis (n.d.) and Jacoby (1994). For a discussion of involuntary experimentation on slaves in the United States, see Washington (2006).
6 It should be noted that adoption of former lab animals is becoming more common. I would also venture to guess that it is more difficult, whether legally or in terms of public relations, to dispose of some species than others.
7 For a discussion on nonhuman resistance, see Kowalczyk's chapter in this volume (Chapter 9).

References

Abadie, R. (2010) *The Professional Guinea Pig: Big Pharma and the Risky World of Human Subjects*, Durham and London: Duke University Press.

Acampora, R. (2006) *Corporal Compassion: Animal Ethics and Philosophy of Body*, Pittsburgh, PA: University of Pittsburgh Press.

Bekoff, M. (2011) "Animal minds and the foible of human exceptionalism", available online at: www.psychologytoday.com/blog/animal-emotions/201107/animal-minds-and-the-foible-human-exceptionalism (accessed 14 December 2013).

Bennett, J. (2010) *Vibrant Matter: A Political Ecology of Things*, Durham and London: Duke University Press.

Benton, T. (1989) "Marxism and natural limits: an ecological critique and reconstruction", *New Left Review*, 178: 51–86.

Benton, T. (1993) *Natural Relations: Ecology, Animal Rights & Social Justice*, London and New York: Verso.

Best, S. (2009a) "The rise of critical animal studies: putting theory into action and animal liberation into higher education", *Journal for Critical Animal Studies*, 7(1): 9–53.

Best, S. (2009b) "Minding the animals: ethology and the obsolescence of left humanism", *The International Journal of Inclusive Democracy*, 5(2): np.

Best, S. and Gigliotti, C. (2007) "Introduction", *Journal for Critical Animal Studies*, 5(2): np.

Birke, L. (2003) "Who – or what – are the rats (and mice) in the laboratory?" *Society & Animals*, 11(3): 207–224.

Birke, L. (2009) "Naming names – or, what's in it for the animals?" *Humanimalia*, 1(1): 1–9.

Birke, L. (2012) "Animal bodies in the production of scientific knowledge: modelling medicine", *Body & Society*, 18(3 and 4): 156–178.

Carney, S. (2011) *The Red Market: On the Trail of the World's Organ Brokers, Bone Thieves, Blood Farmers, and Child Traffickers*, New York: HarperCollins.

Catton, W., and Dunlap, R. (1978) "Environmental sociology: a new paradigm", *The American Sociologist*, 13(February): 41–49.

Cetina, K. Knorr. (1999) *Epistemic Cultures: How the Sciences Make Knowledge*, Cambridge, MA: Harvard University Press.

Cheng, I. (2006) "The beavers and the bees", *Cabinet* 23, available online at: www.cabinetmagazine.org/issues/23/cheng.php (accessed 14 December 2013).

Clark, J. (under review) "A posthumanist perspective on the production of nature thesis".

Cooper, M. (2008) "Experimental labour: offshoring clinical trials to China", *East Asian Science, Technology and Society: An International Journal*, 2: 73–92.

Cooper, M. (2011) "Trial by accident: tort law, industrial risks and the history of medical experiment", *Journal of Cultural Economy*, 4(1): 81–96.

Crist, E. (2002) "The inner life of earthworms: Darwin's argument and its implications", in M. Bekoff, C. Allen and M. Burghardt (eds) *The Cognitive Animal: Empirical and Theoretical Perspectives on Animal Cognition*, Cambridge, MA: MIT Press.

Davis, D. (n.d.). "The problem of slavery", available online at: www.yale.edu/glc/forum/davis.html (accessed 14 December 2013).

Dickens, P. (2004) *Society & Nature: Changing Our Environment, Changing Ourselves*, Malden, MA: Polity.

Dunlap, R. and Catton, W. (1979) "Environmental sociology", *Annual Review of Sociology*, 5: 243–273.

Elliott, C. (2008) "Guinea-Pigging", *The New Yorker*, (January 7): 36–41.

Elliott, C. (2010) "The mild torture economy", *The London Review of Books*, 32(18): 26–27.

Elliott, C. and Abadie, R. (2008) "Exploiting a research underclass in Phase 1 Clinical Trials", *New England Journal of Medicine*, 358(22): 2316–2317.

Elster, J. (1985) *Making Sense of Marx*, Cambridge: Cambridge University Press.

Foucault, M. (1997) "The ethics of the concern of the self as a practice of freedom", in P. Rabinow (ed.) *Ethics (Essential Works of Foucault, 1954–1984, Volume 1)*, New York: New Press.

Foucault, M. (2000a) " *'Omnes et Singulatim'*: toward a critique of political reason", in J. Faubion (ed.) *Power (Essential Works of Foucault 1954–1984, Volume 3)*, New York: New Press.

Foucault, M. (2000b) "The subject and power", in J. Faubion (ed.) *Power (Essential Works of Foucault 1954–1984, Volume 3)*, New York: New Press.

Fracchia, J. (2008) "The capitalist labour-process and the body in pain: the corporeal depths of Marx's concept of immiseration", *Historical Materialism*, 16: 35–66.

Garrard, G. (2012) *Ecocriticism (2nd edition)*, London and New York: Routledge.

Gould, J. R. (2007) "Animal artifacts", in E. Margolis and S. Laurence (eds) *Creations of the Mind: Theories of Artifacts and Their Representation*, Oxford and New York: Oxford University Press.

Gould, J. R. and Gould, C. (2007) *Animal Architects: Building and the Evolution of Intelligence*, New York: Basic Books.

Greenhough, B. and Roe, E. (2011) "Ethics, space, and somatic sensibilities: comparing relationships between scientific researchers and their human and animal experimental subjects", *Environment and Planning, D*, 29: 47–66.

Griffin, D. (2001) *Animal Minds: From Cognition to Consciousness*, Chicago and London: The University of Chicago Press.

Gruen, L. (2011) *Ethics and Animals: An Introduction*, New York: Cambridge University Press.

Guerrini, A. (2003) *Experimenting with Humans and Animals: From Galen to Animal Rights*, Baltimore and London: The Johns Hopkins University Press.

Guthman, J. (2011) "Bodies and accumulation: revisiting labour in the 'production of nature", *New Political Economy*, 16(2): 233–238.

Haraway, D. (2008) *When Species Meet*, Minneapolis and London: University of Minnesota Press.

Harvey, D. (2000) *Spaces of Hope*, Berkeley and Los Angeles: University of California Press.

Harvey, D. (2010) *A Companion to Marx's* Capital, New York: Verso.

Helmreich, S. (2008) "Species of biocapital", *Science as Culture*, 17(4): 463–478.

Hribal, J. (2003) " 'Animals are part of the working class': a challenge to labor history", *Labor History*, 44(4): 435–453.

Ingold, T. (1980) *Hunters, Pastoralists and Ranchers: Reindeer Economies and Their Transformations*, Cambridge: Cambridge University Press.

Ingold, T. (1983) "The architect and the bee: reflections on the work of animals and men", *Man, New Series*, 18(1): 1–20.

Ingold, T. (1986) *Evolution and Social Life*, Cambridge: Cambridge University Press.

Ingold, T. (1987) *The Appropriation of Nature: Essays on Human Ecology and Social Relations*, Iowa City, IA: University of Iowa Press.

Ingold, T. (1988a) "Introduction", in T. Ingold (ed.) *What is an Animal?* London and New York: Routledge.

Ingold, T. (1988b) "The animal in the study of humanity", in T. Ingold (ed.) *What is an Animal?* London and New York: Routledge.

Ingold, T. (2000) *The Perception of the Environment: Essays on Livelihood, Dwelling and Skill*, London and New York: Routledge.

Ingold, T. (2011) *Being Alive: Essays on Movement, Knowledge and Description*, London and New York: Routledge.

Jacoby, K. (1994) "Slaves by nature? Domestic animals and human slaves", *Slavery and Abolition*, 15(1): 89–99.

Jones, O. (2003) " 'The restraint of beasts': rurality, animality, actor network theory and dwelling", in P. Cloke (ed.) *Country Visions*, Harlow: Pearson.

Kim, C. J. (2011) "Moral extensionism or racist exploitation? The use of Holocaust and slavery analogies in the animal liberation movement", *New Political Science*, 33(3): 311–333.

Kohler, R. E. (1994) *Lords of the Fly:* Drosophila *Genetics and the Experimental Life*, Chicago and London: The University of Chicago Press.

Lederer, S. E. (1995) *Subjected to Science: Human Experimentation in America before the Second World War*, Baltimore and London: The Johns Hopkins University Press.

Levins, R. and Lewontin, R. (1985) *The Dialectical Biologist*, Cambridge and London: Harvard University Press.

Lewontin, R. and Levins, R. (2007) *Biology Under The Influence: Dialectical Essays on Ecology, Agriculture, and Health*, New York: Monthly Review Press.

Lukes, S. (2005) *Power: A Radical View (Second Edition)*, New York: Palgrave Macmillan.

Marx, K. (1990) *Capital: Volume I*, New York: Penguin.

Mitchell, R. and Waldby, C. (2010) "National biobanks: clinical labour, risk production, and the creation of biovalue", *Science, Technology, & Human Values*, 35(3): 330–355.

Morgan, L. H. (1986) *The American Beaver: A Classic of Natural History and Ecology*, New York: Dover.

Murray, M. (2011) "The underdog in history: serfdom, slavery and species in the creation and development of capitalism", in N. Taylor and T. Signal (eds) *Theorizing Animals: Re-thinking Humanimal Relations*, Leiden and Boston: Brill.

Nimmo, R. (2011) "The making of the human: anthropocentrism in modern social thought", in R. Boddice (ed.) *Anthropocentrism: Humans, Animals, Environments*, Leiden and Boston: Brill.

Noske, B. (1997) *Beyond Boundaries: Humans and Animals*, New York: Black Rose Books.

Nuffield Council on Bioethics (2005) *The Ethics of Research Involving Animals*, available online at: www.nuffieldbioethics.org/sites/default/files/The%20ethics%20of%20 research%20involving%20animals%20-%20full%20report.pdf (accessed 14 December 2013).

Palmer, C. (2001) " 'Taming the wild profusion of existing things?' A study of Foucault, power, and human/animal relationships", *Environmental Ethics*, 23: 339–358.

Pemberton, S. (2004) "Canine technologies, model patients: the historical production of hemophiliac dogs in American biomedicine", in S. Schrepfer and P. Scranton (eds) *Industrializing Organisms: Introducing Evolutionary History*, New York and London: Routledge.

Petryna, A. (2009) *When Experiments Travel: Clinical Trials and the Global Search for Human Subjects*, Princeton, NJ: Princeton University Press.

Pickering, A. (2008) "Against Human Exceptionalism", available online at: https://eric. exeter.ac.uk/repository/bitstream/handle/10036/18873/XTRwrkshp-250108. pdf?sequence=1(accessed 14 December 2013).

Potts, A. and Haraway, D. (2010) "Kiwi chicken advocate talks with California dog companion", *Feminism & Psychology*, 20(3): 318–336.

Prasad, A. (2009) 'Capitalizing disease: biopolitics of drug trials in India', *Theory, Culture & Society*, 26(5): 1–29.

Rajan, K. S. (2007) "Experimental values: Indian clinical trials and surplus health", *New Left Review*, 45(May–June): 67–88.

Raulerson, J. T. (2010) "Singularities: technoculture, transhumanism, and science fiction in the 21st century", Unpublished dissertation, Iowa: University of Iowa.

Russell, E. (2004) "The garden in the machine: toward an evolutionary history of technology", in S. Schrepfer and P. Scranton (eds) *Industrializing Organisms: Introducing Evolutionary History*, New York: Routledge.

Smith, M. (2001) *An Ethics of Place: Radical Ecology, Postmodernity, and Social Theory*, Albany, NY: SUNY Press.

Tapper, R. (1988) "Animality, humanity, morality, society", in T. Ingold (ed.) *What is an Animal?* London and New York: Routledge.

Thacker, E. (2005) *The Global Genome: Biotechnology, Politics, and Culture*, Cambridge and London: MIT Press.

Tovey, H. (2003) "Theorising nature and society in sociology: the invisibility of animals", *Sociologia Ruralis*, 43(3): 196–215.

Tsing, A. (2012) "Unruly edges: mushrooms as companion species", *Environmental Humanities*, 1: 141–154.

Vint, S. (2009) "Species and species-being: alienated subjectivity and the commodification of animals", in M. Bould and C. Mieville (eds) *Red Planets: Marxism and Science Fiction*, Middletown, CT: Wesleyan University Press.

Wacquant, L. (1995) "Pugs at work: bodily capital and bodily labour among professional boxers", *Body & Society*, 1(1): 65–93.

Waldby, C. (2012) "Medicine: the ethics of care, the subject of experiment", *Body & Society*, 18(3 and 4): 179–192.

Waldby, C. and Cooper, M. (2008). "The biopolitics of reproduction: post-Fordist biotechnology and women's clinical labour", *Australian Feminist Studies*, 23(55): 57–73.

Waldby, C. and Cooper, M. (2010) "From reproductive work to regenerative labour: the female body and the stem cell industries", *Feminist Theory*, 11(1): 3–22.

Warkentin, T. (2006) "Dis/integrating animals: ethical dimensions of the genetic engineering of animals for human consumption", *AI & Society*, 20: 82–102.

Washington, H. (2006) *Medical Apartheid: The Dark History of Medical Experimentation on Black Americans from Colonial Times to the Present*, New York: Doubleday.

Weinstein, M. (2010) *Bodies Out of Control: Rethinking Science Texts*, New York: Peter Lang.

Weisberg, Z. (2009) "The broken promises of monsters: Haraway, animals and the humanist legacy", *Journal for Critical Animal Studies*, 7(2): 21–61.

Whatmore, S. (2006) "Materialist returns: practicing cultural geography in and for a more-than-human world", *Cultural Geographies*, 13: 600–609.

Wilde, L. (2000) "'The creatures, too, must become free': Marx and the animal/human distinction", *Capital & Class*, 72: 37–53.

Wolfe, C. (2003) *Animal Rites: American Culture, the Discourse of Species, and Posthumanist Theory*. Chicago and London: University of Chicago Press.

Wolfe, C. (2008) "Exposures", in S. Cavell, S. C. Diamond, J. McDowell, I. Hacking and C. Wolfe (eds) *Philosophy & Animal Life*, New York: Columbia University Press.

8 The cultural hegemony of meat and the animal industrial complex

Amy J. Fitzgerald and Nik Taylor

We wanted to take the opportunity to write this chapter and contribute to this volume for three main reasons. The first is that we are part of, and witnesses to, the development of a new discipline – Animal Studies – and while we are greatly appreciative of all the work done under this umbrella we are interested to explore what specifically Critical Animal Studies might mean and what it can bring to empirical, as well as theoretical, work. The second is that we feel the belief in the human right to use other animals is so pervasive in modern societies that it is akin to sexism, racism and other forms of discrimination. We also believe that it rests upon similar foundations to other forms of discrimination and thus one cannot successfully counter one without countering the others.

Ecofeminists have historically been at the forefront of interrogating the intersecting subjugation of "nature" and marginalised human groups (see for example, Adams 1991; Gaard 1997; Kheel 1995; Plumwood 1996; Warren 1997). We expand upon this idea later, so suffice to say for introductory purposes that a belief in the need to expand theories of intersectionality to include speciesism motivates us. Finally, we chose to write about the cultural hegemony of animal products as a specific empirical, test-case, because it is here, we believe, that the ultimate expression of human superiority and exceptionalism is made, precisely because it is made normatively and by assumption. This means that challenging it is necessary but also difficult as it requires addressing something so taken-for-granted and embedded that most are unable, or unwilling, to see it. That said, addressing the hidden mechanisms of social life is arguably one of the most important functions of sociology, as Burawoy (2007: 28) points out in his idea of an "organic public sociology", which brings "sociology into a conversation with publics" (see also Twine 2010 and Cudworth, Chapter 1 this volume). One of the projects of such sociology is "to make visible the invisible, to make the private public," which we applaud but would make explicit that achieving such a goal is only possible if we attend to the ways in which power plays out and manifests itself through discourse; hence, our interest in the normative discourses of the acceptability of the human consumption of animal-derived food products.

This chapter, then, will begin to map some of the ways in which entrenched beliefs about animals and humans are created and maintained by different discursive practices. We do this by analysing the various normalisation discourses surrounding the consumption of animal products. Using empirical data drawn from two main fields of practice where animals are constituted simply as consumables (human food and "pet" food products) we seek to begin to uncover the mechanisms whereby the production and consumption of animal products are normalised. We argue that this consumption is so taken for granted in modern industrial societies that there exists a cultural hegemony regarding not just the acceptability, but the *necessity* of animal consumption and seek to deconstruct the specific ways in which this occurs. We consider these issues within a framework that seeks to illuminate the operation of a general speciesist ideology and argue that this is a result of, and at the same time underpins, the "animal-industrial complex" (A-IC) (Noske 1989; Twine 2012). We utilise Twine's (2012: 23) definition of the A-IC as "a partly opaque and multiple set of networks and relationships between the corporate ... sector, governments, and public and private science. With economic, social, and affective dimensions it encompasses an extensive range of practices, techniques, images, identities and markets". In particular, we focus on the interplay between images, identities and markets by analysing the entanglements of animal-derived food production and consumption practices (and in this particular case we mean the literal consumption – ingestion – of animals) as they are sterilised and made palatable for general consumers through various campaigns. This allows us to begin highlighting the interconnected, local and global, character of the animal-industrial complex. We do this in the spirit of Twine's exhortation that in order to make the A-IC an organising concept for CAS we need to continually map the connections between overlapping sectors where animal abuse and exploitation occur. Believing that the obfuscation of meat-as-animal-life is one of the most important cornerstones of the animal-industrial complex under capitalist and neo-liberal regimes, we analyse data where animals are absent, manipulated or otherwise used within a political economy that depends on their very bodies for its maintenance, not in a nutritional sense but in the sense that they are constituted as necessary for maintaining the political economic status quo.

Background and context

While the growing discipline of Animal Studies has legitimised a consideration of animal abuse, much of the focus is on the cruelty perpetrated by individual humans to companion animals. Institutionalised cruelty receives far less attention (for exceptions and more on this see Beirne 2009; Fitzgerald *et al.* 2009; Flynn 2002), perhaps precisely because it is *institutionalized* and therefore subject to both concealment and protection by those with vested interests (see Groling, Chapter 5 in this volume, on this point). Other factors play a role in the concealment of the various animal abuses that take place in the processing of their bodies into products – for example, a desire to maintain the clear boundaries between human and other animals and thus shore up human superiority; a need to "prove" human civility, which necessitates the concealment of anything considered unpalatable and barbaric and in turn led to the moving of slaughter practices from open view to the outskirts of society (Elias 1978; Vialles 1994). As Elias points out,

> It will be seen again and again how characteristic of the whole process of civilization is this movement of segregation, this hiding "behind the scenes" of what has become distasteful. The curve running from the carving of a large part of the animal or even the whole animal at table, through the advance in the threshold of repugnance at the sight of dead animals, to the removal of carving to specialized enclaves behind the scenes, is a typical civilization curve.
>
> (Elias 1978: 99)

However, we believe that these forces legitimate institutionalised animal abuses rather than pre-figure them. By hiding what actually happens – whether this occurs literally through the move to the edges of society of slaughterhouses, or discursively through the constitution of animals as less worthy than humans of moral concern – abusive treatment becomes legitimated and, ultimately, considered to be normal and necessary. Thus we see discourses abound regarding the nutritional necessity for humans to eat meat as well as those that situate animal production as a necessary part of a "thriving" economy. In turn, the abuses that animals suffer as a part of this process are covered up – again discursively and literally – through, for example claims that their welfare is of paramount concern throughout the commodification process, or claims that humans could not live without the nutrition granted to them through the consumption of animal products (see Cole 2011 for a more in depth discussion). This discursive sleight of hand effectively silences any opposition partly because abusive practices are "out of sight, out of mind" but also because they are discursively constituted as normal – "nothing to see here folks, move along".

In our opinion, this is where *Critical* Animal Studies (CAS) comes into its own. Developing, in part, from the rich ideas of ecofeminists from the 1970's onwards, who pointed to the intersected nature of the oppression of nature/ animals/environment and the oppression of women, CAS scholars seek to elucidate the mechanisms of these intersections in order to end them.[1] We would further argue that (some) other feminist approaches have much to offer any analysis of the abuse of animals in modern society. Liz Kelly's concept of a "continuum of violence" (1988) is a particularly useful one here. This concept is based on an acknowledgment that violence against women is far from an aberration and, in fact, is normative. The normative character of violence against women leads to inattention to the ways it plays out on a daily basis because of a pre-occupation with extreme cases. This focus means that the *socially sanctioned* spectrum of male violence against women, which includes, but is by no means limited to, extreme forms of horrific instances of cruelty and abuse, go under-acknowledged and under-analysed. We feel that this is often the case in Animal Studies as well as in society generally, where attention is given to individual instances of cruelty to companion animal species. While worthwhile, this body of scholarship neglects the institutionalised nature of abuse and therefore reinforces a form of speciesism by privileging harms perpetrated against some species over others. As a consequence such theories are bereft of any detailed analyses of power at a structural level, which we feel is necessary.[2]

To better understand how violence against animals at the socially sanctioned end of the "continuum of violence" is normalised and has reached the level of hegemony, our analysis is centred on discourses of the meat production segment of the animal-industrial complex. By using empirical data, we hope to make some of the mechanisms of the A-IC clear and in so doing avoid falling into the trap that Twine identifies when he notes that usage of the concept "may have become simply assumed and almost rhetorical, deployed monolithically to represent, but also to reduce the myriad complexity of the multiple relations, actors, technologies and identities that may be said to comprise the complex" (2012: 15). While it is beyond the scope of this chapter to map all of the interconnections and detail the nuanced interplay between the various arms of the complex, by locating our work empirically we begin to describe some of the concrete mechanisms at play in the A-IC. We focus on the most common themes we observed, but before turning to those findings, we detail our sources of data and analytical method in the next section.

Data and method

We examine two sources of messages about animal-derived food products, targeted at two fairly distinct audiences. The first is websites for red meat exporters, which are aimed at companies looking to secure supplies of red meat from Australia (a major meat exporter). We selected our sample of red meat exporters using an online search function of the Meat and Livestock Australia website. This website (www.mla.com.au/Marketing-red-meat/International-marketing/Red-meat-exporter-database) states that its purpose is that "International meat buyers seeking Australian red meat suppliers can use MLA's Australian Red Meat Exporter Database to find companies that can supply Australian beef, lamb, mutton, goat meat and offal products." The search function has various ways of limiting responses, e.g. "importing destination", or "species type". Leaving all of these fields listed as "any", an open search was performed. This resulted in 136 hits. Of these 110 had active websites broken down by product (beef=94; sheep=74; goat=43; offal=67; other=24), certification (halal=92; organic=37; EQA=41; EU grain fed beef=11), and region (Americas=75; Europe=77; Africa=64; Asia=105; Middle East=80, and Other=27). From these, 55 were selected at random and these websites formed the data for the current analysis. We then analysed the front pages of each website and pulled out various recurring themes, an examination of which occurs below.

The second source is print advertisements published in cooking magazines in North America and aimed at individual consumers. We selected three magazines, *Food and Wine*, *Bon Appétit* and *Cooking Light*, because they are among the top Epicurean magazines according to the circulation averages compiled for the first half of 2012 by the Audit Bureau of Circulations. We analysed all advertisements for products derived from animal sources, including meat, dairy, eggs and "pet" food, appearing in the first seven issues of 2012 for each of these monthly periodicals. These selection criteria resulted in 89 total advertisements (*Food and Wine*=12; *Bon Appétit*=12; *Cooking Light*=65). The removal of duplicate advertisements left us with 74 unique advertisements (63 for human food products and 11 for "pet" food).

The mass media and the power to define

Before moving on to discuss the themes, we want to pause a moment here to reflect on the role of both media and discourse in determining beliefs about, and attitudes toward, non-human animals. Discourse refers to more than simply language:

> Discourse is about the interplay between language and social relationships, in which some groups are able to achieve dominance for their interests in the way in which the world is defined and acted upon.... Language is a central aspect of discourse through which power is reproduced and communicated.
>
> (Hugman 1991: 37)

In modern Western society the media has an enormous power to disseminate ideas about issues, to frame things in certain ways and to determine what is considered normal. As Molloy points out (2011: 1) "animal narratives ... play an essential role in shaping the limits and norms of public discourses on animals and animal issues and so constitute a key source of information, definitions and images". Morgan and Cole (2011) argue that a "flesh-eating hegemony" (p. 120), which is a reflection of dominant social attitudes, permeates the modern media, and it is evidence of this we are keen to analyse. They further point out that "looking at stories relating to other animals ... gives us some idea of what is seen as important or worthy of discussion – and equally importantly, what is *not* worthy of discussion" (ibid.: 120). By expanding this to examine advertisements and websites, which include images relating to the production and consumption of animal products, we can begin to deconstruct how animal production and consumption are normalised in a "carnist" (Joy, 2009), "anthroparchal" (Cudworth (Chapter 1), and Jenkins and Twine (Chapter 11), this volume) culture.

The themes

We analysed our data using Content Analysis, a method that allows the examination of material for themes and meanings. We coded both explicit and implicit messages in the text and images contained in the advertisements and websites. This analysis of the data showed several prominent themes throughout both sources (the magazine advertisements and the websites). We focus on the three most prominent themes here, and while for analytical purposes we tease these themes apart, there are also important interconnections between them which we discuss throughout.

The replacement of "realistic" animals with "happy" animals

One of the themes common to both sets of data is the absence of any acknow-ledgment of the processes used to convert live animals into dead products. In many ways this echoes Adams' argument that,

> Behind every meal of meat is an absence: the death of the animal whose place the meat takes. The "absent referent" is that which separates the meat eater from the animal and the animal from the end product. The function of the absent referent is to keep our "meat" separated from any idea that she or he was once an animal ... to keep *something* from being seen as having been someone.

> (Adams 2010: 13)

This omission of the processes used to produce meat has been documented in other research examining cultural representations of meat (e.g. Heinz and Lee 1998) and particularly in the case of chickens where, as Molloy points out, there is a disconnection between live birds and their "reconfigured" form (2011: 110).

However, while the *processes* that animals are subject to in order to become products – as opposed to whole, living beings – are absent throughout both data sites, actual animals are not. Animals are clearly discursively constituted as "happy" throughout many of the adverts and websites ("happy animals" appeared in 25 per cent of the advertisements and 40 per cent of the meat exporter websites) with the implication that the animals used lived happy and productive lives, presumably until the time came to turn them into products.

Less than one-fifth of advertisements analysed contain images of "live" animals (as opposed to pieces of meat); however, almost half of these depictions are drawings. And only dairy cows are depicted in photographs, which could indicate that pictures of dairy animals (in pastures as opposed to modern dairy farms) are less risky than pictures of animals that are consumed as meat. As Molloy (2011: 114–115) notes,

> much of the popularly available imagery of dairy farming has been gener-ated by the dairy industry, where advertising ... has continued to deploy culturally specific visions of contented cows in rural landscapes. As a result the publicly available meanings of dairy farming and cows have been refracted through readings of nature as "environment" and "landscape".

In turn, Molloy argues, this is used to suggest "natural" and "healthy" and is reflected back upon dairy cows and their produce, thus suggesting healthy products obtained within a welfare-friendly environment (Molloy 2011: 114, 120). Thus, even though some companies use images of animals in their advertisements, they employ idyllic images of dairy cows in pastures and even cartoon depictions of happy, appeasing, docile animals. This docility is central to the concept of human rights to "control" other species (Cole 2011). The ads more commonly rely on text and cartoon depictions to convey their message that the animals used in these products at least lived happily (no mention was ever made of happy deaths) and were even used willingly. Others have also pointed to the use of these techniques to sell animal products to consumers (e.g. Cole 2011). The use of cartoon/animated animals and the construction of suicidal animals willingly giving up their bodies constitute genres in the advertising of animal products (see also Griffin, Chapter 6 in this volume). Advertisements for Starkist tuna are illustrative of these two genres. These ads present a cartoon depiction of Charlie the Tuna, who is gesturing with his fin as though he is proudly presenting the product – tuna flesh. The three different ads for Starkist tuna included in our sample all contained Charlie the Tuna and a print statement thanking Charlie. Another ad for Skinny Cow ice cream includes an illustration of a cow lying horizontally (and even provocatively) with a smile on her face and a measuring tape around her thin midsection. These depictions not only communicate implicit claims about the animals being kept in such humane conditions that they are happy about it, they also imply that aware of their own limitations, these animals give themselves over to more able, human hands. These messages are reminiscent of historic claims about slaves and those imprisoned in concentration camps benefitting from these arrangements because they are being "taken care of" (see Patterson 2002; Spiegel 1996).

We note that the claims about the happiness of animals are also devoid of content about the actual welfare of the animals being used, which is also complemented by the absence of visual depictions of "realistic" animals – those that actually are or resemble those used by industrial animal agriculture. The notion of happy agricultural animals no doubt mitigates angst among some consumers. As Cole (2011: 84) points out in his application of Foucauldian ideas regarding "pastoral power" and "happy animals", such discourses "attempt to remoralize the exploitation of 'farmed' animals in such a way as to permit business as usual, with the added 'value' of ethical self-satisfaction for the consumer of 'happy meat.'"

On the production side, claims about the "naturalness" of the living conditions of animals were linked with claims about their "happiness". However, it is evident that there is some selectivity used in visually representing the naturalness of these conditions and the happiness of animals. For instance, while one meat exporter website provided video of the lifespan of cattle and sheep from birth through to arrival at the slaughterhouse, the footage ended when the gate the animals were herded through at the slaughterhouse was closed. Thus the last image we have of these animals is one of them as whole and healthy, and specific links between these animals and the images of plated meats are left unmade and subject to a "symbolic distancing" (Hamilton and Taylor 2013). Similarly, less extensive visuals were provided of the processes that chickens were subject to, presumably because they tend to be raised in more concentrated and confined facilities. It is also worthy of note that this was the only website that included any reference whatsoever to the slaughterhouse.

Thus, both meat exporter websites and the magazine adverts market a sanitized version of production, not only through the happy animals discourse but also through the visual images they choose to use, as well as those they omit and conceal. It is not surprising that those marketing animal products to individual end-users or consumers would not provide realistic visuals of the way that the majority of animals used for human consumption are raised and killed. Doing so might risk turning individual consumers off consuming these products (although as Salih, Chapter 3 this volume, points out, exposure to information about the production process does not necessarily translate into changes in consumptive behaviour). What is more surprising is that marketing through the meat exporter websites, which targets business interests, is similarly sanitised, although not to the same degree. The marketing here does contain images of live animals, and one even includes video footage of some animals. However, the websites are also saturated with images of idealised rural agriculture. Modern industrialised and concentrated animal agriculture is not represented here, even though the target audience is businesses and not individual consumers. This can be analysed in several different ways. First, the meat exporter websites could be cognisant of the chance that the general public could happen upon their publicly available marketing materials, although animal product consumers are not known to go searching for information about how the animals they consume live and die; most consumers do not want to know how the meat, dairy and eggs they consume are produced (Foer 2009). Second, it could also mean that they are concerned with a form of corporate "impression management" (Goffman 1959). If, for instance, they depicted how these animals are actually raised, the importers they seek to do business with may opt to take their business elsewhere – to a company that is providing a more sanitized view of the industry. This possibility seems more likely than the former, although it may not tell the whole story. Irrespectively, as Twine points out in his discussion of the discourses surrounding genetically modified animals, the language used is 'uncanny' in that it "combines instrumental language related to profitability, efficiency and death, with the subjectification of [particular animals] as a 'she'" (2012: 110). This "uncanny" subjectification flies in the face of the realities of the slaughter at the heart of the business of these corporations. Finally, it is also possible that sanitised depictions of the industry have become so hegemonic that companies simply reproduce the same message and strategies over and over: the animals used are happy and only those that appear that way (i.e. in pastures instead of in confinement barns) are to be depicted. This would indicate the substantial power of the ideology promulgated by the A-IC to continually reproduce itself, as well as demonstrating "how capitalism creatively commodifies its own excess" (Twine 2012: 19).

The relative absence of live animals (and the total absence of "realistic", industrially-produced animals) in these ads stands in stark contrast to what we found in the 11 additional advertisements we analysed that were marketing animal products in the form of pet food. Of those advertisements, all but one (91 per cent) contain images of animals, but not the species of animals contained in the food; instead these images are of "pet" or companion animal species (i.e. cats and dogs) who will presumably be consuming the food. They are pictured eating, posing for the camera, or interacting lovingly with human companions. Consistent with the majority of advertisements for human food, none of these advertisements for pet food feature images of the species of animals contained in the food. Thus, some species of animals are actively rendered invisible as part of the commodification process (see Shukin 2009), while other species are intentionally foregrounded and constructed by corporations as family members that deserve the best that money can buy. These companies are exploiting and benefitting from the existence of a sociozoologic scale (Arluke and Sanders 1996), which privileges some species of animals (e.g. pets) over others (e.g. livestock and poultry). The A-IC, as a whole, relies upon the continued cultural salience of this sociozoologic scale, which is in turn reaffirmed by the presumption of its naturalness.

Romanticisation of "naturalness"

The "happy animals" narrative was frequently connected with a discourse about naturalness. We noted the romanticisation of nature (in 42 per cent of the meat exporter websites) and of relationships between human farmers and the animals they keep (in 8 per cent of meat exporter websites). The advertisements were also replete with messages about the naturalness of animal-derived products. For instance, the copy in an ad for Kerrygold's cheese and butter uses the word "natural" three times in the span of two sentences: "We don't mind telling you what goes into our pure, all natural Irish cheese and butter. It's pure, all natural Irish milk that comes from cows that graze on pure, all natural Irish grass." Another ad for dairy products asserts the "naturalness" of their product based on their claims that they do not use artificial growth hormones.[3] Interestingly, the small print under the claim about growth hormone-free cows states "Our farmers pledge not to use artificial growth hormones. No significant difference has been shown in milk from cows treated with the artificial growth hormone rbST and non rbST treated cows." Despite this interesting qualification, which we can only assume is in place to protect against lawsuits, we find the word "natural" is used frequently in both samples to imply that a product is healthy.

We suggest that this discourse of naturalness serves to normalize the production and consumption processes of animal-derived food products. The romanticisation of nature and of human relationships with both it and the animals considered to be an integral part of it, conceptually associates food animal production with nature and the idea of the "natural". This implicitly legitimates human use of animals based on rhetoric of *necessity* (nature-as-resource) and of control, which, in turn, rests upon post-Enlightenment beliefs in human exemptionalism and the supremacy that technological advances are held to give us as a species (Castree and Braun 1998). Research has found that the presumed naturalness of the consumption of meat is so pervasive that even those organisations that challenge the nature-as-resource perspective (i.e. environmental organisations), generally fail to question the human "need" to eat meat (Packwood Freeman 2010).

It is worthy of note, however, that the discourse regarding the "naturalness" of these "products" is actually one of a carefully controlled nature. These are not images of a wild, windswept nature that defies human control and dominance. Rather they are images of a romanticized, yet domesticated, nature, which is again linked to human dominance through a civilising discourse. As Ingold points out (1994: 6), "man's rise to civilization was conceived to have its counterpart in the domestication of nature".

This "natural order" discourse was both gendered and speciesist. In our analysis, sexism was apparent in the reification of traditional gender roles in both the magazine advertisements and through The Red Meat Exporters' websites. Perhaps not surprisingly, some of the advertisements for animal products were clearly directed at women, presumably as the primary shoppers. On the flip side, the materials of The Red Meat Exporters depicted farmers as exclusively male. This resonates with Cudworth's (2008: 43) arguments that the process of "becoming-meat" in an anthroparchal (i.e. human dominant) system is a gendered one. As she explains:

> As a complex social system, anthroparchy is intersectionalised ... [where] the intersection of capitalist and patriarchal relations is particularly marked.... The object of domination in the manufacture of meat is patriarchally constituted. As such animals are largely female and are usually feminized in terms of their treatment. Farmers disproportionately breed female animals so they can maximize profit via the manipulation of reproduction. Female animals that have been used for breeding can be seen to incur the most severe physical violence within the system, particularly at slaughter. Female and feminized animals are bred, incarcerated, raped, killed and cut into pieces, and this tale of becoming-meat is very much a story of commodification. Yet whilst the production of meat is shaped by relations of capital and patriarchy, it is most clearly a site in which anthroparchal relations cohere as certain kinds of animals are (re)constructed as a range of objects for human consumption.

Linked to the gendered nature of oppression in animal product representation is speciesism, particularly in the ways that (hegemonic) masculinity is linked to meat eating (Adams 2010; Rogers 2008). As Rogers demonstrates in his analysis of television advertisements, failure to consume meat, particularly among men, is framed as challenging the "natural balance" between genders and species. Meat consumption is linked with "natural" predation, particularly by men. The materials we analysed also naturalize speciesism and this occurs on a few different levels. On one level they support the underlying speciesist assumption that humans have the right to enslave and commodify animals in the name of profit. This assumption supports the constructed dichotomy between human and non-human animals. However, we also find another level of speciesism at play here that points to the existence of a continuum between human and non-human animals. As discussed earlier, in the magazine advertisements it was rare to see photographs of live animals; however, we find that in the advertisements for pet food products, photos of companion animal species are ubiquitous, while the animals contained in the advertised products are once again omitted. This finding illustrates the power of speciesism: some animal species are foregrounded in photographs and referred to as family (i.e. companion animals), while others are literally and symbolically annihilated (i.e. animals used as food).

Sexism and speciesism sit happily side by side in the materials we analysed and this again demonstrates the utility of a critical animal studies approach that is grounded in intersectionality. Speciesism and sexism (as well as other forms of domination) are intertwined precisely because they originate from, and within, a particular system of oppression. As Nibert (2002) points out:

> the oppression of various devalued groups in human societies is not inde-pendent or unrelated; rather, the arrangements that lead to various forms of oppression are integrated in such a way that the exploitation of one group frequently augments and compounds the mistreatment of another.
>
> (Ibid.: 4)

The hegemony of these forms of oppression makes them appear natural.

The naturalness discourse in the materials we analysed most explicitly nor-malise the current mode of industrial animal agriculture. On the production side, both the red meat exporter websites and the magazine advertisements make claims about the animals being raised and managed in "natural" conditions. These claims divert attention away from the industrial conditions that the vast majority of animals are raised within in developed countries. They can also simultaneously normalise these conditions by implicitly framing them as "natural". This is possible because there is no objective standard for "natural-ness", which also makes it a useful and inexpensive marketing tool, akin to "tradition".

The websites framed messages of "naturalness" within the larger theme of happy animals and romanticized agriculture. Within this larger theme, science and technology were also apparent and seemed to coexist comfortably with mes-sages of naturalness. It seems that Western cultures have accepted science and technology as a natural part of meat and dairy production. This acceptance speaks to the power of the A-IC to frame their increasingly technologically advanced practices not only as harmless, but even as natural. It also enables them to capitalise on the cultural authority of science and illustrates the interconnect-edness of actors within the A-IC, such as agribusiness companies, marketing companies, companies and governmental agencies invested in new scientific and technological developments and the pharmaceutical industry.

Good health and good taste

In addition to normalising the ways in which animal-derived food products are produced, we suggest that the "naturalness" discourse more generally legitimises the consumption of animal-derived products and even assists in constructing them as healthy and smart choices (i.e. if it's "natural", it must be healthy). In turn, this also plays out through discourses that suggest that animals reared in natural, healthy ways ("happy animals") will taste better and thus, as Cole (2011: 93) notes, the intermingling of animal welfare friendly and "happy meat" discourse "always posits a win-win scenario: happy animals taste better". By implication those who choose these products demonstrate their own good taste, ethically and gastronomically. By far the most common message conveyed in the advertisements analysed is that consuming animal products is a healthy and therefore smart decision. This finding is consistent with Heinz and Lee's (1998) observation that the concerns raised about the healthfulness of meat in the 1980s and 1990s caused the industry to begin "aggressively promoting meat as a natural, healthy part of daily food intake" (ibid.: 86). Seventy per cent of the advertisements explicitly or implicitly claim that consuming animal products is healthy. One of the more explicit messages appears in an ad for Campbell's Chunky Healthy Request Chicken Noodle Soup. The focal point of the ad is a heart-shaped bowl, which presumably contains the chicken noodle soup being advertised. Above the bowl the headline reads: "The flavour that captured your heart, made heart healthy." To reaffirm the message being communicated about the healthfulness of the product, in the lower left hand corner there is a heart with a check mark through it and a statement indicating that the product is certi-fied by the American Heart Association because it is low in saturated fat and cholesterol. The copy at the bottom of the advertisement states "It's amazing what soup can do", which in combination with the rest of the advertisement seems to imply that this soup may actually even be able to reverse dangerous health conditions. At the very bottom of the ad, in small print, the following dis-claimer appears: "While many factors affect heart disease, diets low in saturated fat and cholesterol may reduce the risk of heart disease." Similarly, the large print banner at the top of an ad for pork by the National Pork Board in the US, reads simply: "Be healthy", and the ad also contains the American Heart Associ-ation checkmark for being "Extra Lean", which in combination implies that it is a healthy food choice.

While Heinz and Lee (1998), in their analysis of cultural representations of meat, found that messages about the healthfulness of meat raised critical ques-tions about whether it could be healthy and taste good, we find in the ads we analyse that claims about the healthfulness of the products are accompanied by claims about the tastiness of the product two-thirds of the time. Illustrative of the frequent tethering of good taste with healthfulness, the headline of an advertise-ment for Al Fresco chicken sausage and Classico creamy Alfredo sauce urges readers to "Lighten up, in the most delicious way possible." Two different ads by the Beef industry – specifically by The Beef Checkoff programme of the US

Cattlemen's Beef Board, whose logo reminds one of the American Heart Association's 'heart check mark' programme – emphasize the leanness and good taste of their product. The tag line of one ad refers to it as "Lean and delicious, with endless possibilities." Exploiting the traditional tension between good taste and healthfulness, which Heinz and Lee (1998) identify, while claiming a confluence of the two in their product, one series of ads by TruMoo Chocolate Milk includes pictures of two men: one as an angel (dressed like a milkman of years gone by) and the other (dressed in a black delivery uniform) as the devil. The banners on their ads include the following statements: "Purely good and devilishly delicious" and "Good nutrition meets great taste." According to the copy in the ad, "TruMoo is a delicious chocolate milk you can feel good about giving your kids, every day."

While taste was not necessarily specifically referred to on the meat exporter websites it was implied through images of meat-based food, plated and looking appetising. Fifty two per cent of websites analysed had this imagery. Both those marketing to individual consumers and to businesses seeking to import meat therefore emphasise the good taste of the product through text and this is often accompanied by images of pieces of meat or of dairy products. These products are framed as being about more than meeting nutritional needs; flavour is a critical component. However, we find an important difference between the claims about taste made in the magazine advertisements and by The Red Meat Exporters. In the magazine advertisements the claims about the good taste of the products commonly appear with reference to the healthfulness of the product. This coupling is not observed in the marketing materials of The Red Meat Exporters. Those purchasing their products – businesses that will then sell to consumers – do not need to be convinced that the product is not only tasty but also healthy. They simply need to be convinced that the product is tasty because their ability to modify taste itself is constrained. They can, however, subsequently add claims about the healthfulness of the product in their own marketing, which, in turn, has an ethical implication – in order to be "tasty", these animals had to be reared in a welfare-friendly manner. These interconnected discourses (happy animals/happy meat, animal welfare, good taste, ethical responsibility) ultimately reinforce the idea that farm animals are no more – or less – than meat-in-waiting, which is illustrative of the ways in which discourses communicated from within the A-IC are tailored to the specific audience of interest:

> "Happy meat" discourses therefore posit that happiness becomes an adjunct of meat (or for that matter eggs), something to be consumed along with the muscle fibres, fat and blood. The "ethical consumer" is morally satiated by consuming the happiness of the animals at the same time as her or his belly is filled with their corpses or secretions. The juxtaposition of "welfare" and "quality" is therefore more significant than a legitimation of exploitation.
>
> (Cole 2011: 94)

We therefore suggest that the happy animal/animal welfare discourse is making it possible to reconcile claims about healthfulness and tastiness that previously would have seemed incongruent. For instance Franklin (1999) has asserted that as health concerns have been raised about meat, it has come to be "marketed for its taste, flavours and versatility rather than its essential health and strength giving properties". (ibid.: 174). We find, however, that these two claims are not necessarily mutually exclusive, and are likely being brought together by the development of a happy meat/animal welfare discourse.

The coupling of the messages about taste and healthfulness in the magazine advertisements could also be read as an acknowledgement that individual consumers are sufficiently concerned about the healthfulness of these products that they require reassurance even in the presence of great taste. It may also serve to assuage those consumers who are uneasy about consuming these products in two ways. First, one is made to feel less guilty about consuming an indulgently tasty product if it is also said to contain health benefits. Second, and more implicitly, the pairing of claims of health benefits with claims of good taste can help to normalise the consumption of animal products. Most people are of the opinion that meat products taste good. That alone, however, is not necessarily justification for consuming them. Combining claims of taste with claims of health benefits can be more convincing and reassuring for individual consumers. It serves to reinforce the notion that consuming animal products is necessary for health reasons, which is also connected to the theme of naturalness, as discussed earlier. The coupling of these themes may belie an evolving environment wherein the agricultural arm of the A-IC cannot rely simply on the taste of their product and cultures of consumption, and instead needs to leverage additional justifications. Further research exploring animal agriculture discourses over time could usefully map this trajectory

This discourse played out slightly differently in the meat exporter websites. Claims about the healthfulness of animal products were not entirely absent in the marketing materials of The Red Meat Exporters we analysed; however, they took a backseat to the messages they chose to foreground. Rather than stressing benefits to individual health of meat consumption, these sites pointed to their clean, green, safe production (18 per cent), scientific/quality control management (22 per cent) practices and the advantages of their national or regional location (62 per cent). We point to the audiences being targeted by both sets of actors to explain this divergence. The Red Meat Exporters are targeting businesses looking to import red meat. They are selling a brand and do so by drawing on the image of Australia to sell products, pointing to the country's history of meat production and the physical and cultural environment and messages of safety and scientific rationality. On the other hand, the advertisements we analysed are aimed at end-users – the individuals purchasing food for themselves and/or their family; cooking it and consuming it. Here the most prevalent message is the alleged healthfulness of the products. It appears that healthfulness is not necessarily a quality inherent to animal products: it is a discourse aimed at individual consumers in response to relatively recent questioning of the effects of the consumption of animal products on human health. Our analysis of these two players in the A-IC (advertisers and red meat exporters) therefore illuminates how messages are shaped by economic objectives: the advertisements in the cooking magazines are aimed at increasingly health conscious consumers, whereas the Red Meat Exporter websites are aimed at importers interested in a reliable and safe supply of meat.

Conclusion

Our aim in this chapter was to outline the strengths of a specifically *Critical Animal Studies* analysis of the ways in which animal products form a corner-stone of the A-IC and to outline some of the mechanisms by which this occurs. We wanted to do this, in part, to facilitate Twine's call to make the A-IC an organising concept for CAS, agreeing with him that the strength in this approach is that it "contextualizes the use of animals as food not primarily within a rubric of inadequate ethical frameworks but as part of the wider mechanisms of capit-alism and its normalizing potential" (Twine 2012: 15). We point to three main themes within our samples that normalise the consumption of animal products and the practices of industrial animal agriculture.

The discourses emanating from the A-IC are of particular importance today. The relationship between people and their food used to be immediate and direct. A growing divide between production, processing and consumption has been promulgated by the A-IC, which has positioned itself to be provisioner of information about animal-derived food products that consumers no longer have first-person information about. People now receive mediated messages about their food. As Heinz and Lee remark, "We do not so much eat meat as we consume socially produced meaning" (1988: 98). This therefore makes the mes-sages the A-IC communicates to potential consumers particularly important. We have sought to illuminate these messages here, paying particular attention to how they normalise consumption of animal-derived products and therefore protect their own financial interests and the status quo of a capitalist system grounded in the industrialized production and processing of animals for human (and "pet" animal) consumption more generally.

Our analysis also illustrates how the continuum between humanity and ani-mality intersects with the continuum of violence, introduced by feminists and applied to violence against women (e.g. Kelly 1998), which we seek to apply to violence against animals. The consequence of this intersection is that the percep-tion of harm against animals depends upon the context within which the harm takes place and the species of the animal involved. To illustrate the importance of context, consider the death of a cow by having his/her throat slit. If under-taken within a slaughterhouse by an employee, it is deemed socially acceptable, even beneficial. The same act undertaken by someone wandering by a feedlot would be considered socially unacceptable, and indeed deemed criminal. Our analysis indicates that the A-IC is working skilfully to normalise and rationalise the violence inflicted upon certain species of animals in order to make them con-sumable, while simultaneously admitting the existence of an animality–humanity continuum instead of a strict dichotomy between the two. This is observed in the elevation of pet animals to the level of family, while simultaneously subjugating other species of animals for their consumption (see Herzog 2011). The salience of this demarcation is being witnessed at the time of this writing vis-à-vis the outrage inspired by the discovery of horse meat in various beef products consumed by unsuspecting consumers, primarily in Europe. The violence visited upon those horses is being denounced not because they are slaughtered any differently than any other animals (they are not), but because as a species they occupy a liminal space between pet and consumable livestock. Ironically, the A-IC may have a vested interest in allowing some animals to be closer to the humanity side of the animality–humanity continuum because the existence of carnivorous "pet" animals provides an important market for the industry's (by-)products.

There are few things presumably as private and intimate as the consumption of food. We offer up this analysis as one instance of how a sociological perspective can and ought to demonstrate how the private is publicly mediated and individual experiences are in fact social (Burawoy 2007; Mills 1959/2000). We do so through employing a CAS perspective that elucidates the larger social structures that the discourses promulgated by the A-IC are embedded within. This analysis also suggests that as CAS scholars continue to investigate the A-IC, it will be imperative to count marketing and public relations as critical, constitutive components.

Notes

1 There are other factors that demarcate a CAS approach, but our purpose here is not to list them in order to place rigid demarcations re what can, and cannot be considered CAS. Rather, it is to apply key CAS principles to an empirical test case.
2 We want to clarify that in drawing this distinction between Animal Studies and Critical Animal Studies we are not implying that there is some sort of impermeable divide between the two. We recognise that there is considerable overlap between the two fields and that scholars frequently contribute to both bodies of scholarship. Our point is that the goals of this chapter require a critical engagement with the animal-industrial complex and the forms of power that it is built upon and reproduces, which necessitates grounding it in Critical Animal Studies.
3 For a detailed examination of how milk came to be constructed as a pure and natural product, see DuPuis (2002).

References

Adams, C. (1991/2010) *The Sexual Politics of Meat: A Feminist-Vegetarian Critical Theory*, New York: Continuum.

Arluke, A. and Sanders, C. (1996) *Regarding Animals*, Philadelphia: Temple University Press.

Beirne, P. (2009) *Confronting Animal Abuse*, Lanham: Rowman & Littlefield Publishers, Inc.

Burawoy, M. (2007) "For public sociology", in D. Clawson, R. Zussman, J. Misra, N. Gerstel, R. Stokes, D. L. Anderton and M. Burawoy (eds), *Public Sociology: Fifteen Eminent Sociologists Debate Politics and the Profession in the Twenty-first Century*, Berkeley: University of California Press.

Castree, N. and Braun, B. (1998) *Remaking Reality: Nature at the Millennium*, London: Routledge.

Cole, M. (2011) "From 'animal machines' to 'happy meat'? Foucault's ideas of disciplinary and pastoral power applied to 'animal-centered' welfare discourse", *Animals*, 1: 83–101.

Cudworth, E. (2008) "Most farmers prefer blondes: The dynamics of anthroparchy in animals' becoming Meat", *Journal for Critical Animal Studies*, 6(1): 32.

Cudworth, E. (2011) *Social Lives with Other Animals: Tales of Sex, Death and Love*, London: Palgrave.

Du Puis, E. M. (2002) *Nature's Perfect Food: How Milk Became America's Drink*, New York: New York University Press.

Elias, N. (1978) *The Civilising Process*, trans. by Edmund Jephcott, Oxford: Blackwell.

Fitzgerald, A., Kalof, L. and Dietz, T. (2009) "Spillover from 'The Jungle' into the larger community: Slaughterhouses and increased crime rates", *Organization and Environment*, 22: 158–184.

Flynn, C. (2002) "Hunting and illegal violence against humans and other animals: exploring the relationship", *Society & Animals*, 10: 137–154.

Foer, J. S. (2009) *Eating Animals*, New York: Penguin Books.

Franklin, A. (1999) *Animals and Modern Cultures: A Sociology of Human-Animal Relations in Modernity*, London; Thousand Oaks; New Delhi: Sage Publications.

Gaard, G. (1997) "Ecofeminism and wilderness", *Environmental Ethics*, 19(1): 5–24.

Goffman, E. (1959) *The Presentation of Self in Everyday Life*, New York: Doubleday Anchor.

Hamilton, L. and Taylor, N. (2013) *Animals at Work: Identity, Politics and Culture in Work with Animals*, Boston and Leiden: Brill Academic Publishers.

Heinz, B. and Lee, R. (1998) "Getting down to the meat: the symbolic construction of meat consumption", *Communication Studies*, 49(1): 86–99.

Herzog, H. (2011) *Some we Love, Some we Hate, Some we Eat*. New York: Harper-Collins.

Hugman, R. (1991) *Power in the Caring Professions*, Basingstoke: Macmillan

Ingold, T. (1994) "From trust to domination: an alternative history of human animal relations", in A. Manning and J. Serpell (eds) *Animals and Human Society: Changing Perspectives*, London: Routledge.

Joy, M. (2009) *Why We Love Dogs, Eat Pigs, and Wear Cows: An Introduction to Carnism*, Massachusetts: Conari Press.

Kelly, L. (1988) *Surviving Sexual Violence,* Minneapolis: University of Minnesota.

Kheel, M. (1995) "License to kill: an ecofeminist critique of hunters' discourse", in C. Adams and J. Donovan (eds) *Animals & Women: Feminist Theoretical Explorations*, Durham: Duke University Press.

Mills, C. Wright (1959/2000) *The Sociological Imagination*, Oxford University Press.

Molloy, C. (2011) *Popular Media and Animals*, London: Palgrave.

Morgan, K. and Cole, M. (2011) "The discursive representation of non-human animals in a culture of denial", in B. Carter and N. Charles (eds) *Human and Other Animals: Critical Perspectives*, London: Palgrave Macmillan, p. 122.

Nibert, D. (2002) *Animal Rights/Human Rights: Entanglements of Oppression and Liberation*, Maryland: Rowman and Littlefield.

Noske, B. (1989) *Humans and Other Animals; Beyond the Boundaries of Anthropology*, Montreal: Black Rose Books.

Packwood Freeman, C. (2010) "Meat's place on the campaign menu: how US environmental discourse negotiates vegetarianism", *Environmental Communication: A Journal of Nature and Culture*, 4(3): 255–276.

Patterson, C. (2002) *Eternal Treblinka: Our Treatment of Animals and the Holocaust*, New York: Lantern Books.

Plumwood, V. (1996) "Nature, self, gender: feminism, environmental philosophy, and the critique of rationalism", in K. Warren (ed.) *Ecological Feminist Philosophies*, Bloomington, Indianapolis: Indiana University Press.

Rogers, R. (2008) "Beasts, burgers and hummers: meat and the crisis of masculinity in contemporary TV advertisements", *Environmental Communication*, 2(3): 281–301.

Shukin, N. (2009) *Animal Capital: Rendering Life in Biopolitical Times*, Minneapolis; London: University of Minnesota Press.

Spiegel, M. (1996) *The Dreaded Comparison: Race and Animal Slavery*, Philadelphia: New Society Publishers.

Twine, R. (2010) *Animals as Biotechnology: Ethics, Sustainability and Critical Animal Studies*, London: Earthscan.

Twine, R. (2012) "Revealing the 'animal-industrial complex' – a concept and method for critical animal studies?" *Journal for Critical Animal Studies*, 10(1): 12–39.

Vialles, N. (1994) *Animal to Edible*, Cambridge: Cambridge University Press.

Warren, K. (1997) (Ed.) *Ecofeminism: Women, Culture, Nature*, Bloomington; Indianapolis: Indiana University Press.

9 Mapping non-human resistance in the age of biocapital

Agnieszka Kowalczyk

Marx held not a labour, but a class, theory of value. The basic commodity form of which value is the expression is the class struggle itself which (...) is over the imposition of that form. The substance of value is abstract labour but that concept denotes not the essence of humanity but a social process – the substitutability of labour, its mobility which derives from the struggle over the division and re-division of the working class.

(Cleaver 1976: 9)

Introduction

It has become commonplace to insist that Marxism does not carry any political weight for animal liberation or environmental issues (Eckersley 1992; Ferkiss 1993; Toledo 1996). Although the problem of the place of "Nature" within a Marxian framework has been broadly (even if not conclusively) discussed among Marxist scholars (see, for example Benton 1996; Burkett 1999a, 2006; Foster 2000; Grundmann 1991; Parsons 1977; Schmidt 1971), their findings – as opposed both to ecocentric and mainstream environmental approaches – can hardly be considered as influential. Similarly, much more dispersed and incidental Marxist-based elaborations on animal liberation (see, for example Hribal 2003, 2010; Perlo 2002; Torres 2007), with their critical orientation towards animal rights and animal welfare discourses, can be viewed as marginal within animal studies. Simultaneously, some kind of revival of interest in Marx's analysis can be observed in regard to debate over the emergence of the concept of "Biocapitalism"[1] (Cooper 2008; Helmreich 2008; Rajan 2006; Thacker 2005; Waldby and Mitchell 2006). Although these kind of theoretical debates are sometimes considered as unimportant for actual praxis, the objective of this chapter is to show how a specific re-reading of Marxist theory can inform the struggle against the exploitation of non-human animals.

Traditionally, Marxist thought has subscribed to the belief of "species imperialism"[2] (Wilde 2000: 38). Yet, a political reading of Marx reveals this understanding to be anachronistic. This mode of reading, introduced by Harry Cleaver (2000) and enriched by findings of socialist ecofeminism and critical posthumanism, shows that Marx's concept of labour under capitalism is much wider than the human realm. If this is the case, is it possible for workers and animal activists – and non-human creatures as well – to unite in the common struggle? Can non-humans resist exploitation under the rule of (bio)capital? Are animals a part of the working class? I claim that answering these questions requires rethinking both the labour-value nexus (along with the dichotomy of production and reproduction) in Marx's writings and the notion of resistance. I assume that extending Marxist theory to the non-human perspective can be understood as simply drawing final conclusions from Marx's work itself. Moreover I attempt to defend the view that it can equip us with tools useful for mapping human as well as non-human resistance. By engaging a political reading of *Capital* (Cleaver 2000), Autonomist Feminist and ecofeminist works on (re)production and a critically posthumanist perspective, I attempt to establish the set of theoretical-practical tools for localizing places of potential resistance to the overwhelming exercise of power of (bio)capital. The aim is to show that the process of broadening the influence of (bio)capital is accompanied by a simultaneous process of changing the character of resistance, which enters into the non-human realm.

Does critical animal studies need Marx?

Ariel Salleh's assertion that the "materialist restatement of the humanity-nature interface may be used to enhance political alliancing between ecology, gender, postcolonial, and worker struggles in an era of globalisation" (Salleh, 2001: 448) raises the question of what kind of materialist theory can help us understand the dense net of human–animal–capital relations in a way that advances the struggle for "more-than-human justice" (Mallory 2008). In this context, I attempt to build an argument based on a premise that there can be no animal liberation as long as large-scale global capitalism persists. Although human–animal dualism certainly is not an invention of capitalism, we should acknowledge, first, the influence of the capitalist mode of production on animal exploitation and, second, the complexity of connections between a number of modern dichotomies and particular stages of capitalism.[3]

This intertwining between theory and practice is a crucial element in the definition of Critical Animal Studies (CAS) anticipated in my further considerations. Assuming a difference between animal studies and CAS, we can state that the latter:

> argues that "the animal" includes all sentient beings, including humans, and thus "animal liberation" cannot be properly formulated and enacted apart from "human liberation" and vice versa; (...) thus animal, human, and Earth liberation are inseparably intertwined in the politics of "total liberation".
>
> (Best 2009: 15)

I argue that the re-reading of Marx's work presented in this chapter can contribute to this project. It should be emphasized that, in my view, CAS needs Marxism to remain "critical" as much as Marxism needs CAS to confront the rule of capital. Moreover, it is not possible to determine any starting point or prevailing direction of this urgent theoretical and practical exchange. What I propose here is rather a simultaneous reworking of both these approaches in order to establish theoretical-practical tools for recognizing the significance of non-human resistance.

Although I partly agree with Salleh when she writes that "a generic notion of labor is essential to ground and integrate worker, women's, peasant, indigenous, and ecological politics" (Salleh 2010: 206), it should be noted that the objective of this chapter is not to unconditionally adopt the labour theory of value to show its ongoing relevance. Rather, the following analysis is aimed at challenging Marxist heritage in regard to establishing a theoretical tool-kit, with a considerable explanatory force in relation to the contemporary plane of resistance to capitalism. In these attempts to overcome the limits of labourism, I anticipate some of Harry Cleaver's argument. In his classical introduction to *Capital*, Cleaver (2000) identifies general modes of reading Marx's essential work. According to Cleaver, during the last 80 years, *Capital* has become an object of theoretical study rather than a political tool with possible practical applications. Therefore, Cleaver distinguishes two main interpretations of Marx: ideological and strategic. Ideological readings perceive his work as a base for ideological critique or for the critical interpretation of capitalism, whereas a strategic approach to *Capital* assumes that it serves not only as a basis of a critique of ideology but, above all, it gives strategic tools for mapping class struggle. However, a further distinction must be made as to whether the strategic mapping is from the point of view of capital or from that of the working class. In this perspective, the latter informs and helps to develop working-class struggle. Cutting across this distinction between ideological and strategic readings of Marx is another breakdown between reading *Capital* as philosophy, as political economy and reading it politically. To read Marx philosophically means reading his works as critical interpretations, as a form of ideology. Reading Marx as political economy can include both elements of ideology (critique of capitalism) and elements of a strategic reading in the interests of capital. What, then, does Cleaver mean as political reading? Of course, he is aware that all readings are political regarding their practical applications and influence on class relations. However, in his theoretical frame, he reserves the term "political" for a "strategic reading of Marx which is done from the point of view of the working class" (Cleaver 2000: 30). Cleaver points out that there are always two opposite perspectives in analyses of the capitalist mode of production: those of capital and those of the working class. Therefore, class analysis requires a two-sided investigation, which explores the definition and importance of each category from the differing perspectives of those two classes. Such an approach reveals not only that capital imposes certain conditions on the working class, but that the latter also takes an active part in co-shaping those conditions by struggling against them.

In summary, one line of striking convergence between a political reading of *Capital* and CAS perspectives especially requires accentuation, and it concerns a kind of bottom-up approach. CAS and the mode of political reading (of Marx) share the similar methodology of the theoretical-practical inquiry. They both begin their analysis from examination of the actual situation of, respectively, non-human animals and the working class. In order to grasp the complex power relations in which those groups are entangled, CAS (Twine 2010a: 9–14) – as well as a political reading of Marx – engage intersectionality, acknowledging its importance for mapping the possibilities of struggle against those power relations. Therefore I argue that when the "animal standpoint" (Best 2009: 16) of CAS and "the working class perspective" inherent to the mode of reading (*Capital*) politically are brought together, they can mutually render each other more intelligible with regard to an intersectional analysis of exploitation under capitalism. Moreover, as Castree rightly notes, a Marxist critical analysis of capitalism's impact on nature helps to go beyond two extreme approaches (Castree 2000), which CAS appears to avoid as well. In the first place, a Marxist perspective allows us to escape the trap of specific "bourgeois technocentrism", which assumes the possibility for the successful management of nature within existing social and economic conditions. Second, a Marxist framework eludes any specific kind of "radical ecocentrism", based on the premise that, in the precapitalist mode of production, we were dealing with "Nature in itself", thus returning to something that can build a foundation of post-capitalist reality. Since the aim here is to keep the analysis symmetrical, the modernist assumptions of Marxist thought should be now brought to light, together with the possibilities of overcoming them through a critical, posthumanist re-reading. Understanding of some of the most disturbing contradictions of Marxism requires acknowledging both the unavoidable class character of every analysis and its entanglement with modern hierarchies.

Marxism and its discontents

My elaboration particularly focuses on the issue of animal liberation in a Marxist context but, in assuming the intersection between the exploitation of non-human animals and the environment, it is necessary to briefly reconstruct the discussion concerning Marxism and Nature in general. Marx's remarks on Nature are rather occasional and dispersed, although some attempts for a more consistent exposition of his theory have been offered by numerous scholars (Burkett 1999a, 2006; Foster 2000; Grundmann 1991; Parsons 1977; Schmidt 1971). When it comes to the Marx–Nature–Capitalism complex, "there is still much disagreement over the meaning and ecological significance of even core categories (like value)" (Castree 2000: 6). It is not my intention to resolve these theoretical disagreements. Rather, the question posed here concerns the possibility – given the goal of a political reading of *Capital* – of taking the nineteenth-century modernist burden off Marx's *oeuvre* in order to use his theory as a tool for understanding various forms of contemporary social struggles. In this regard, my argument proceeds in two stages: first, the necessity of historicizing Marx's work will be claimed as a logical consequence of his own theoretical assumptions; second, the modern – in Latour's sense (Latour 1993) – character of the majority of Marxist elaborations on Nature will be problematized (Castree 2000).

The ecofeminist contribution to this theoretical debate to radically historicize Marxian notions is crucial. As Salleh points out, "In fact, the constructionist aspect of ecofeminism interrogates the very foundations of historical material-ism, with its supposedly transhistorical concepts of history, nature, and labor" (Salleh 2003: 64). Possibilities for recognizing the significance of reproduction and Nature were undermined by, among other things, Marx's own Enlighten-ment conviction that reason in tandem with technology might shape the way towards a post-capitalist future. In other words, we can say that the scholars working at the intersection of Marxism and Feminism historicize the categories of Nature and reproduction in Marx's theory to the extent of arguing not simply that they undergo important variation through time, but that they historically have been absent as an effect of the ideological blindness of the author of *Capital*.

However, we should do justice to Marx and acknowledge that throughout *Capital* he challenges a number of modern ideological assumptions, which work for the benefit of capital. Nevertheless, the sphere of reproduction/Nature remains a blind spot for him. Yet, Salleh argues: "Had he been writing in another era, he might well have developed different vantage points – this is certainly implied by his dialectic of internal relations." (Salleh 1997: 70). This kind of approach – focused on Marx's method rather than on a literal and static under-standing of his work shared by "orthodox Marxism" (Lebowitz 2003: 25; Lukács 1966) – is another premise in the following investigations.

Moreover, with regard to the unfinished character of Marx's manuscripts, it should be noted that Marx's essential book, not without reason, was named *Capital*. Michael Lebowitz is one of the scholars who suggest that Marx had no opportunity to write "the missing book(s)" of his project, which were supposed to have been devoted to wage-labour. *Capital* is one-sided in its findings, which are formulated from capital's point of view and this is why it "looks upon workers (...) not as subjects for themselves" (Lebowitz 2003: 119). This per-spective not only resonates with Cleaver's claims but can also be traced in Bur-kett's line of argument, developed against the most common ecological criticisms of Marxism: "(...) while value inadequately represents natural con-ditions, this is a contradiction of capitalism, not of Marx's analysis" (Burkett 1999b: 89).

In contrast with the standard vision in which Marx's very work is depicted as being irremediably interlaced with modern premises, it is argued here that instead the modes of re-reading his work are problematic. Castree (2000) employs Latour's framework of the "Modern Constitution" (Latour 1993) in order to describe those difficulties with Marxist approaches towards "Nature". He claims that at a higher level of abstraction – which should be applied to Marxist theories before they could usefully inform practical action – many of them are trapped within Modern Constitution dualisms. Of course, in the spec-trum of apparent modern dichotomies with which Marxist work has been charged, the charge of maintaining the Nature–Society dualism can be viewed as primary and structuring. Castree argues that viewing the Marx-Nature-Capitalism node through the lens of the Modern Constitution acts against achieving essential Marxist goals.

The mode of reading Marx that is proposed here can be situated at the intersection of political and critically posthumanist (non-modern) modes of reading. While a variety of definitions of "posthumanism" have been suggested, this paper will use the meaning of this term presented in the work of Donna Haraway, combined with some insights from Bruno Latour (Latour 1993). I argue that in order to face the challenges of contemporary (bio)capitalism, we have to move beyond anthropocentrism[4] and the Modern Constitution, as Latour calls it. One question that needs to be asked, however, is how to avoid anthropocentric prejudices in those multi-faceted encounters between human and non-human creatures. Donna Haraway (2008) shows that this task requires more than refiguring the rules the distribution of attributes between beings ascribed to Nature or culture (and consequently to spheres of reproduction and production). Rather, it is the issue of rethinking the very definitions of those attributes. In my opinion, this "posthumanist gesture", as Papadopoulos (2010) calls it, constitutes the necessary step towards a truly political reading of *Capital*.

Teresa Brennan's (1997, 2000) strategy of overcoming the subject/object distinction in the labour theory of value significantly contributes to this project of undermining the modernist assumptions that underlie Marx's theory. When Marx assumes that "[p]rofit is made by selling a commodity at its value" (Marx 1975: 44), he opposes specific explanations of the origin of profit in the capitalist mode of production, which are exclusively derived from market phenomena or particular capitalist abilities and attributes (for example entrepreneurial skills). Therefore, Marx's indication of labour as a source of value is a theoretical proposal of great political and ethical significance.[5] Acknowledging that profit does not come from the sphere of circulation means that capitalists do not sell commodity above its value but, rather, that they do not pay for the part of the value created by the labourer that is above what is required for the labourer to live. Therefore, necessary labour time is that portion of the working day during which reproduction of the labourer takes place. Surplus labour time is imposed on the labourer by capitalists, so that surplus value can be created. Brennan (1997, 2000) formulates this process in terms of subjective and objective factors in production. Objective factors (in Marx's terms fixed capital/dead labour) do not add value, while subjective factors (in Marx's terms variable capital/living labour) are the source of surplus value. For Marx, both Nature and technology are a part of fixed capital. In an attempt to overcome this subjective/objective dichotomy, Brennan argues that not only labouring human subjects add energy to the process of production, but Nature does as well. In order to reject Marx's typically modernist objectification of Nature, Brennan proposes rebuilding his theory on a different premise related to notions of time and energy. Assuming that there are no legitimate reasons to exclude nature from living labour,[6] she argues "that all natural sources of energy entering production should be treated as variable capital and sources of surplus-value" (Brennan 1997: 179). If labour power is treated as an energy source, Marx's dichotomy of subjective and objective aspects of production can be replaced with the opposition of living energy (both human and non-human) and dead technology. Nevertheless, keeping technology in the realm of "dead labour" goes counter to Latour's formulation of the Non-modern Constitution (Latour 1993). It is also debatable regarding the above-formulated critically posthumanist premises. It seems that, at this point, Brennan's conception should go even further in the reinterpretation of the very categories of dead and living labour rather than expanding the domain of one of these concepts. This raises the question of whether focusing on the energetic aspect of the Labour Theory of Value can provide common ground for both organic and technological non-human entities.[7] We should acknowledge the complexity of such a task since, in Marxian analysis, treatments of nature, non-human animals and technology, even though they mainly overlap, have a distinc-tive character and need separate analysis before building commonalities. So although problematic in the context of the general non-modernist claims of this chapter, this shift provides (at least) a possibility for extending the theory of surplus value towards the organic non-human realm.[8] However, this conclusion would have no significance if it were not for being intertwined with the issue of resistance to this overwhelming capitalist imposition of work. The following two sections will attempt to reveal a further basis for expanding the potential alliance between various human and non-human beings subordinated to capital, pre-paring us to establish a set of conditions for a common site of resistance.

More-than-social factory

The starting point for the Italian Autonomist Marxists is a re-reading of Marx's *oeuvre* from the perspective of remarks included in *Grundrisse* (Marx 1973) and *Results of the Immediate Process of Production* (Marx 1992a: 949–1084). Their research concentrates around the question of how the capitalist mode of produc-tion increasingly subordinates social relations as a whole. Mario Tronti's (Tronti: 2006) analysis of capitalism is focused on the total process of accumulation. He traces how the circulation and reproduction examined in Volume II of *Capital* (Marx: 1992b) also involves the reproduction of classes. The recognition of this fact leads him to conclude that, in spite of assumptions of traditional Marxist political economy, we cannot simply identify capital with the "factory". Acknowledging that the reproduction of the working class involves not only work in the factory but also work performed in households and communities, Tronti introduces the notion of the "social factory" (Cleaver 2000; Thoburn 2003). According to the "social factory" thesis, the whole of society becomes simply an articulation of production. As capitalist development advances, the relationship between the factory and society becomes increasingly organic. We are witnessing such a moment of capitalist development in which social relations have become moments of the relations of production. Life itself emerges as a plane of capitalized activity.[9] The subsumption of society (or rather "naturecul-ture") under capitalism proceeds in two stages. We are dealing with formal sub-sumption when labour and relations created outside capitalist production are incorporated into its reign; preserved, however, in an unchanged form (Marx 1992a: 645, 944, 1992c: 1019–1023). In this case, capital simply absorbs pre-existing labour practices. Subsumption becomes real if capital creates new, properly capitalist forms of labour (Marx 1992c: 1023–1025; 1034–1038). Then, when formal subsumption occurs, the division between the inside and outside of capital is maintained whereas, in the case of the real subsumption,[10] this division is blurring and:

> it is no longer the distinct individual entities of the productive workers that are useful for capitalist production, nor even their 'work' in a conventional sense of the word, but the whole ensemble of sciences, languages, knowledges, activities, and skills that circulate through society.
>
> (Thoburn 2003: 81)

This raises the question of redefining the working class in order to include non-factory (unwaged) workers. As Cleaver notices: "To say that the working class sells its labour-power to capital must be understood broadly: the working class includes those who work for capital in various ways in exchange for a portion of the total social wealth they produce." (Cleaver 2000: 84). In the same spirit, Negri encourages the broadening of the definition of class unity and the creation of a conception of class more adequate to new forms of the reign of capital – as he emphasizes: "the potentiality of a new working class now extended throughout the entire span of production and reproduction – a conception more adequate to the wider and more searching dimensions of capitalist control over society and social labour as a whole" (Negri 1988: 209). I will argue that this theory not only informs Marxist analyses of worldwide struggles involving students, women, the unemployed or peasants in the Third World (Costa 1972: 27–32), but can also provide a point of departure for incorporating a more-than-human perspective into Marxist thinking.

"Natural" alliance

Before we can proceed with an analysis of the Autonomist Feminist thesis developed through Tronti's work, the clarification of the assumed relation between "Nature" and "Women" is necessary. In this chapter, it is definitely not argued that women are "closer to nature" in any ontological sense. However, we cannot deny the existence of a complex relation between the oppression of "Women" and "Nature" under capitalism. Moreover, we have to acknowledge the consequences of discourse, which places Women and Nature opposite to a presumed rational male culture. The most interesting aspects for the current argument are the political and materialistic dimensions of such an opposition since "the material-discursive link between 'women' and 'nature' can be used productively to destabilize relations of oppression, through a coalitional politics that creatively 'plays with' these constructed connections and deploys them strategically" (Mallory 2008: 3).

Autonomist Marxist Feminists contribute to this project by taking further the analysis of the social factory sketched above. The first serious discussions of reproductive work as labour emerged during the 1970s within the Italian Autonomist Marxist movement. On a theoretical level, this vastly expanded Tronti's work on the non-factory part of the working class, which focuses on the key role of the wage in hiding – not only the unpaid part of the working day in the factory, but also unpaid work outside the factory. As Federici and Cox emphasize:

> Since Marx, it has been clear that capital rules and develops through the wage, that is, that the foundation of capitalist society was the wage laborer and his or her direct exploitation. What has been neither clear nor assumed by the organizations of the working class movement is that precisely through the wage has the exploitation of the non-wage laborer been organized.
>
> (Federici and Cox 2012: 28)

This perspective draws on Marx's work on the reserve labour army, yet it goes beyond it in seeing the reproduction of labour-power as crucial for the accumulation of capital. As a result of acknowledging that Marxism failed to recognize the role of reproductive labour in the capitalist mode of production, the "Wages for Housework" campaign emerged. Although it assumes that capitalism uses the patriarchal structures that preceded it and developed alongside it, "Wages for Housework" emphasizes that domestic work, as a socially important experience, is directly connected to the struggles of the working class. It makes clear the way in which the wage divides the class hierarchically into wage (factory) sector and unwaged sector, in the manner that the latter appears to be outside of the working class. Some ecofeminist scholars come to similar conclusions. As Ariel Salleh notices:

> These externalised groupings constitute an unspoken 'class', even though their reproductive labours are essential to capitalist economies. I use the term meta-industrial class to describe such people whose 'labour' is considered 'outside' of capitalism and is untheorised in Marxist class analysis, for example, domestic care-givers, peasant farmers, and indigenous hunter gatherers.
>
> (Salleh 2001: 448)

In summary, thinkers such as Mariarosa Dalla Costa (1972), Silvia Federici (2004) and Leopoldina Fortunati (1995) show how Marx's notion of labour is used by capital in order to create divides within the working class. And, although focused on housework as a paradigmatic instance of this mechanism, their analysis serves as the basis of a conceptualization of Nature as a productive force rather than passive "resource" or "free gift". This gesture creates an important link between ecofeminist theory and Marxist thought and from this perspective constitutes another plane of potential intersection between a political and critically posthumanist reading of Marx.

The struggle against capital has gone to the dogs

In classical Marxist terms, workers as an historical subject form a true "class for itself" only when they engage in struggle against the imposition of work. Thus, in order to map the position the non-human occupies in the capitalist mode of production we should first pose the question of the possibility of resistance in the non-human realm. This leads us to the problem broadly discussed in CAS – can animals exercise agency? As Jason Hribal notices, presenting his theoretical-practical perspective on this issue:

> Agency is discussed as a theory, but it is not applied in practice. The agents (i.e. the animals themselves) dissipate into a vacant, theoretical category. This is a view from above. History from below is not a theory. It is a methodology or form of analysis, which can be applied to the study of historically un/underrepresented groups.
>
> (Hribal 2007: 102)

Although Hribal refers here to theoretical investigations in general, his argument can be perceived as particularly adequate in respect to Marxism and its deep internal tension caused by its dual character as both a theoretical and a political project. Clearly, his attitude resonates with Cleaver's political reading of *Capital* from the point of view of the working class. This mode of thinking, enriched by Haraway's insights mentioned earlier, provokes closer investigation of the notion of agency itself before denying or attributing agency to non-human animals. Thus, this argument consists of two parts. First, I will focus on the intersection between resistance and agency by exploring the possibilities for deriving both of these from the corporeal realm rather than (only?) conscious subjectivity. Then, by claiming the importance of the collective aspect of "agentival capacity", I will insist on the importance of alliances across species boundaries for struggling against capitalism.

There has been little discussion about non-human resistance to capitalist oppression. Given this, it is hard to overestimate Jason Hribal's contribution to unveiling the history of violent resistance to human ill-treatment (Hribal 2003, 2010). In *Fear of the Animal Planet* (Hribal 2010), he sketches a history of animal struggles against exploitation, mainly focusing on zoos and circuses. In this investigation he extends the notion of agency to animals, identifying them as a kind of political agent revolting against human domination:

> Captive animals escaped their cages. They attacked their keepers. They demanded more food. They refused to perform. They refused to reproduce. The resistance itself could be organized. Indeed, not only did the animals have a history, they were making history. For their resistance led directly to historical change.
>
> (Ibid.: 29–30)

In the history of these rebel acts, he highlights their non-contingent character. Hribal argues that denying agency in the case of animal resistance is one of the key features of the strategy, which is adopted by circuses and zoos in order to convince the public that there is no reason for those animals to attack. As he emphasizes, everyone tries to ignore the fact that usually animals target concrete individuals and then only after experiencing long-lasting exploitation. Moreover, "(...) the standard operating procedure is to deny agency. The key words to remember are 'accident', 'wild', and 'instinct'". (ibid.: 24). In this case, the dichotomy of nature and culture is strongly supported by the discourse of biological "reaction" rather than a conscious "response", which would be worth taking into account. In contrast to that narrative, Hribal points out that captive animals, because of their experience and training, are aware of the consequences of their behaviour. However, despite this, they still carry out actions against their oppressors. According to Hribal, this proves that these acts of animal violence (or in many cases, simply the refusal of work) can be understood as a true form of resistance. As he states: "These animals (...) are rebelling with knowledge and purpose. They have a conception of freedom and a desire for it. They have agency." (ibid.: 26). As an answer to potential accusations of anthropomorphism, Hribal claims that the idea of anthropomorphism as such is entangled with various political and economic interests as well as cultural and social contexts. The separation between humans and non-humans is absolutely compatible with the capitalist imperative of efficiency and growth, which, for those being subordinated, means exploitation. Although in the context of the aforementioned re-readings of Marx, the last part of Hribal's argument appears valid, still his conception of resistance raises questions.

Therefore, we should explore how the notion of agency and the assumptions underlying this notion can be an obstacle for the exploration of non-human resistance and what kind of theoretical approach can be introduced as an alternative. We need a conceptualization of agency that can constitute a theoretical shift away from treating the "male waged worker" mode of resistance as universal and as having the inherent characteristics of struggle against exploitation. Narrowing the notion of resistance to the conscious actions performed by labourers simply legitimizes the existing system of capitalist power relations rather than undermining it.

Foucault's remarks on resistance can provide a starting point for disrupting those modernist notions of who, or what, counts as a political subject. Beginning with his famous statement that "Where there is power, there is resistance" (Foucault 1978: 95), he elaborates on plural forms in which resistance can be shaped, including spontaneous acts "inflaming certain points of the body, certain moments in life, certain types of behaviour." (Foucault 1978: 96). I will not engage here in a discussion on the concept of active/reactive resistance in Foucault's work (see, for example: Toymentsev, 2010) – which although important is not fundamental at this point to my argument. Rather, a Foucauldian perspective serves here as a proposal, opening spaces for the revalorisation of a body in the capitalist nexus of power. In this regard the potential of Foucault's writing to mobilize thought rather than his exact findings, is crucial as Foucault himself ignores the process of reproduction and specificity of female experience of the body. Federici, when analysing the significance of the "rebel bodies" in the transition from feudalism to capitalism, notices:

> the body has been for women in capitalist society what the factory has been for male waged workers: the primary ground of their exploitation and resistance, as the female body has been appropriated by the state and men and forced to function as a means for the reproduction and accumulation of labor.
>
> (Federici 2004: 16)

Even though Foucault's formulation of resistance is far from flawless, a number of feminist scholars draw from it to reveal complex ways of intersection between the different modalities of agency (Diamond and Quinby 1988; Faith 1994; McLaren 2002; Oksala 2004; Sawicki 1991). Challenging the limits of Foucauldian concepts and extending the reach of its explanations, feminist scholars make us more nuanced about various forms of struggle. Moreover, the analysis of experiences of women in different cultural contexts forces us to pay attention to more subtle signs of resistance. On the one hand, relations of power are inscribed in the body, but the body can also become the plane of resistance to those relations by transforming oppressive practices: "Wherever power is infused across the range of disciplinary sites, there it simultaneously intersects with the force of resistance, even at the most microscopic, cellular and capillary levels of existence" (Faith 1994: 38). This conceptualization enables us to recognize acts of resistance even at an unconscious biological level (Coppin 2003), and this is what opens the possibility for establishing a notion of non-human resistance, even to those who deny consciousness to certain non-human animals. Coppin has no doubts that:

> (…) a nonhuman animal is involved in both the discipline and the resistance that previously had been the sole province of human beings. Certainly the form that porcine agency takes has altered over the years and conditions – sows in tightly confined farrowing crates have far less opportunity to act than their foremothers who farrowed in forest nests – but they remain agents of resistance and change.
>
> (Coppin 2003: 614)

Acts of resisting exploitation performed by non-human bodies do not necessarily have to be thoughtful (although such characteristics cannot be denied to them in specific situations) to be recognized as significant. Even biologically-based acts of survival should not be reduced to "merely instinctive" or meaningless in relation to resistance to the reign of capital. Whether or not a response or a reaction of the exploited to exercised power is planned or intentional, it is within this realm of the body that the individual non-human animal becomes a part of the political collectivity. These are necessary steps in establishing an embodied and non-deterministic materialism which, as I argue, can lay the foundation in organizing the common struggle of "(...) all these embodied subjects [which] are bound together in the material relations of white supremacist anthropocentric capitalist patriarchy" (Mallory 2008: 9).

Critters of the world, unite!

We should emphasize that such resistance cannot simply overthrow the power exercised by capital over non-human animals. However, these kinds of acts of resistance may refocus the debates on the role played by non-human animals in the capitalist mode of production. For instance, acknowledging non-human resistance weakens processes of victimization, consequently generating specific political empowerment through the blurring of hierarchical divisions imposed by capital on the working class. Moreover, like in the case of struggle over recognizing reproduction activities as work, the demand for acknowledging the importance of animals in producing surplus value "(...) is a revolutionary demand not because by itself it destroys capital, but because it attacks capital and forces it to restructure social relations in terms more favourable to us and consequently more favourable to the unity of the class." (Federici 1975: 5).

Then, in light of foregoing explorations, what kind of perspective on organizing struggle emerges? Answering this question requires further elaboration on resistance, with emphasis on its collective and relational aspect. Carter and Charles (2011) offer us a thorough analysis of the concept of agency in selected recent sociological theories in the context of the question of animal agency. In general terms, they define "agency as the relationally generated capacity for social action" (ibid.: 253). In the context of the present analysis, we can find their reconstruction of Archer's (2000, 2003) concepts of Primary and Corporate Agency particularly interesting, although our conclusion may differ. Primary Agency is derived from the involuntary occupation of a position in the distribution of resources in society. Primary Agency as "a property of collectives sharing similar life chances" (Carter and Charles 2011: 253) is exercised by everyone, including non-human animals, who gain this characteristic on the basis of being embedded in social relations. However "(t)his agency does not enable a transformation of an anthropocentric system. For this to happen, what would be required is human Corporate Agency" (Carter and Charles 2011: 265), defined as explicit reflection upon one's position within given social relations. This ability is connected with "reflexive embodiment", understood as the capability to recognize oneself as a subject and object. It emerges both from animals with "practical consciousness" of their environment and from language, which is unique to humans. As the result of being deprived of Corporate Agency, animals "cannot organise collectively to resist the anthropomorphic relations of power and domination within which they are enmeshed. They can, however, act individually to avoid particular effects of these relations (...)" (ibid.: 258). Carter and Charles admit that animals can struggle over conditions of their existence but their efforts cannot be considered political as they are not mediated by the symbolic realm of representation. In this context, non-human animals are incapable of collective political struggle. As these authors argue, this emphasis on the distinctiveness of humans, based on language, is an effect of the sociological approach they choose to assume.

However, considering that Carter and Charles do not deny the existence of communication between human and non-human animals, could Corporate Agency be established across species boundaries if we sacrifice elements of this sociological perspective in favour of a class-based approach? It seems that Mallory, when writing about acting together in the chain of equivalencies, could provide a basis for this kind of "Trans-species Corporate Agency". Based on her observations of the Seattle anti-WTO protests, she notes that:

> the protest speech of the sea turtle voiced through human actor/agents who felt this performative affinity with such creatures was no less real than that of the union workers who were engaged in self-representation and enactment of a collective subjectivity
>
> (Mallory 2008: 8)

For Mallory human and non-human, materially-based agency should be perceived as "complexly co-emergent" (ibid.: 11). In this perspective, there is no doubt that the possibility exists for a common site of resistance for labour movements, "new social movements", animal liberation movements and non-humans themselves. Overall, it is in the actual practice of resistance where "humans and the more-than-human world make embodied political subjectivities (…) together" (ibid.).

Conclusion

This chapter has given an account of, and the reasons for, the critically posthumanist rereading of Marxian notions in the context of building a common site of resistance to capitalism. First, conditions and a claimed mutually desired relation between CAS and Marxism were established. Then, using theoretical tools from both these perspectives, I explored problems arising at the intersection of "Nature" and labour, production and reproduction in order to recognize the modernist burden of previous analyses of capitalism, which work in the interests of capital. This theoretical framework enabled a critical resituating of class struggle within the broader view of the variety of capitalist power relations. It gave focus to forms and patterns of exploitation that cannot be easily explained by labour theory. This included the recognition that resistance to power relations based on anthropocentric assumptions is both integral to, and distinct from, all other resistances to capitalism. However, it is not sufficient to argue that although the emancipation of the working class is a key facet of liberation struggles, it does not encompass other sites of resistance. Rather, the issue is how to achieve solidarity in the process of bringing those fragmented resistances together. In general, therefore, it seems that CAS and Marxist efforts should be combined to expose how hierarchies built on the division of labour and corresponding to discursive modern great divides, have destructive effects upon solidarity among those subordinated to capitalism.

Notes

1 The notion of "biocapitalism" can be misleading because it suggests some kind of quality change within the capitalist mode of production in the context of biotechnological development (Helmreich 2008). I would rather view it as a change in the intensity, since the biological factor has always been important for capitalist accumulation, even if its significance was not fully recognized by Marx. So I rather use this term to emphasize what previously has been a blind spot for Marxist analysis, that is, an essential role of the biological in creating a surplus value.

2 Wilde uses this notion commenting on Benton's analysis of Marx's approach towards human–animal relations (Benton 1993). Benton, focusing on Marx's *Economic and Philosophic Manuscripts of 1844* (Marx 1992d), states that besides the apparent humanism of Marx's early writings, his description of human nature is naturalistic (Benton 1993: 23–57). However, due to his speciesism Marx fails to recognize that human and non-human animals suffer similar abuse under capitalism. Benton concludes that "Marx's contrast between the human and the animal cuts away the ontological basis for such a critical analysis of forms of suffering shared by both animals and humans who are caught up in a common causal network" (ibid.: 42). In Wilde's perspective, this can undermine the communist project as a whole.

3 What makes the overcoming of the Nature/Culture dichotomy in anti-capitalist thinking all the more urgent is the fact that capitalism itself has gone beyond that dichotomy long ago. Although in its origin capitalism was based on the separation of human from Nature so that it could impose wage-labour, it seems that in its further development, the very initial separation allowed even more intensive capitalist production of hybridized entities.

4 For an interesting mapping of the concept of "posthumanism(s)", see (Braidotti 2006) and (Twine 2010b). In this chapter, I try to use the notion "anthropocentrism" both in "a theoretical ontological sense" and "a more political sense related to critiques of human supremacism" (Twine 2010b: 184).

5 Katherine Perlo (2002) argues that assuming the necessity of dialectical progress within Marxism itself, the same – what she calls – "quality of sympathy" (ibid.: 303) that caused Marxism to devaluate animals – in order to highlight suffering of the working class – should now be applied towards nonhuman beings as well.

6 Marx argues that the ability to exercise one's will is what distinguishes a human labour from an animal's activity (Marx, 1992a: 283–284). However, under the reign of capital, human will is increasingly suppressed. Moreover not every worker can act willingly and consciously – indeed, just "a labour aristocracy" appears to have this privilege (Brennan 2000: 85–90).

7 For the interesting elaborations of Marx's theory of machines drawn from different perspectives, see (Caffentzis 2013: 127–202) and (Mirowski 1991).

8 I would like to thank the editors for their valuable comments on this issue.

9 We can observe an interesting convergence between Tronti's thesis and some ecofeminist findings. As Vandana Shiva claims,

> With engineering entering the life-sciences, the renewability of life as a self-reproducing system comes to an end. Life must be engineered now, not reproduced. A new commodity set is created as inputs, and a new commodity is created as output. Life itself is the new commodity.
>
> (Shiva, 1988, p. 84)

10 For an interesting elaboration on real and formal subsumption in regard to the place of non-human animals in the capitalist mode of production, see Shukin (2009).

References

Archer, M. S. (2000) *Being Human: The Problem Of Agency*, Cambridge: Cambridge University Press.

Archer, M. S. (2003) *Structure, Agency And The Internal Conversation*, Cambridge: Cambridge University Press.

Benton, T. (1993) *Natural Relations: Ecology, Animal Rights and Social Justice*, London: Verso.

Benton, T. (ed.). (1996) *The Greening of Marxism*, London: The Guilford Press.

Best, S. (2009) "The rise of critical animal studies: putting theory into action and animal liberation into higher education", *Journal for Critical Animal Studies*, 7(1): 9–53.

Braidotti, R. (2006) "Posthuman, all too human: towards a new process ontology", *Theory, Culture & Society*, 23(7–8): 197–208.

Brennan, T. (1997) "Economy for the Earth: The labour theory of value without the subject/object distinction", *Ecological Economics*, 20(2): 175–185.

Brennan, T. (2000) *Exhausting Modernity. Grounds for a New Economy*, London: Routledge.

Burkett, P. (1999a) *Marx and Nature. A Red and Green Perspective*, New York: St. Martin's Press.

Burkett, P. (1999b) "Nature's 'free gifts' and the ecological significance of value", *Capital & Class*, 68: 89–111.

Burkett, P. (2006) *Marxism and Ecological Economics: Toward A Red And Green Political Economy*, Leiden: BRILL.

Caffentzis, G. (2013) *In Letters of Blood and Fire: Work, Machines, and the Crisis of Capitalism*, New York: PM Press – Autonomedia.

Carter, B. and Charles, N. (2011) "Conceptualising agency in human-animal relations: a sociological approach", in J. Bull (ed.) *Animal Movements. Moving Animals. Essays on Direction, Velocity And Agency In Humanimal Encounters*, Uppsala: Uppsala University.

Castree, N. (2000) "Marxism and the production of nature", *Capital & Class*, 24(3): 5–36.

Cleaver, H. (1976) "Internationalisation of capital and mode of production in agriculture", *Economic and Political Weekly*, XI(13).

Cleaver, H. (2000) *Reading Capital Politically* (2nd edn, p. 192), London: Anti-Theses/ AK Press.

Cooper, M. (2008) *Life as Surplus: Biotechnology And Capitalism In The Neoliberal Era*, Seattle: University of Washington Press.

Coppin, D. (2003) "Foucauldian hog futures: the birth of mega-hog farms", *The Sociological Quarterly*, 44(4): 597–616.

Costa, M. D. (1972) *Women and the Subversion of the Community: The Power of Women and the Subversion of the Community*, Bristol: Falling Wall Press.

Diamond, I. and Quinby, L. (1988) *Feminism and Foucault: Reflections on Resistance*, Boston: Northeastern University Press.

Eckersley, R. (1992) *Environmentalism and Political Theory*, Albany: SUNY Press.

Faith, K. (1994) "Resistance: lessons from Foucault and feminism", in H. Stam and L. Radtke (eds) *Power/Gender: Social Relations in Theory and Practice*, London: Sage.

Federici, S. (1975) *Wages Against Housework. Revolution at Point Zero. Housework, Reproduction, and Feminist Struggle*, Bristol: Falling Wall Press.

Federici, S. (2004) *Caliban and the Witch*, New York: Autonomedia.

Federici, S. and Cox, N. (2012) *Counterplanning from the Kitchen: Revolution at Point Zero. Housework, Reproduction, and Feminist Struggle*, New York: PM Press – Autonomedia.

Ferkiss, V. (1993) *Nature, Technology, and Society*, New York: New York University Press.

Fortunati, L. (1995) *The Arcane of Reproduction: Housework, Prostitution, Labor and Capital*, (H. Creek, Trans.), New York: Autonomedia.

Foster, J. B. (2000) *Marx's Ecology: Materialism and Nature*, New York: Monthly Review Press.

Foucault, M. (1978) *The History of Sexuality. Volume I: An Introduction*, New York: Pantheon Books.

Grundmann, R. (1991) *Marxism and Ecology*, Oxford: Oxford University Press.

Haraway, D. (2008) *When Species Meet*, Minneapolis: University of Minnesota Press.

Helmreich, S. (2008) "Species of Biocapital", *Science as Culture, 17*(4): 463–478.

Hribal, J. (2003) "'Animals are part of the working class': a challenge to labor history", *Labor History, 44*(4): 435–453.

Hribal, J. (2007) "Animals, agency, and class: writing the history of animals from below", *Human Ecology Review, 14*(1): 101–112.

Hribal, J. (2010) *Fear of the Animal Planet. The Hidden History of Animal Resistance*, Petrolia/Oakland: Counter Pounch/AK Press.

Latour, B. (1993) *We Have Never Been Modern*, Cambridge, MA: Harvard University Press.

Lebowitz, M. A. (2003) *Beyond Capital*, Basingstoke: Palgrave Macmillan.

Lukács, G. (1966) "What is orthodox Marxism?" *International Socialism*, (24): 10–14.

McLaren, M. A. (2002) *Feminism, Foucault and Embodied Subjectivity*, New York: SUNY Press.

Mallory, C. (2008) "Ecofeminism and a politics of performative affinity: direct action, subaltern voices, and the green public sphere", *Ecopolitics Online Journal, 1*(2): 2–13.

Marx, K. (1973) *Grundrisse*, New York: Random House, Inc.

Marx, K. (1975) *Value, Price and Profit*, New York: International Publishers.

Marx, K. (1992a) *Capital. A Critique of Politial Economy. Volume One*, New York and London: Penguin Books and New Left Review.

Marx, K. (1992b) *Capital. A Critique of Political Economy. Volume Two*, London: Palgrave Macmillan.

Marx, K. (1992c) "Results of the immediate process of production", in B. Fowkes, *Capital. A Critique of Politial Economy. Volume One*, London: Penguin Books and New Left Review.

Marx, K. (1992d) "Marx's economic and philosophic manuscripts of 1844", in R. Livingstone and G. Benton (Trans.), *Early Writings*, London: Penguin Books and New Left Review.

Mirowski, P. (1991) *More Heat than Light: Economics as Social Physics, Physics as Nature's Economics*, Cambridge: Cambridge University Press.

Negri, A. (1988) "Archaeology and project: the mass worker and the social worker", in *Revolution Retrieved: Writings on Marx, Keynes, Capitalist Crisis and New Social Subjects (1967–83)* London: Red Notes.

Oksala, J. (2004) "Anarchic bodies: Foucault and the feminist question of experience", *Hypatia, 19*(4): 97–119.

Papadopoulos, D. (2010) "Insurgent posthumanism", *Ephemera: Theory and Politics in Organization*, *10*(2): 134–151.

Parsons, H. (1977) *Marx and Engels on Ecology*, Westport: Greenwood Press.

Perlo, K. (2002) "Marxism and the underdog", *Society & Animals*, *10*(3): 303–318.

Rajan, K. S. (2006) *Biocapital: The Constitution of Postgenomic Life*, Durham, NC: Duke University Press.

Salleh, A. (1997) *Ecofeminism As Politics: Nature, Marx and the Postmodern*, London: Zed Books.

Salleh, A. (2001) "Sustaining Marx or sustaining nature?" *Organization & Environment*, *14*(4): 443–450.

Salleh, A. (2003) "Ecofeminism as sociology", *Capitalism, Nature, Socialism*, *14*(1): 61–74.

Salleh, A. (2010) "From metabolic rift to 'metabolic value': reflections on environmental sociology and the alternative globalization movement", *Organization & Environment*, *23*(2): 205–219.

Sawicki, J. (1991) *Disciplining Foucault: Feminism, Power, and the Body*, London: Routledge.

Schmidt, A. (1971) *The Concept of Nature in Marx*, London: New Left Books.

Shiva, V. (1988) *Staying Alive. Women, Ecology and Development*, London: Zed Books.

Shukin, N. (2009) *Animal Capital. Rendering Life in Biopolitical Times*, Minneapolis: University of Minnesota Press.

Thacker, E. (2005) *The Global Genome: Biotechnology, Politics, and Culture*, Cambridge: MIT Press.

Thoburn, N. (2003) *Deleuze, Marx and Politics*, Abingdon, UK: Taylor & Francis.

Toledo, V. M. (1996) "The ecological crisis: a second contradiction of capitalism?" in T. Benton (ed.) *The Greening of Marxism* London: The Guilford Press.

Torres, B. (2007) *Making a Killing. The Political Economy os Animal Rights*, Edinburgh: AK Press.

Toymentsev, S. (2010) "Active/reactive body in Deleuze and Foucault", *Journal of Philosophy: A Cross-Disciplinary Inquiry*, *5*(11): 44–56.

Tronti, M. (2006) *Operai e capitale*, Roma: Derive Approdi.

Twine, R. (2010a) *Animals as Biotechnology. Ethics, Sustainability and Critical Animal Studies*, London: Earthscan/Routledge.

Twine, R. (2010b) "Genomic natures read through posthumanisms", *The Sociological Review*, *58*: 175–195.

Waldby, C. and Mitchell, R. (2006) *Tissue Economies. Blood, organs And Cell Lines In Late Capitalism*, Durham, NC: Duke University Press.

Wilde, L. (2000) "'The creatures, too, must become free': Marx and the animal/human distinction", *Capital & Class*, *24*(3): 37–53.

Part IV

Contesting the human, liberating the animal

Veganism and activism

10 'The greatest cause on earth'

The historical formation of veganism as an ethical practice

Matthew Cole

Introduction

From 1948–1951, *The Vegan*, the quarterly journal of The Vegan Society (the world's first, founded in the UK in November 1944), bore the strapline, 'Advocating living without exploitation' on its front cover. That ambition to live without exploitation is arguably fundamental to contemporary Critical Animal Studies (CAS), especially when we consider that to live without exploitation entails *active engagement* with what were in 1944, and remain today, brutally exploitative social systems, for many humans as well as for other animals. In this chapter, I draw on a Foucauldian reading of early Vegan Society publications to argue that modern veganism (that is, veganism since the formation of The Vegan Society) was a critical enterprise at birth, in a way that anticipated CAS in some respects: from its inception, veganism was engaged in a revolutionary transformation of human relationships with other animals, with other humans, and with vegans themselves. This is evident in *An Address on Veganism*, delivered to the eleventh Congress of the International Vegetarian Union (IVU) by Donald Watson, co-founder of the Vegan Society, and coiner of the word 'vegan', in 1947:

> The vegan renounces it as superstitious that human life depends upon the exploitation of these creatures whose feelings are much the same as our own, and by comparison [...] wishes to see a world in which animals will live healthily, free from exploitation, mutilation, and slaughter.
>
> (Watson 1947a: 2)

Beyond this, veganism also began with an engagement, *through* this foundational transformation of human–other animal relationships, with the reconfiguration of intra-human relationships:

> The vegan considers the abolition of the slaughter-house as a reform of the first importance in the reconstruction of human society, for so long as it remains society has no morality [...] Once [we] accept that the strong have the right to exploit the weak, [...then] the basis upon which decent society can be built is destroyed.
>
> (Ibid.)

Finally, veganism instantiated a process of transformation of vegans themselves, as individuals:

> The first part of the task is to pull down within ourselves the old life based on exploitation [...] it is well that we examine our lives and ask whether creatures are living and dying in hell that our needs may be supplied.
>
> (Ibid.: 2, 3)

The challenge of this radical dawn of veganism has always been in danger of being eclipsed by its stigmatization and distortion by a largely hostile discursive environment in the modern West. One ambition of this chapter, therefore, is to continue a programme of work that contests the misrepresentation of veganism by discursive authorities, including the mass media (Cole and Morgan 2011) and (non-CAS) academia (Cole 2008). There is no space to fully recapitulate these arguments here but, briefly, that misrepresentation centres on marginalizing and ridiculing vegans as aloof eccentrics, tainted by elitism, anachronistic puritanism, and therefore irrelevance. This marginalization reduces veganism to the third, personal, aspect of transformation alone and by doing so also misrepresents that personal transformation itself as solipsistic self-indulgence rather than the *consequence* of a revolutionized relationship with animal others. This therefore functions to obscure that foundational re-ordering of human–other animal relationships, which provides the moral impetus of veganism, and also denies the relevance of veganism to re-ordering society and forms of intra-human exploitation. In other words, this individualizing form of marginalization distracts attention from the embedded structures of oppression highlighted by Watson. In contradistinction, this chapter aims to foreground the inter-relationship between these three aspects of transformation that gave veganism its initial critical vigour and radicalism, summed up in Donald Watson's characterisation of veganism as 'the greatest Cause on earth' (1947b: 7).

As indicated by the quotations above, this understanding of early modern veganism will be elaborated in this chapter, from an analysis of the first publications of The Vegan Society, from its formation in 1944 to the first adoption of a formal definition of veganism in 1950. The principal source is *The Vegan* magazine (the quarterly journal of the Society from 1946 until the present day, which began publication as *The Vegan News* in 1944). This source is supplemented with pamphlets published by the Society in its first seven years, such as Watson's *An Address on Veganism*, and with later writings by Watson, which reflected on the early history of The Vegan Society. The reason for this specific historical focus is that it was over this period that the hitherto dispersed practices of de facto 'vegans' coalesced into a purposive programme for inculcating those practices throughout society: 'Our purpose in founding The Vegan Society was not to claim the discovery of a new ethic, but to show how this ethic could be implemented' (Watson in Farhall 1994: ii).

As Leah Leneman has documented, the formation of the Vegan Society at this point in time was stimulated by the refusal of the UK Vegetarian Society to allow a 'non-dairy' section in its journal (1999: 227), in response to letters in the early 1940s from vegan pioneers, including Leslie J. Cross and Donald Watson. Leneman (1999) also documents that now familiar arguments for veganism, on the grounds of the inherent cruelty and exploitation of the 'dairy' industry, were already being vigorously debated, and won, within the Vegetarian Society in the early twentieth century. It is important to note here that what was denoted by 'veganism' since 1944 already existed in prototype among the first self-defined 'vegetarians' in the early 1800s (Davis 2012: 32–34). The dilution of this foundational radicalism of vegetarianism in the nineteenth century, which was contested at various points by attempted moves towards 'veganism', is documented by John Davis, the historian of the IVU (2012). This chapter does not at all claim that human plant-based lifestyles – voluntary, or involuntary – for reasons of privation, commenced in the nineteenth century, let alone in 1944. This was also recognized by Donald Watson, who, in recalling the formation of The Vegan Society, commented in 1989 that, '[t]he early vegans were probably a more numerous group consisting of most of the human race – before humankind invented tools to subdue and kill animals' (1989b: 21). Here then, Watson exemplifies CAS' contemporary emphasis on social transformation through destabilizing the hierarchical dualism of 'human' over 'animal' (see Twine 2010: 12 and critiques of human exceptionalism by Clark (Chapter 7), Peggs (Chapter 3), and Jenkins and Twine (Chapter 11) in this volume). Watson's evocation of a fall from vegan grace rhetorically foregrounds the historical contingency of human domination, and therefore makes thinkable a vision of its reversal. The particular importance of the foundation of The Vegan Society, compared with earlier periods of debate within the vegetarian movement, inhered the moral codification of veganism as distinct from the, by then, confused identity of vegetarianism. Simultaneously, The Vegan Society developed shared ethical practices, through which 'vegans' self-consciously constituted themselves and each other, as an organized body of people committed to radical social and personal transformation.

Veganism and Foucauldian ethics

In this light, Michel Foucault's later work on ethics offers a fruitful way to understand veganism, as it was experienced and constituted by pioneering vegans themselves. In his analysis of ancient Greek ethical practices in *The Use of Pleasure* (1992), Foucault distinguished between more or less fixed moral codes and lived ethical practices that are orientated, in different ways, towards those codes:

> For a rule of [moral] conduct is one thing; the conduct that may be measured by this rule is another. But another thing still is the manner in which one ought to 'conduct oneself' – that is, the manner in which one ought to form oneself as an ethical subject acting in reference to the prescriptive elements that make up the code.
>
> (1992: 26; see also Foucault 1983: 263)

This distinction is reflected in the way that veganism was first codified, and the way that Society members were enjoined to relate to themselves in light of that code. The first formal definition of veganism was proposed by Leslie Cross in 1949; 'the principle of the emancipation of the animals from exploitation by man'[1] (Cross 1949b: 16 – emphasis in original). This proposal was modified to become the first 'official' definition of veganism adopted by the Vegan Society, at its AGM on 11th November 1950: 'the doctrine that man should live without exploiting animals' (The Vegan Society 1954: 1). This eleven-word distillation of the discursive labour of the first seven years of the Vegan Society contains the essence of the new 'moral code' of veganism. But over the same period, the Society was engaged in forging a new ethics; 'a member [of the Society is ...] one who undertakes to live out vegan principles as far as he or she is able according to circumstances' (Cross 1951: 3). The *practice* of living out vegan principles was therefore the focus of The Vegan Society, more so than was the elaboration of moral *arguments* in favour of veganism and in opposition to the exploitative status quo. Leneman argues that:

> No correspondent [to *The Vegetarian Messenger*] in any period after World War I attempted to argue that cruelty was not a necessary component of the dairy industry; by then, the ethical argument for what became known as veganism had been won.
>
> (1999: 226)

Instead, Leneman points out that correspondence in this period tended to focus on *how* to live as a 'vegan' (ibid.: 227), or in other words, on ethical practices, rather than on moral arguments. This focus is germane to contemporary CAS, which emphasizes activist praxis over rarefied academic debate (Best *et al.* 2007; ICAS 2012 and see Glasser, Chapter 12, in this volume).

Chloë Taylor argues that this focus on ethical practice in Foucault's analysis of ancient Greek ethics, can help the contemporary 'animal liberation movement', overcome the impasse of a (victoriously) concluded moral debate, which nevertheless fails to change the behaviour of those who are intellectually convinced by the arguments (2010: 81–83).[2] That is, ethical practice can work directly on transforming the habits inculcated through speciesist disciplinary regimes, which shape human relationships with other animals in an exploitative direction. It is important to note, though, that Foucault did not seek to authorize a return to Greek ethics as such, which he characterized as 'disgusting' for its masculinist and nonreciprocal preoccupation with virility (1983: 253). That is, it was only open to an elite class of privileged men to engage in solipsistic ethical practice, in the context of a society organized around a patriarchal moral principle of virility, which objectified women, boys, and slaves.

As Taylor argues, the same pattern can be found in Foucault's analysis of ancient Greek dietetics, 'in so far as it is solely concerned with the effects of foods on the eating subject, and not at all with where that food came from' (Taylor 2010: 79). Taylor goes on to characterize contemporary non-vegan practices, specifically 'meat'-eating, as analogously 'disgusting' to the patriar-chal ethics of the Greeks: 'In the virile pleasures of eating chickens, cows, ducks, turkeys and lambs, we do not think about the pleasure of the other – the pleasure of non-human animals' (2010: 80). The use of Foucault's analysis therefore is not through any access it gives to an ethical 'golden age', which we might rein-state. Instead, it offers a model for thinking through how veganism, from its 1944 foundation, critiqued the speciesist status quo through ethical practice, more than it did through moral argument (although the latter remains important). In light of Leneman's point above, the focus on ethical practice over moral argu-ment offers an important rebuttal to the enduring disparagement of veganism as a 'lifestyle'. For the early vegans, the robustness of the moral reasoning for veganism meant that the pressing problem was the practical problem of how to embed veganism as a lived reality throughout society; in other words, emphasiz-ing *doing* veganism rather than *exhorting* veganism. Therefore, the pioneers of modern veganism arguably intuited Joseph Tanke's formulation of Foucault's work on ethics; 'the attempt to think tactically in an effort to develop possibil-ities of resistance to enforced forms of normalcy through the conscious re-constitution of the self' (2007: 89). While Taylor argues that Foucault's ethics hold lessons for the contemporary 'animal liberation movement' (2010: 81–86), the argument developed in this chapter is that The Vegan Society already fol-lowed those lessons, or rather, it formulated and promulgated them itself, through its discourse and practice.

The complexity of this parallel reformulation of moral principles and the assembling of ethical practices in the early years of The Vegan Society can therefore be illuminated by working through Foucault's four-stage analysis of ancient Greek ethics, contained in *The Use of Pleasure* (1992: 25–32). This 'ethical fourfold' is summarized by Rabinow as 'ethical substance, mode of sub-jectivation, ethical work, and telos' (2000: xxvii). Ethical substance refers to 'the aspect or the part of myself or my behaviour which is concerned with moral conduct' (Foucault 1983: 263). Mode of subjectivation refers to 'the way in which people are invited or incited to recognize their moral obligations' (ibid.: 264). Ethical work refers to the techniques through which the ethical substance is worked on, or in short 'what are we to do' (ibid.: 265). Finally, telos refers to 'the kind of being to which we aspire when we behave in a moral way' (ibid.). The following four sections of this chapter consider each of these aspects in the context of the early Vegan Society, by analysing exemplary quotations from the archival sources outlined above.

Compassionate non-exploitation and desire: the ethical substance of veganism

For Foucault, modern Western society emphasizes 'feelings' towards others (especially sexual desire) as the primary ethical substance to be transformed through ethical practice (1983: 263). Feelings towards others of a very different kind are evidently an aspect of the ethical substance of veganism, but a more precise characterization might be the way in which vegans position themselves in relation to an extant moral hierarchy of species. Through this hierarchy, (non-vegan) humans, then and now, come to habitually reproduce themselves as the rulers of all other species and of all other individual animals through the social-ization process (see Stewart and Cole 2009; and Cole and Stewart forthcoming 2014). Exiting, and dismantling this moral hierarchy and replacing it with a rela-tion of compassion towards other animals was therefore central to the ethical substance of veganism: 'The vegan ethic has developed apace because there is no logical point where it can be stopped short of a new relationship with the rest of sentient creation' (Watson 1989c: 21). But crucially, that repositioning was bound up with an aestheticization of the vegan self through the reconstruction of desire – away from the bodies and reproductive products of the murdered, muti-lated, imprisoned masses of other animals, and towards the 'fruits of the earth' (Henderson 1946a: 19).

This repositioning of humans in relation to other animals is manifested expli-citly in the abdication of the 'privileges' of exploiting other animals: 'The time has come for us to boldly renounce the idea that we have the right to exploit animals' (Watson 1947c: 12). In turn, that abdication of privilege is founded on a compas-sionate other-animal orientation, as in this example from Leslie Cross, who argues that veganism is based on, 'the emancipation of the animal world from man-imposed cruelties and unnecessary exploitation' (1947: 4). Reflecting on the Soci-ety's fiftieth anniversary, Donald Watson wrote that: 'Strong as the health and ecological reasons for veganism are, they came second to compassion as the force that drove us on to start the Society at that most difficult time of food rationing' (in Farhall 1994: ii). In one example, compassion extends to the empathic adoption of the position of the animal other: 'You kidnap my children, steal my milk, kill me and eat me, and then strut about in my hide' (anon, in Watson 1945c: 6). An other-orientation extends to specifically condemning the 'disgusting virility' commented on by Taylor, as in Alec Martin's damning of 'dairy' and egg industries for their 'diabolical sex exploitation of the animals' (1945: 6). Martin later added, 'surely the exploitation of sex in animals is the most debased form of all' (1947: 8). The violent practices involved in the sexual exploitation of cows are described thus:

> The mating of cattle is nothing but prostitution [...]. Generally speaking, a great many cattle have no normal sex life. They mate not because they will it, but because man wills it, and when he wills it [...]. The cow must be bulled, and bulled she is by fair means or foul [...] some cows remain obstinate and they are placed in a framework, their feet tied, and the bull led up to them.
>
> (Barlow 1945: 9)

Barlow's invocation of prostitution rhetorically positions pimping other animals as one of the 'privileges' that the vegan renounces. The ethical ugliness of pimping is therefore implicitly counterposed to the ideological idyll of 'dairy farming' as a beneficent or benign form of human relation with other animals:

Cool dairies and dainty dairymaids have been in the literary picture from time immemorial, but we must not allow their charm to lull our critical faculty. Behind the dairy lurks the inevitable slaughter-house with all its horrors.

(Reid 1948: 9)

These kinds of rhetorical tactics work to construct this aspect of the ethical substance of veganism – human relationships with other animals – as something that has been debased (albeit that that debasement has been ideologically masked for the majority of non-vegans), by inculcation into the privileges of domination. In remedy for that debasement, Cross argued that 'the impelling element [of veganism] is compassion for animals' (Cross 1949b: 16), illustrated by the writing of Martin and Barlow above, and more generally through repeated references to the cruelty of exploitation in The Vegan Society's early publications. Exemplary is this rejoinder, which explicitly links veganism as a *practice* with the *feelings* of compassion and empathy for other animals:

One of our critics informs us that the reform we advocate is 'so very difficult'. The cow too must not find it easy when her successive calves are taken away from her, but she can do nothing about it. We can.

(Watson 1945a: 2)

Compassionate, non-exploitative relations with other animals are constructed as a re-aestheticization of human existence, in that the vegan good life is also the beautiful life. This is illustrated, for instance, by a metaphor of non-veganism as parasitism: 'We seek to emancipate ourselves from parasitism on animals' (Watson 1945a, 1; see also Watson 1944: 1) or in Watson's reflection that drinking cow's milk, '...was certainly wrong aesthetically, and we could conceive of no spectacle more bizarre than that of a grown man attached at his meal-times to the udder of a cow' (1988: 11). This ethico-aesthetic reworking of the relation of human vegans to other animals is complemented by the reconfiguring of corporeal desires, principally through new dietary practices, but also through using plant-based or synthetic clothing, toiletries, cosmetics, and so on. New dietary practices were emphasized from the beginning of The Vegan Society, not least due to the catalytic discussion of the cruelties of 'dairy' in the pages of *The Vegetarian Messenger* in the early 1940s; 'man's food should be derived from fruits, nuts, vegetables, grains and other wholesome non-animal products and [...] it should exclude flesh, fish, fowl, eggs, honey, and animals' milk, butter and cheese' (The Vegan Society 1944: 1). Even in the choice of the word 'vegan', Donald Watson made clear the importance of the aesthetic appeal of new dietary practices: 'We need a name that suggests what we do eat, and if possible one that conveys the idea that even with all animal foods taboo, Nature still offers us a bewildering assortment from which to choose' (1944: 2).

The elaboration of this 'bewildering assortment' is pursued in the context of the 'ethical work' of The Vegan Society below, but this emphasis on the alimentary abundance offered by plant foods is also significant as a marker of the compassionate relation to self that veganism instantiated. This is brought out in Donald Watson's self-deprecating comment in the first issue of *The Vegan News*: 'We can hardly wish to be classed as moral giants because we choose to live on a diet so obviously favouring self-preservation' (1944: 3).[3] This theme of good health through vegan diet plays a number of roles. First, from the beginning, it was recognized that good vegan health was crucial to counteract anti-vegan discourse:

> We may be sure that should anything so much as a pimple ever appear to mar the beauty of our physical form, it will be entirely due in the eyes of the world to our own silly fault for not eating 'proper food'.
>
> (Watson 1944: 3)

Second, it provided a component of veganism-as-care of the self, to borrow the title of Foucault's third volume of *The History of Sexuality* (1990). In other words, to be compassionate towards the vegan self included a concern with one's own corporeal well-being. One way in which this was emphasized was in the publication of vignettes about thriving vegan children in *The Vegan*:

> Several people have said to me lately, 'I think you are jolly lucky, you must be the only mother in Streetly whose children haven't been ill this winter', and I answer, 'It all depends where you live; we live at the house where the milkman never calls'.
>
> (Mayo 1946b: 12)

The importance of this witness to vegan flourishing cannot be underestimated, in a hostile climate in which animal foods were widely thought to be essential for children's development. Watsons's 1947 IVU address stressed that governmental and scientific authorities, together with public opinion, opposed veganism on health grounds (a point which, despite the accumulation of copious evidence to the contrary in the intervening years, remains salient today).

Finally, good vegan health, especially that of children, was a crucial component of compassionate concern for other humans. Anxiety about 'imposing' veganism in case it proved risky for health was evident in the early issues of *The Vegan*. To assuage that anxiety, the Society posted a questionnaire for readers who were parents of vegan children in 1947. The results were reported in the spring 1948 issue:

> [The questionnaire] confirms the effectiveness of a vegan diet in building up healthy children, for, with the exception of an occasional cold, no vegan child appears to have had any serious illness. This is grand testimony indeed to the vegan way of living.
>
> (Mayo 1948: 16)

Although a self-report health survey like this might have little academic credibility, it must be remembered that at this point, in the absence of research, the evidence for the health effects of veganism were limited to anecdotes and the pioneering work of a handful of sympathetic nutritionists such as Dr. Cyril Pink (whose testimony about children thriving on plant-based diets under his care regularly featured in *The Vegan*). Compassion for other humans also emerges in arguments for the role veganism could play in what would now be called global food security: 'The competition between animals and man for the world's grain must end before the problem of famine can be solved' (Watson 1947a: 12). This is discussed further below as an aspect of the telos of veganism.

Taken together then, the ethical substance of veganism centred on compassionate non-exploitation: For other animals, for the self, and for other humans. This was complemented by the reconfiguration of desire away from the bodies and products of other animals, and towards a plant-based cornucopia. This stood in contradistinction to the extant exploitation of other animals, which was constructed as condemning humans to dubious 'privilege' at the price of ethical, aesthetic, and physical corruption.[4]

Proselytizing, exemplification, and collaboration: modes of subjectivation to veganism

Anticipating the first 'official' definition of veganism, Leslie Cross wrote that it could offer 'a definite, clear and precise "veganism," which shall have the consent and allegiance of every person who joins the Society' (1949a: 15). In his article, Cross expresses concern that a constitutional and therefore 'binding' definition is needed to prevent fragmentation of the vegan movement, and furthermore, that a definition ought to be 'a principle, from which certain practices logically devolve' (1949a: 15). This formulation might almost have been written with Foucault's ethical fourfold in mind. Foucault argues that modes of subjectivation are ways, 'in which the individual establishes his relation to the rule and recognizes himself as obliged to put it into practice' (1992: 27). The reconfiguration of dietary (and other) practices discussed in the last section therefore connects the ethical substance of veganism with the mode of subjectivation. That is, it raises the question of *how* to enact the principle of compassionate non-exploitation, and the new alimentary desires that flow from it. In short, how were early vegans to become and remain vegan in a non-vegan world? There are three aspects to the mode of subjectivation constructed by the early Vegan Society, which will now be considered in turn: a proselytizing mode, an exemplary mode, and a collaborative mode.[5]

The establishment of the Society at all, and the coining of 'veganism' immediately established the organization as a proselytizer of the new moral code and how to enact it. The first aims of the Society, written in its 1944 Manifesto and formalized in its first rules in 1947, emphasized its advocacy role for veganism (The Vegan Society 1944; 1947). The very effort to keep *The Vegan* in circulation, often on a financial shoestring, is testament to the imperative to proselytize. Stalwarts of the early Society also worked with tremendous enthusiasm and energy to promote veganism among non-vegans. *The Vegan*, for instance, documents numerous efforts to persuade others of the merits of veganism, ranging from (sometimes hostile) vegetarians to members of Parliament. Of particular note is the work of Fay K. Henderson, wife of Watson's successor as the editor of *The Vegan* in 1947, G. Allen Henderson. Fay Henderson embarked on a tour of the UK and Ireland in the winter of 1946, promoting veganism through public talks and cookery demonstrations, thereby pioneering a consciousness-raising model for vegan activism, which endures in countless vegan fairs and street stalls to this day. To this end, the Society positioned itself as an educator about veganism,[6] in order to equip readers with the tools for proselytization:

> It is our duty to recognise the obligation we owe to these creatures and to understand all that is involved in the consumption and use of their live and dead products. Only thus shall we be properly equipped to decide our own attitude to the question and explain the case to others who may be interested but who have not given the matter serious thought.
>
> (Henderson 1947b: 2)

Much of the content of *The Vegan* is also distinguished by a passionate, exhortative style, as illustrated by the title of this chapter, and elsewhere within it. However, reflecting in 1989, Watson reported that, '[w]e agreed that conscience is not a mere evolutionary accident and that [...] "the still small voice within" was the most persuasive orator we were ever likely to hear' (1989a: 13). So, the exhortative style of some of The Vegan Society's output was intended to enable its audience to hear their own inner voices of conscience, voices habitually silenced by the disciplinary power of human species privilege: 'The Vegan Society here is built up of individuals who have acted on inner conviction. Each has chosen the vegan way as an expression of personal belief' (Henderson no date: 15).

This segues into the second mode of subjectivation, to conduct oneself as an exemplar of veganism. The emphasis on 'inner conviction' lent pioneering vegans an authority based on attesting to lived, experiential vegan truth, rather than as hectoring mouthpieces of dogma. To exemplify the principles of veganism was therefore partly to embody its principles. This was pursued in a variety of forms, not least in the vignettes of thriving vegan children discussed above. For example, the father of 12-year-old David Sharpe wrote:

> His health is simply taken for granted – he is always tip-top! [...] He is contented, happy and interested – deeply in love with life [...] David went vegan of his own volition: it gradually dawned on him about two years ago that it was wrong for us to handle the eggs of fowls and to drink the milk of cows. Instinctively he refused to take part in the acts of robbery. [...] He is sure proof of the fact that to be a Vegan is right.
>
> (Watson 1946a: 10–11)

Other examples include Frank Needham's article, 'Coal mining on a vegan diet': 'I can unreservedly testify not only to the adequacy but to the entire suitability of the Vegan diet for the arduous and unpleasant work on which I am engaged' (1947: 10). Then as now, demonstrating the adequacy of plant-based diets to sustain intense physical effort was rhetorically important in the face of scepticism. As well as personal testimonies, *The Vegan* also promoted others as exemplars, as in this report on 'Vegans Overseas': 'This little survey shows us that champions of our cause are setting an example to the rest of mankind in many parts of the world' (specifically USA, South Africa, and Singapore in this article) (Smith 1947: 20). However, vegan exemplarity was embedded in a challenging social context, acknowledged in *The Vegan*:

> From the big correspondence I received as Secretary it is clear that the problem most pressing to the Vegan is not whether the diet is adequate, but the social implications arising from practising it strictly on all occasions [...] the Vegan is confronted with these rival claims of personal consistency and of accommodating himself constructively and sociably in a wicked world with as little violation of conscience as possible.
>
> (Watson 1947c: 12)

Watson stressed the difficulty of vegan perfection in the context of a society that was absolutely dependent on the exploitation of other animals, including in areas that individual vegans would find it very difficult to avoid unwilling complicity, notably the use of slaughterhouse by-products in crop growing (1945c: 7). Exemplarity was therefore constructed not as the cultivation of personal purity, but in working towards eliminating exploitation, i.e. an engaged, active veganism, rather than an aloof, self-satisfied veganism: 'The truest consistency must surely be the effort to clean up the mess which at present *dirties us all*' (Watson 1945c: 7 – emphasis added). The particular problem of slaughterhouse by-products inspired the development of a regular column in *The Vegan*, 'Horti-Vegan Notes', by Alec Martin, which confronted and attempted to tackle this central issue of the possibility of vegan consistency; 'if a system of agriculture which does not rely upon the exploitation of animals cannot maintain a healthy soil and a healthy people [...] then our whole vegan case and ideal must fall down' (Martin 1949: 18). Martin's column anticipated the later development of the Vegan Organic Network (founded in Manchester in 1996 as the Vegan-Organic Horticultural Agricultural Network), which continues to address this issue (Vegan Organic Network 2013).

In terms of the social difficulties of veganism, *The Vegan* published models of vegan exemplarity in trying circumstances. For instance, in an article on hiking in remote areas of Scotland, J. R. Clark recommended self-catering as the best solution to maintaining a vegan diet, but noted that:

> There is, of course, almost always to be met in these [youth] hostels the morbid smells of cooking animal flesh. [...] My own diet on tour consisted basically of raw greens, bread and potatoes, which I bought on my way, and dried fruit, nuts and peanuts, which I took with me. I also took a camper's

butter pot filled with soya butter and a tin (with pressed-down lid) filled with peanut butter.

(Clark 1946: 13)

Nevertheless, the social difficulty of veganism was acknowledged as an often more troubling problem than practical issues, such as finding suitable food. At the same time as the questionnaire for vegan parents discussed above, the Society distributed a general questionnaire for all members. Reporting on the findings in 1948, Watson wrote that:

The main problem confronting the vegan is not one of health, nor expense, nor domestic labour; it is the social upheaval necessary to practice strict veganism [...]. The Vegan Society, realising the great difficulties of achieving its ideal in a world organised to cater for orthodoxy, does not ask members to pledge themselves to any degree of consistency, save that of serving in the way they think best under prevailing conditions.

(Watson 1948: 13)

This apparent concession to 'inconsistency' may seem surprising, even uncomfortable, with hindsight (the Society now requires a minimal pledge to adhere to a diet suitable for vegans, as a criteria of full membership). However, Watson was concerned with maintaining consistency among those who claimed it for themselves:

We can hope to present our case strongly only so long as those who form the nucleus of Full Members are faithful. In any Movement there must be a most dramatic loss of spiritual power when signatures are not honoured.

(Watson 1945b: 3)

Furthermore, Watson's response to the questionnaire's finding is indicative of the tone of *The Vegan*, which encouraged the flourishing of that 'inner voice' in favour of castigating inconsistency among aspiring vegans. In Foucauldian terms, this might also be interpreted as connoting the difference between a self-transformative vegan ethic and veganism-as-disciplinary power, enforced by vegan 'authorities'.

The exemplary mode was, therefore, not intended to foster icons of vegan dogmatism, but to encourage collegial and empathetic guides through the quagmire of exploitation that 'dirties us all'. The exemplary mode therefore also necessitated a collaborative mode, addressed towards overcoming social (and other) obstacles: 'Veganism will lose if its missionaries become solitary ascetics' (Watson 1945c: 8). Ethical transformation, therefore, was orientated towards the collaborative solution of the obstacles that vegans and veganism faced: 'We have the faith that *collectively* we can solve the problems arising from our desire to be humanely fed, clothed, and otherwise provided for' (Watson 1945b: 2 – emphasis added). The 'Vegans Overseas' article added that vegan exemplars, 'are conducting a lonely struggle, and in all pioneering work it is of tremendous help to know that others are working on similar lines', and to that end the Society offered to put overseas vegans in touch with each other by post (Smith 1947: 20). This encouragement of a collaborative mode of subjectivation is of course evident in the organization's name; The Vegan *Society*, and this was pursued further through the establishment and encouragement of local vegan groups, precisely in order to provide mutual support in the face of social difficulties (the first stage in the development of a network of Vegan Society Local Contacts, which endures to this day). In recognizing the social as well as personal aspects of veganism, G. Allen Henderson wrote:

> ... there is the social attitude which is concerned with the procedure to be adopted when entertaining guests or when having meals prepared by others. It is to assist the solution of the problems arising under this heading that the formation of Local Groups and Discussion Circles is being encouraged.
>
> (1947b: 1)

While exemplarity also extended to appeals within the pages of *The Vegan* to establish ideal vegan communities, the emphasis was on the integration of proselytization, exemplification, and collaboration as modes of subjectivation that orientated vegans towards engagement with the non-vegan world. To be vegan in the 1940s was therefore to be actively attempting to re-shape both self and society. Precisely how this was done was a matter of the cultivation of ethical techniques, or practices.

The transformation of exploitative habits: the ethical work of veganism

Ethical practice builds on the mode of subjectivation, that is, on how vegans were enjoined to relate to the vegan principle of compassionate non-exploitation and its complementary reconfiguration of desire. As Foucault puts it, ethical work is pursued 'not only in order to bring one's conduct into compliance with a given rule, but to attempt to transform oneself into the ethical subject of one's behavior' (1992: 27). Unlike Greek ethics however, the ethical work of veganism was not exclusively inward looking, but was directed towards self-transformation through and because of a transformed relationship with other animals. The elaboration of the ethical work of veganism occupies more space than the other aspects of the ethical fourfold in early Vegan Society publications, which therefore exemplified Watson's stress on showing how to *implement* vegan ethics, discussed in the introduction to this chapter. The central theme of this ethical work is the transformation of the habitual practices that flow from and reproduce the disciplinary regime of species exploitation.

Although the Vegan Society was always concerned with more than diet, transforming food habits was nonetheless the central plank of the ethical work it elaborated. This is most obvious in the publication of vegan recipes and tips on vegan food preparation, beginning with a recipe for vegan rice pudding in the second issue of *The Vegan News* in 1945. The recipe was followed with the caveat that, '[f]or those of us who lose our childish instincts slowly, the great drawback with this pudding is that it has no skin!' (Watson 1945a: 4). But later in the same issue, the theme of parasitic dependence on the exploitation of cows was invoked:

> When I hear of the business man in the prime of life ordering a pint of milk to be delivered at his office daily, I feel I would like to say to him 'Don't you think at your age it is time you were weaned'?
>
> (Watson 1945a: 9–10)

Choosing the skinless rice pudding therefore already works as a re-aestheticisation of self, contrasted with the infantilized image of the habitual dependence on exploitation. The meaning of vegan food choices, then, was always to be found in the renunciation of speciesist privilege, more than a hedonic calculation of gustatory pleasure between vegan and non-vegan foods. Among the earliest literature distributed by The Vegan Society was Margaret Rawls' pamphlet, *Vegetarian Recipes Without Dairy Produce* (no date) initially produced by the Leicester Vegetarian Society, of which Donald Watson was the secretary at the time of the formation of The Vegan Society. The first book published by the Society was Fay Henderson's *Vegan Recipes: Amplifying the non-dairy diet* (1946b). This book therefore bears the distinction of being the first cookbook to include the word 'vegan' in the title, although de facto 'vegan' recipe books were published as early as 1874 (Davis 2012: 49–50).

Notwithstanding the rice pudding example cited above, a key theme of early vegan recipes, and more general discussions of food, was on the pleasures of vegan eating. For example, in response to a *Daily Mail* article by a doctor advocating increased rations of animal products, Watson replied that:

> Your view that monotony in diet may be the cause of indisposition is a sad reflection on the lack of knowledge in the use and preparation of natural foods that are available. It is not difficult to name a hundred different foods that have been available throughout the war.
>
> (Watson 1946e: 9)

Watson therefore asserted the pleasures of vegan food in spite of the challenging context of rationing, which hit vegans harder than most:

> The vegetarians had been successful in obtaining an additional cheese ration in lieu of meat, but all my approaches to the Ministry of Food to obtain some comparable concessions for vegans failed.
>
> (Watson 1988: 11)

The elaboration of a new vegan cuisine, while occasionally conceding that old habits die hard, as in the rice pudding example, aimed to transform those habits.

For example, in response to a letter requesting a recipe for a vegan equivalent of Welsh rarebit (a dish based on butter and cheese), Fay Henderson wrote that '[t]his request reminds one again how very much we are inclined to live by habit. Actions, thoughts, and even tastes are influenced by the habits of our childhood'. Henderson then reasserted the connection between ethical work and the new relation with other animals that gave it its meaning: 'One result of the acceptance of veganism is the breaking down of old habits and the building up of new ones based on an awareness of life as a whole' (1948: 18).[7] In a similar vein, Watson argued, in the first issue of *The Vegan News* that, 'once the [vegan] diet has been arranged the sight and smell of dairy produce is soon forgotten' (1944: 4).

The publication of vegan recipes also played an educative function, feeding the proselytizing mode of veganism, in that it equipped vegans with a response to incomprehension at what constituted a vegan's diet: 'One is inclined to weary of the oft-repeated exclamation: "Good gracious! No eggs or milk or even cheese! Whatever do you live on?"' (Henderson 1946a: 19). As well as the re-education of gustatory desire, discussions of food also fed the exemplary mode of veganism in that they emphasized the importance of a *healthy* vegan diet. The deleterious effects of cow's milk on human health were regularly discussed in the early publications of the Society, notably the pamphlet, *Is Milk a Curse?* (Goodfellow 1945). The counterpoint was provided in public lectures (for instance, one of Fay Henderson's 1948 talks was entitled 'Veganism: Health Without Animal Food') and articles in *The Vegan* exploring the health benefits of a vegan diet, particularly for children:

> Dear Vegan Mothers [...] I have every confidence that a Vegan diet can be adapted to suit a baby after breast feeding is finished, so, in spite of adverse criticism, have confidence and what is morally right will be the best.
>
> (Mayo 1946a: 11)

The parenting of vegan children was an especial focus of attention, through Kathleen Mayo's regular 'Vegan Baby Bureau' column, and the contributions of Dr Cyril Pink and others. This concern with infant feeding played a large part in stimulating interest in developing plant-based alternatives to cow's milk, especially for vegan women unable to breastfeed (Watson 1946c: 2). That initial concern later developed into the formation of the separate Plantmilk Society; 'although the Society's task is a practical one, involving hard facts of science and commerce, the force which brought it into being was the moral force of compassion' (The Plantmilk Society 1957: 1). This became the Plamil Foods company, founded in the 1960s, which is still in business as an entirely vegan company at the time of writing (Plamil Foods Ltd. no date).

The transformation of habitual practices was a response to existing knowledge about exploitation, as in the examples above, or the exploration of the possibilities for vegan horticulture, discussed earlier. It was also manifested in a concern to unmask the extent of, thus far, hidden dependence on exploitation of other animals. To that end:

> Mr. Drake and Mr. Spencer have undertaken the work of making extensive enquiries to ascertain whether foodstuffs and commodities conform to Vegan requirements. The Reports will appear in due course and will subsequently be published in the form of a Handbook which will be revised annually.
>
> (anon in Watson 1946a: 16)

This effort continued and intensified throughout the Society's history, manifested in recent publications such as *The Animal Free Shopper* (The Vegan Society 2010) and the Vegan Society's current Trademark scheme, which acts as a vouchsafe that products are free from ingredients taken from other animals and have not been tested on other animals (The Vegan Society no date). Vegans were enjoined to collaborate in the work of verifying the vegan status of products: 'Please send in your discoveries and help to make the Vegan Trade List comprehensive. Any commodities you are doubtful about will be gladly investigated' (Spencer 1946: 7); 'Supporters are [...] strongly urged to make a habit of reading the ingredients' (Drake 1946: 8). Likewise, vegans were encouraged to assess the extent to which they could be catered for; 'a survey of vegan facilities in London restaurants is already being made' (Sowan 1946: 16). Echoing the discussion above of the non-disciplinary character of the exemplary mode, Spencer addressed readers thus:

> You are *aiming* at ultimate consistency. The commercial world is not out to help you except by chance, and you have to succeed gradually, demonstrating what can be done to eliminate connivance at animal exploitation in spite of adverse circumstances.
>
> (Ibid.: 5 – emphasis added)

Ethical work further developed the collaborative mode of subjectivation through the encouragement of social contact among vegans, not least through the establishment of the Local Contacts network and local vegan groups, discussed above. This demonstrated the adoption of a new official aim of The Vegan Society in 1947: 'To extend and organise Veganism nationally and internationally, and to facilitate contacts between those endeavouring to follow the Vegan Way of Life' (The Vegan Society 1947).

Returning to Spencer's aim for 'ultimate consistency', the ethical work of veganism therefore *aimed* at something beyond the present conditions of life. The transformation of exploitative habits provided the means through which vegans could develop their proselytizing, exemplary, and collaborative modes of subjectivation to the vegan principle of compassionate non-exploitation. While problematizing any claims to vegan purity, the Society nonetheless orientated the ethical work of vegans towards a transformed social context in which vegan 'purity' would not be an effortful achievement of a select few, but the social and experiential norm for all. The next section therefore considers the telos of early veganism as the normalization of veganism, for the self and for society.

Ending exploitation for society and self: the telos of veganism

Foucault's understanding of ethical telos inheres in the way in which specific ethical actions gesture beyond themselves and towards a 'mode of being characteristic of the ethical subject' (1992: 28). So, while each vegan act is ethically significant in itself, it also commits the actor to 'being vegan' through repeated ethical work. Through this process, veganism might therefore be understood to become a durable, 'normal' identity for the individual vegan. But, as stressed throughout this chapter, early vegan ethics were concerned with the transformation of society *through* the transformation of self. This theme has been reiterated throughout this chapter, so this section provides a brief recapitulation of this symbiotic personal and social telos.

As with Leslie J. Cross earlier, Dr Cyril Pink could almost have had in mind the ethical fourfold, in this case its personal telos as a 'normalized' vegan identity, when he wrote:

> First must come the vision – we see the goal, we appreciate a principle, and recognise it as Truth. Next, perhaps after a long delay, the principle 'works through' into daily life, and man must become vegan. He would say that it comes naturally to him, to be so, then he is indeed vegan and will never change.
>
> (Watson 1948: 17)

The personal telos of vegan ethics was typically expressed as an emancipation of the self from a position of privilege, which is aesthetically transfigured from a complacent supremacy to complicity with horror:

> Human existence does not depend upon the inconceivable tyranny now existing against animals, in fact progress is impeded enormously by it. To renounce this tragic heritage is to be born again, to a life sometimes more difficult, but always of clearer conscience and more satisfying conclusion.
>
> (Watson 1946d: 2)

Watson's negative telos, i.e. the description of what veganism moves away from, is complemented by the positive telos articulated in Marion Reid's description of the re-aestheticized vegan self: 'Vegans believe in beauty so strongly that they are determined to make their lives beautiful from the foundations upwards by banishing all exploitation of other lives as far as possible.' Reid takes this further by, once again, insisting on the embedding of vegan ethics in a compassionate other-orientation: 'Veganism is not based on a negative law of non-killing only, but rather on a warm, vibrant sentiment of goodwill and brotherhood towards all living creatures' (Reid 1948: 2).

That other-orientation therefore segues into a consideration of the social telos of veganism. This was already envisioned from the first issue of *The Vegan News*: 'The Vegan Society seeks to abolish man's dependence on animals, with its inevitable cruelty and slaughter, and to create instead a more reasonable and humane order of society' (Watson 1944: 1). The practical vision of a re-ordered human society came in the form of arguments against the injustice of feeding plant food to 'domesticated' animals, in the context of human inequality in access to food, reiterated several times by Donald Watson (and others) throughout the 1940s, for example:

> To meet the problem of famine, the earth must be made to produce the foods which man needs [...] Of the many factors affecting the problem of food shortage none is more pressing than the competition for grain between man and his domesticated animals.
>
> (Watson 1946b: 1, 2)

Typically, Watson foregrounded the mutuality of exploitation of other animals and human self-destruction: 'In enslaving the animals man has enslaved himself to the laws governing scarcity' (1947a: 13). Analysing this intersection of the exploitation of nonhuman animals with human impoverishment has been a key theme of CAS scholarship (see Twine 2010). CAS has also focused on exposing the ideological obfuscation of exploitation (for example, see Nibert 2002). This exposure is manifested repeatedly in the early writings of The Vegan Society, and forms a key aspect of its social telos. For instance, this was expressed in G. Allen Henderson's characterization of non-vegan society as a 'dark citadel', which inhibited recognition of the beneficence of veganism, 'obscured by walls of prejudice, habit, apathy, perverted tastes, faulty reasoning, vested interests [...] We are determined that these walls shall be breached' (Henderson 1947a: 2).[8]

As suggested in the introduction to this chapter, the vegan telos therefore combines compassionate non-exploitation of other animals with an emancipated vegan self *and* a more compassionate human society. Vegan ethics, from the beginning, was directed towards these interconnected goals of transforming human beings and transforming human society, with both flowing from the foundational reconfiguration of human–nonhuman animal relations.

Conclusion

This chapter has shown how Foucault's work is useful for interpreting the social construction of early veganism as an ethical practice, which always already transcended a preoccupation with solipsistic dietetics. The application of Foucault's 'ethical fourfold' has suggested that the 'ethical substance' of veganism inhered in a re-evaluation of the relationship between humans and other animals, and the replacement of human species-based privilege with compassionate non-exploitation. Derived from this is a reconfiguration of desire, including, but not limited to, alimentary desires for plant-based food and drink. Individual vegans, through membership of the Society, were enjoined to cultivate a tripartite 'mode of subjectivation', as proselytizers and exemplars of veganism, and as collaborators in ethical communities that sought to resist and transcend the disciplinary regime that reproduced species privilege. To that end, the 'ethical work' of veganism centred on the transformation of exploitative habits that reproduced human domination of other animals and mutual support in formulating tactics to engage with a sceptical, intransigent, or oblivious social context. That ethical work was in turn directed towards a symbiotic personal and social telos, in which vegan identity becomes embedded as an experiential norm of compassionate, non-exploitative living, as the basis of a more compassionate and peaceable society.

This analysis suggests that vegan ethics anticipated some core concerns of contemporary CAS, especially the inseparability of personal ethical commitment to opposing exploitation on the one hand and social transformation towards a peaceable future on the other (see Best *et al.* 2007). Contrary to stereotypes of vegan elitism, the formation of a new vegan ethics was intrinsic to a utopian social movement dependent on the *repudiation* of human superiority over others – whether other humans or other species of animals. Vegan ethics animated a socially transformative endeavour, which foregrounded the intersecting exploitations of humans and other animals, at least in some respects.[9] The breath-taking scope of the transformative vision of the vegan pioneers outlined in this chapter may inspire a re-centring of vegan ethics in the practice and advocacy of all those who oppose exploitation in its myriad and pernicious forms: 'Details of origin should always be interesting, and vital if a movement is to remain faithful to its precepts' (Watson 1989a: 13).

Notes

1 The gendered language of *The Vegan* was conventionally sexist for the time, and is reproduced here unedited. Despite the overt challenge to species privilege offered by early veganism, conventional species hierarchy was also reproduced linguistically, in formulations such as '...lower forms of animal life' (Watson 1944: 1). However, there was also already a questioning of what we would now call the speciesism embedded into conventional language. For example, Leslie Cross wrote of the 'enslavement' of other animals, and made use of denaturalizing quotation marks around 'domestication' (1949b: 17), in a rhetorical technique that is commonly used in contemporary CAS writing (see for example Nibert 2002). Similarly, Donald Watson recognized the need to de-normalize the use of other animal's products through the choice of language; 'we should realise the obeisance implied by the term "nut-milk". Is there no more vegan word or phrase?' (Watson 1945d: 11).

2 See Salih's discussion of the failure of moral argument alone to precipitate vegan transitions in Chapter 3 of this volume.

3 It should be noted that at the time of the formation of The Vegan Society, vitamin B_{12} had not yet been discovered, and therefore its dietary importance had not been recognized or promoted. Some vegan pioneers suffered ill-health as a result of this lack of knowledge at the time (Watson 1988: 11). Ensuring that vegans are adequately provided with sufficient B_{12} and are informed about its importance remains a core concern of the contemporary Vegan Society, arguably for a similar set of reasons as those detailed here for the early Society.

4 Spiritual corruption could also be included here; 'if the curse of exploitation were removed, spiritual influences, operating for good, would develop conditions assuring a greater degree of happiness and prosperity for all' (The Vegan Society 1944).

5 Although with different emphases, strikingly, the '10 principles of CAS' (Best *et al.* 2007) also incorporate elements of proselytization, exemplarity and collaboration into a mode of subjectivation; that is a way of *doing* CAS that enacts its transformative aims for the CAS practitioner as well as for wider society.
6 The present day Vegan Society is an educational charity.
7 In Chapter 11 of this volume, Jenkins and Twine echo Henderson, by critiquing the contemporary mistaking of *automatic* food choice for *autonomous* food choice, as a defence of non-vegan mainstream food practices that are socialized and culturally reproduced ad nauseam.
8 The theme of exposing the 'dark citadel', via diverse academic and activist tactics, is central to CAS praxis. For instance, see Fitzgerald and Taylor in Chapter 8 of this volume for a contemporary analysis of the discursive normalization of the Animal-Industrial Complex, or Glasser's discussion of the media impact of Radical Direct Action, in Chapter 12 of this volume.
9 For fuller discussions of intersectionality and its importance to CAS, see the contributions by Cudworth (Chapter 1) and Peggs (Chapter 2) in this volume. Perhaps the most notable omission from early Vegan Society discourse vis-à-vis contemporary CAS is an explicit critique of capitalism, which is present in many of the contributions to this volume.

References

Barlow, E. G. (1945) 'Dairy farming', in *The Vegan News*, 5: 9.
Best, S., Nocella II, A. J., Kahn, R., Gigliotti, C. and Kemmerer, L. (2007) *Introducing Critical Animal Studies*. Available at: www.criticalanimalstudies.org/wp-content/uploads/2009/09/Introducing-Critical-Animal-Studies-2007.pdf (accessed 19 May 2013).
Clark, J. R. (1946) 'A vegan in the Cairngorms', in *The Vegan*, 2(4): 13–15.
Cole, M. (2008) 'Asceticism and hedonism in research discourses of veg*anism', *British Food Journal*, 110(7): 706–716.
Cole, M. and Morgan, K. (2011) 'Vegaphobia: derogatory discourses of veganism and the reproduction of speciesism in UK national newspapers', *British Journal of Sociology*, 61(1): 134–153.
Cole, M. and Stewart, K. (forthcoming 2014) *Our Children and Other Animals: The Cultural Construction of Human-Animal Interaction in Childhood*, Farnham: Ashgate.
Cross, L. J. (1947) 'Veganism defended', in *The Vegan*, 3(2): 4.
Cross, L. J. (1949a) 'In search of veganism – 1', in *The Vegan*, 5(2): 13–15.
Cross, L. J. (1949b) 'In search of veganism – 2', in *The Vegan*, 5(3): 15–17.
Cross, L. J. (1951) 'The new constitution', in *The Vegan*, 7(1): 2–3.
Davis, J. (2012) *World Veganism – past, present and future*. Available at: www.world-vegfest.org/index.php/blogs/john-davis/55-world-veganism-free-e-book (accessed 6 January 2013).
Drake, B. (1946) 'Food findings', in *The Vegan*, 2(2): 7–8.
Farhall, R. (1994) 'The first fifty years: 1944–1994', in *The Vegan*, Autumn: i-xlii.
Foucault, M. (1983) 'On the genealogy of ethics', in P. Rabinow (ed.) *Ethics: Subjectivity and Truth*, London: Penguin.
Foucault, M. (1992) *The Use of Pleasure, The History of Sexuality: Volume Two*, London: Penguin.
Foucault, M. (1990) *The Care of the Self, The History of Sexuality: Volume Three*, London: Penguin.

Goodfellow, J. A. (1945) *Is Milk a Curse?* Leicester: The Vegan Society.

Henderson, F. K. (no date) *Vegan Viewpoint*, Ambleside, Westmorland: The Vegan Society.

Henderson, F. K. (1946a) 'Vegan fare', in *The Vegan*, 2(4): 19–21.

Henderson, F. K. (1946b) *Vegan Recipes: Amplifying the non-dairy diet*, Leicester: The Vegan Society.

Henderson, F. K. (1948) 'Pot pourri', in *The Vegan*, 4(1): 18.

Henderson, G. A. H. (1947a) 'Editorial', in *The Vegan*, 3(1): 1–2.

Henderson, G. A. H. (1947b) 'Editorial', in *The Vegan*, 3(3): 1–2.

Henderson, G. A. H. (1947c) 'Editorial', in *The Vegan*, 3(4): 1–2.

Institute for Critical Animal Studies (ICAS) (2012) *About Us*. Available at: www.criticalanimalstudies.org/about/ (accessed 19 May 2013).

Leneman, L. (1999) 'No animal food: the road to veganism in Britain, 1909–1944', *Society and Animals*, 7(3): 219–228.

Martin, A. (1945) 'Points from letters received', in *The Vegan News*, 3: 5–6.

Martin, A. (1947) 'On milk and eggs', in *The Vegan*, 3(2): 8–9.

Martin, A. (1949) 'Horti-vegan notes', in *The Vegan*, 5(2): 18–19.

Mayo, K. V. (1946a) 'Formation of a baby bureau', in *The Vegan*, 2(2): 9–12.

Mayo, K. V. (1946b) 'Vegan baby bureau', in *The Vegan*, 2(4): 11–13.

Mayo, K. V. (1948) 'Vegan baby bureau', in *The Vegan*, 4(1): 16–17.

Needham, F. (1947) 'Coal mining on a vegan diet', in *The Vegan*, Spring 1947.

Nibert, D. (2002) *Animal Rights/Human Rights: Entanglements of oppression and liberation*, Lanham, MA: Rowman and Littlefield.

Plamil Foods Ltd. (no date) *About us*. Available at: www.plamilfoods.co.uk/about/ (accessed 14 January 13).

Plantmilk Society, The (1957) *Plamil Newsletter 1*, Uxbridge: The Plantmilk Society.

Rabinow, P. (2000) 'Introduction: the history of systems of thought', in R. Rabinow (ed.) *Ethics: Subjectivity and Truth*, London: Penguin.

Rawls, M. B. (no date) *Vegetarian Recipes without Dairy Produce*, Leicester: The Leicester Vegetarian Society.

Reid, M. (1948) 'Good and evil', in *The Vegan*, 4(3): 2–4.

Smith, E. T. (1947) 'Vegans overseas', in *The Vegan*, Autumn 1947.

Sowan, F. A. (1946) 'Notes on the London group', in *The Vegan*, 2(3): 15–16.

Spencer, P. (1946) 'Vegan commodities', in *The Vegan*, 2(2): 5–7.

Stewart, K. and Cole, M. (2009) 'The conceptual separation of food and animals in childhood', *Food, Culture and Society*, 12(4): 457–476.

Tanke, J. (2007) 'The care of the self and environmental politics: towards a Foucaultian account of dietary practice', *Ethics and the Environment*, 12(1): 79–96.

Taylor, C. (2010) 'Foucault and the ethics of eating', in *Foucault Studies*, 9: 71–88.

Twine, R. (2010) *Animals as Biotechnology: Ethics, Sustainability and Critical Animal Studies*, London: Earthscan.

Vegan Organic Network (2013) *About Us*. Available at: http://veganorganic.net/about-us/ (accessed 14 January 2013).

Vegan Society, The (no date) 'Is this product vegan?' Available at: www.vegansociety.com/lifestyle/is-this-product-vegan.aspx (accessed 14 January 2013).

Vegan Society, The (1944) *The Vegan Society Manifesto*, Leicester: The Vegan Society.

Vegan Society, The (1947) *Proposed Rules of The Vegan Society*, Ambleside, Westmorland: The Vegan Society.

Vegan Society, The (1954) *Rules of the Vegan Society*, 4th edn, Reigate, Surrey: The Vegan Society.

Vegan Society, The (2010) *The Animal Free Shopper*, 9th edn, Birmingham: The Vegan Society.

Watson, D. (1944) *The Vegan News*, 1.

Watson, D. (1945a) *The Vegan News*, 2.

Watson, D. (1945b) *The Vegan News*, 3.

Watson, D. (1945c) *The Vegan News*, 4.

Watson, D. (1945d) *The Vegan News*, 5.

Watson, D. (ed.) (1946a) *The Vegan*, 2(3).

Watson, D. (1946b) 'Editorial', in *The Vegan*, 2(2): 1–2.

Watson, D. (1946c) 'Editorial', in *The Vegan*, 2(3): 1–3.

Watson, D. (1946d) 'Editorial – The case for veganism', in *The Vegan*, 2(1): 1–7.

Watson, D. (1946e) 'Starving in the midst of plenty', in *The Vegan*, 2(1): 8–9.

Watson, D. (1947a) *An Address on Veganism*, Ambleside, Westmorland: The Vegan Society.

Watson, D. (1947b) 'An appeal by the president', in *The Vegan*, 3(4): 7–8.

Watson, D. (1947c) 'The president's log', in *The Vegan*, 3(1): 12–13.

Watson, D. (1948) 'The president's log', in *The Vegan*, 4(2): 12–13.

Watson, D. (1988) 'Out of the past', in *The Vegan*, Summer 1988: 11.

Watson, D. (1989a) 'Retrospect', in *The Vegan*, Spring 1989: 13.

Watson, D. (1989b) 'Retrospect', in *The Vegan*, Autumn 1989: 21.

Watson, D. (1989c) 'Retrospect', in *The Vegan*, Winter 1989: 21.

11 On the limits of food autonomy

Rethinking choice and privacy

Stephanie Jenkins and Richard Twine

Introduction

Those familiar with traversing the still nascent field of animal studies at the ever increasing number of seminars and conferences will have, in all likelihood, encountered the elephant in the room of whether those present consume animals. Instead of being ill received as an awkward presence, animal studies scholars, of all people, ought to welcome this particular metaphorical animal not least because the questions raised are some of the most important to the future direction of the field. Yet questions around whether animal studies scholars should be consuming meat and other animal products are also germane to posing that same question to broader populations in society. For the first time, academia provides a host (if precarious) to a self-identified group of vegan scholars under the banner of critical animal studies (CAS). Simultaneously, contemporary Western (perhaps especially these given their histories of accumulated consumption) societies are facing (yet are largely facing *away* from) fundamental questions about normalised practices in domains such as energy generation, transport, but also food consumption and production.

Given that radical infrastructural change in food practices, inclusive of a turn away from animal consumption, could have important mitigation effects against climate change (e.g. Berners-Lee *et al.* 2012; Berners-Lee and Clark 2013: 158) it seems incumbent to better understand social inertia around these practices. Whilst such an understanding may give us some purchase on the internal debate within the field of animal studies this, whilst important to those of us involved, is parochial in the wider context of societal change or stasis. In this chapter we consider some of the conceptual framings that serve to reinforce hegemonic meat/dairy centric eating practices with the hope of evaluating key components of food norms. In particular we focus on the interconnected ideas of autonomy, privacy and choice. We see these as pivotal in perpetuating the depoliticization of human–animal relations and to the deflection of critical thinking around food practices. We offer this critique for the purpose of expanding areas of potential empirical research and for suggesting that, politically, it is these discourses of food that should be directly targeted.

As many people in animal studies have argued, the consumption of animals as food is so taken for granted, normalised and habituated that it simply is not reflected upon as a *relationship* between humans and other animals. There is potentially an analogue with other practices here, such as the everyday routinized use of electrical items not being associated with energy consumption per se, with methods of re-association being the hallmark of climate awareness campaigns. We are interested in the sort of discursive strategies that are brought into play when this routine objectification of animals is disturbed and questioned, be that for direct animal ethics reasons or for arguments associated with ecology (although we also suggest that projected impacts of climate change on all species shows this to be a false distinction). This entails an interest in those meanings and forms of background knowledge that people bring, but do not necessarily reflect upon, to their food practices.

Bound up in the assumption that consuming animals is just consuming food rather than a particular relationship with another animal is, we argue, an assumption of human autonomy from 'nature', which downplays the possibility that our food practices have ecological impacts that in turn shape the material context of human beings. Questioning this assumption of autonomy has been a central labour of environmental philosophy (especially those intersectional works that focus on dualism, e.g. Plumwood 1993) and the posthumanist contestation of human–animal dualism. Assumptions of human autonomy in the context of food practices are an example whereby these aforementioned ontological challenges (which are also political) pertain to the everyday of embodied social life. In seeking to better understand critiques of autonomy we opt at this juncture to turn to critical work in a largely non-food domain.[1] Whilst aware of traditional sociological work that has critiqued autonomy we turn to, for the most part, different iterations of feminist thought as they pertain to this area. We make this choice because we see in the feminist focus – on bodies, interdependency and anti-dualistic ontology – heuristic tools for approaching our food-related critique of autonomy. In drawing especially upon feminist bioethics, ecofeminism and posthumanist feminism, our objective is to contribute to the literature on food meanings, stressing the interplay of autonomy, privacy and choice in maintaining the normality and dominance of animal consuming food practices.

Feminist bioethicists' challenge to autonomy

In mainstream bioethical discourse, autonomy emerged as one of the main ethical principles – especially in the context of healthcare and medicine. The feminist critique of bioethics has argued that the choice of core principles such as autonomy has been shaped by the 'masculine marking of the generic subject conventionally assumed in ethics' (Rawlinson, 2001: 340–1). Empirical examination of the experience of differently positioned people reveals a much more complex picture that may question the value of autonomy and instead stress social interdependencies and, for example, values of care. This is not to detract from the value of autonomy in the sense of a check on medical authority and the empowerment of recipients of medical provision. The feminist critique is more concerned with the way in which a particular construction of masculinity has been taken as representative of the human. This, it is argued, is a classic notion of the masculine as abstract, asocial, individualistic and largely unencumbered by the messiness of affective social relationships (see Donchin 2001: 366). This tends to downplay the way in which every human is, at some point in their lives, likely to be responsible for or dependent upon others, and continually dependent on ecology and nonhuman others. Our social embeddedness calls into question the ability and desirability (in all contexts) of autonomy, instead stressing our inescapable vulnerability and contra-traditional conceptions of masculinity. One area where feminist bioethicists have been critical of the uses of autonomy has been within discourses around maternal/foetal relationality. Concern has been expressed over the way the foetus is sometimes subjectified as autonomous from the body of the mother, who may herself be subjected to a process of comparative objectification (Donchin 2001: 371). This could even be said to echo historically entrenched views of the female body as an instrument for the perpetuation of the royal line and the association of women as 'body people' (see Twine 2001), vis-à-vis the 'rational male'.

Consequently feminist bioethicists have wanted to explicate the context around a particular assumption of autonomy. Although they have, for the most part, yet to meaningfully extend this to the more-than-human[2] (Twine 2010a) or to the area of food practices, there is obvious relevance to rethinking assumptions of autonomy in such domains in a way that makes clear that bioethics must better incorporate environmental and public health considerations into its model of health (see Twine 2010b). Assumptions of autonomy in the food system clearly take place in the commodity fetishism of contemporary global consumerism, wherein we typically assume our well-being to be autonomous from the relations and conditions of production. Moreover autonomy, we argue, also inhabits the dualism between culture and nature, and between human and animal in the context of food consumption. It is obvious to say this, but very seldom do bioethicists make attempts to try and think about how autonomy could apply to nonhuman animals. Analogous to the critique from feminist bioethics, we argue that this sense of autonomy is similarly shaped by assumptions of a specific masculinity being generalised to the 'human' and that it similarly speaks to disavowed responsibility and denied interdependency (Donchin 2001: 375) and vulnerability. This assumption of autonomy is materially secured and reinforced by the social and spatial sequestration of animal killing and processing under the auspices of the animal-industrial complex (Pachirat 2011; Twine 2012).

Some of those writing in radical feminism and ecofeminism – important precursors to critical animal studies – also noted the role of the principle or norm of autonomy in naturalising the killing of animals for food consumption. In her important 1993 paper *The Feminist Traffic in Animals*, Carol J. Adams considers some of the strategies that have been used by some feminists to deflect attention away from the possibility of animal exploitation constituting an important part of the feminist intersectional vision (see also Twine 2010c). Indeed, in discussing the politicization of food practices at feminist conferences Adams shows that the CAS elephant in the room has some lineage. Moreover, of special interest to us is that Adams also highlighted autonomy, privacy and choice (she uses the term pluralism) as important discursive strategies used for depoliticising food practices, and, therefore, human–animal relations. Adams argues that, 'The invocation of autonomy – the insistence that enforcing vegetarianism at a conference restricts an individual's autonomy – presumes that no one else's liberty is at issue in food choices' (1993: 210). She further argues that a claim of autonomy in this context can be seen as a discursive strategy for renaming privilege and power. Yet non-vegan delegates at a conference who want to underline some sort of right to consume animals in that context are, we might assume, simply giving voice to a broader social norm that shapes one's food practices as largely autonomous from much in the way of critical thought about power, relationality and exploitation. Furthermore, we argue, and this ought to be seen as a peculiar practice for animal studies scholars, this perpetuates a view of human practices as autonomous from the more-than-human and fails to fully see the consumption of other animals as the reiteration and performance of human–animal relations. In order to understand inertia in food practices, we wish to explore this normative framing further by exploring related discourses of privacy and choice.

Eating practices as private

Eating in Western cultures is often deemed a private matter almost to the extent that it is seen as impolite, and taboo to question what another eats. Private here may mean that the topic is personal, but it may also pertain to its association with the domestic sphere assumed to be private. Even though a substantial amount of eating now takes place outside the home, and often in public places, food practices seem to partly retain their association with the private sphere. The privacy of the gendered familial sphere continues to hold sway over assumptions around food practices wherein their personalisation continues to assume to erect a boundary from the political. Feminists have obviously had a lot to say about the public/private divide and the depoliticisation of the private sphere, most notably in issues such as the social construction of gender and sexuality, domestic violence and the inequitable domestic division of labour (usually in heterosexual couples). How gender and other such inequities shape food purchasing and consumption practices is very complex. It has been suggested that women may be more likely than men to change eating practices according to ethical arguments and health advice, and that they tend to do most of the shopping (Beardsworth *et al.* 2002), that women may defer to their husbands' food preferences (Lynn Brown and Miller 2002) and that heterosexual coupling provides a context for varied and sometimes competing scripts of masculinity performed through food practices (Sobal 2005). Food privacy norms provide part of the context to such politics of the 'private' sphere. What CAS can add to feminist analyses of this sphere as a site of power is to point to its role in the normalisation of various human–animal relations. This undoubtedly includes the socialisation of young children into a culture's dominant ethical relation to other animals (Stewart and Cole

2009), and the intersection of human–animal relations with other constructions of difference such as gender or class.

Ecofeminist writers Carol J. Adams (1993) and Val Plumwood (1995), who have been influential upon CAS, have written on privacy, themselves influenced by earlier feminist work on the public/private distinction by, for example, Nancy Fraser (1987) and Carole Pateman (1983). Thus Adams talks of the way in which the 'purchasing, preparing and eating of food has been cast as a private-domestic matter' (1993: 199). Following Fraser, she also points to the economisation of aspects of the animal-industrial complex as another means by which they are depoliticised, for example, an understanding of the growth of factory farms in terms of the 'demands of the market' (ibid.). This mirrors contemporary assumptions about the so-called livestock revolution, which is often framed in terms of the growing prosperity of 'developing' nations entailing increased meat/dairy consumption, apparently without any influence from governmental and corporate actors in the political economy.

Plumwood's focus on the role of privacy in liberal democracy adds to this analysis. Her interest is specifically in the way in which norms of privacy serve to curtail democratic oversight. Drawing upon the work of Carole Pateman and others, Plumwood explores how privacy has, in certain contexts, come to be understood as non-interference. She writes:

> As a claim to exclude others from participation in the relevant sphere of decision-making, the liberal account of privacy has been used both to restrict the scope of the political, to effect the exclusion of women and other groups (slaves, servants, domestic animals) counted as part of the household, and also to relegate most economic affairs to the realm of the private, to exclude them from democratic control and prevent their realisation as political, as subject to social discussion and determination.
>
> (1995: 107)

As she makes clear, the understanding of privacy – be that to the invariably male head of a household or company – has traditionally been seen as a freedom from interference. We can see this in a central claim of neoliberalism, the desire for an unregulated 'free' market. Yet here, a conflation of privacy with freedom is masking of the protection of male economic and familial power. Privacy can be oppressive as much as it may be freedom enhancing. For Plumwood, this liberalist conception of privacy is a significant factor in both perpetuating social inequality and the human control over nonhuman nature. Demarcating a boundary of depoliticised privacy around the family and a substantial part of the economic sphere[3] exempts from oversight the two main sites of production and consumption, perhaps where human practices have most environmental impact (Plumwood 1995: 108). This calls into question the ability of contemporary, nominally democratic, societies to respond to the challenge of climate change.

What is also interesting, at least in relation to the domestic sphere, is that in spite of norms of privacy, the family as a social institution has obviously *also* been the object of governmental biopolitical scrutiny. This encompasses a wide range of attempted interventions – many centred on children, but also more generally, in a Foucauldian sense, on the regulation of the health of the population. Any account of the privacy of the domestic sphere must be qualified by this history of biopolitics. Moreover this has included governmental interest in the diet of populations. However, in spite of the nascent discourse of sustainable diets or ecological public health (Rayner and Lang 2012) that thinks critically about the ecological impact of what we eat, most biopolitical interest in Western diets has centred on particular interpretations of 'healthy eating', nutrition and, more recently, obesity. There has been very little focus on the carbon intensity of diets at the consumption end and official support for vegetarian or vegan diets, in spite of their lower carbon intensity, has been, as far as we know, close to non-existent. Promoting veganism is left to vegans themselves – obviously still a small, grassroots and under-resourced minority.

Within the politics of privacy, various politically diverse groupings attempt to police the boundary of acceptable intervention. Thus, from some libertarian and conservative perspectives, accusations of paternalism and nanny-state[4] criticisms attempt to protect what is framed as an 'attack on freedom'. In a different manner some on the left have framed such biopolitics, especially those pertaining to obesity or smoking, as being couched in classist assumptions and devoid of any questioning of the consumption habits of the wealthy. If the focus is on the energy intensity of different patterns of consumption, then this is a good point because high consumption tends to correlate with high incomes. However, there is a disconnect between most versions of 'sustainable diet' discourse, which usually shies away from fundamentally challenging norms of animal consumption in favour of reduction arguments, and CAS or vegan critiques that make it clear that autonomy and privacy norms are protective of abusive and exploitative practices, and that, fundamentally, animals should not be viewed as either property or food.

Consuming animals as 'freedom of choice'

As well as these arguments for autonomy and privacy, a third related discursive strategy, which similarly serves to depoliticise food practices, is to speak about food in terms of *freedom of choice*. This plays off autonomy in its assumption that the choices of the individual are somehow sovereign and free from much in the way of social or ecological consequence. In the smoking debate opponents have been able to rather successfully counter this in terms of a discourse of passive smoking. Whilst there is as yet no concept of 'passive eating', the way in which diets high in animal products (or involving a large degree of transportation and processing) disproportionately contribute to the cumulative carbon in the atmosphere, makes clear the strong potential for specific modes of eating to impact on others, human or otherwise. Various attempts have been made by bioethicists to move towards less individualistic understandings of the human consumption of other animals, including arguments for the taxation and banning of such consumption (see Deckers 2013; Nordgren 2012; Wirsenius *et al.* 2011). A classic defence by meat eaters is to declare that their food practices are a personal choice. Furthermore, vegans are often asked to 'respect' the choices of others as a means of closing down critical conversations.

However, what is lost from such a request is the recognition that for most consumers of animal products no choice as such has been made. Consuming animals is a dominant cultural practice, and so it is part of the set of normalised values and ontological distinctions of the culture we are born into. Vegans and, to an extent, vegetarians represent resistant socialisation 'failures', and it is typically these people (unless, for example, they have been raised in a vegan sub-culture) who have exercised agency and decided to go against the dominant norm. Discourses of choice de-socialise and personalise eating practice as a means of attempting to remove them from the political.

A posthumanist critique of food autonomy

Examining posthumanist critiques of autonomy, particularly in the work of Judith Butler, reveals two normative assumptions that undergird the norm of food privacy. First, food autonomy is a vital expression of personal freedom, encompassing a wide range of morally acceptable choices. Second, the norm of food privacy extends the right to be left alone in food selection; freedom from interference means not only protection from governmental regulation, but also isolation from social judgment, even when eating in a public space.

A critical animal studies analysis reveals both of these assumptions to be false. Not only is the omnivore's 'freedom' to eat bacon a perversion of the concept of freedom itself, but meat eating enacts a double assault on autonomy. While we will hold off on an in-depth analysis of food autonomy in general, we believe that a 'freedom' that requires the suffering and slaughtering of an animate creature cannot be considered a freedom in any sense, because it inherently conflicts with a moral other's rights to life, bodily integrity and autonomy. In short, eating bacon harms the pig. Furthermore, given the ubiquity of compulsory anthropocentrism, omnivores do not exercise autonomy in any meaningful way when rehearsing its habits or consuming its unintelligible violence.[5] Although the limited ability of consumers to ensure fairness in global food networks cuts across dietary practices, the consumption of animals for food differs morally from the use of products with controversial production practices such as bananas and palm oil. While it may be possible to cultivate and harvest plant products through fair and ecologically sound methods, eating animals is inseparably intertwined with the suffering, death and exploitation of sentient life.

Appeals to food privacy function to impede reflective engagement and deflect moral criticism back to the interlocutor; questioning dietary habits exceeds the boundaries of acceptable political discussion. As a pleasure of the (eater's) body, food choice is safeguarded from surveillance, interference and judgment by a zone of privacy. The norm of food privacy enacts a double assault on autonomy:

first in the ontological exclusion of animals from the moral community, and second, through compulsory participation in anthropocentric violence.

Claims to food privacy require that nonhuman animals are recognized only as property, not as living, embodied creatures demanding our respect, protection and solidarity. Food autonomy is believed to be immune from public criticism (e.g. 'don't judge my diet') because 'no one' is harmed in its exercise. This perspective, of course, presumes that nonhuman animals do not count as 'someone' and that their pain and deaths do not rise to the level of morally impermissible harm (see Jenkins 2012). The prosaic instrumentalisation and commodification of nonhuman corpses as meat conceals anthropocentric violence because its harms are not perceived as violations; animal pain does not count as suffering, and mechanised 'rendering' of animals is not recognised as mass murder.

Human exceptionalism, as a worldview distributing moral worth only within the human 'family', perpetuates itself through the unquestioned and widespread treatment of animals as raw materials for our entertainment and consumption. The incessant repetition of speciesist privilege through 'harmless' daily practices like shopping and eating normalises the devaluing of, and disregard for, nonhuman creatures. As Nicole Shukin argues in *Animal Capital* (2009), the animal-industrial complex commodifies, disassembles and fetishises animal bodies into discrete consumable products; we do not purchase corpses, but meat, leather, fur and dairy.

Food privacy reinforces hegemonic animal-centric eating practices under the pretence of a moral right that systematises the exclusion of nonhuman animals from the moral community. Not only are animals' interests excluded from the evaluation of rights claims, but also the biopolitical commodification, management and industrialisation of animal lives are constitutive of the normative apparatus itself. Consequently, moral discourse reinforces the exploitation of animals because their bodies become a material condition for human autonomy. When animals are deemed unworthy of moral consideration, calls to protect their well-being are interpreted as threats to human autonomy. Simply stated, social acceptance of rights for nonhuman animals would delegitimize the right to consume their flesh. The exclusion of animals from moral consideration is therefore a material condition for food autonomy and privacy. Because the consumption of animals is marked as a personal choice, food privacy has 'been immunized against political challenge' (Butler 1998: 523).

CAS demonstrates that the decision to be vegan or to remain complicit with anthropocentric dietary norms cannot be described as an amoral, personal lifestyle choice. Deploying privacy as a strategy for deflecting criticism incorrectly assumes that such food choices fall within the boundaries of acceptable autonomous actions. As the taboos against eating other humans and companion species demonstrate, food autonomy does not translate into the license to eat whatever (or whomever) one wants. The prohibition against cannibalism is not encountered as a limitation on human freedom, because this food 'choice' is incompatible with the ownership, exploitation and murder of fellow human beings. Humans are not food and the desire to consume the flesh of *homo sapiens* is pathological.

In her books *Precarious Life* (2006) and *Frames of War* (2009), Judith Butler refocuses ethics to consider 'who' counts as a morally considerable life. Her social ontology of the 'war on terror' provides a model for an other-focused ethics that problematises our norms for moral perception and performance. Examining the frames of moral responsiveness reveals how relations of power, regimes of knowledge and modes of subjectivation distribute others on a continuum of attention, response and concern. Following Butler, we understand moral responsiveness as the affective capacity to respond to others that is conditioned, shaped and sustained by historically contingent frames differentially valuing others as respond-able or as others as *moral* others, who demand consideration. This methodology helps to articulate how human exceptionalism operates to prevent encountering animals as morally considerable others. Because a universalised concept of humanity governs entrance into the moral community, Butler argues that norms 'have developed historically in order to maximize precariousness for some and minimize precariousness for others' (Butler 2006: 12–13).

Specifically, if nonhuman beings are 'not conceivable as lives within certain epistemological frames', then their lives do not count as worthy of protection and their deaths are not recognised as grievable (Butler 2009: 1). The boundary between moral and amoral others determines whose interests receive normative protection through the rights to life, autonomy and privacy. This means that, first, the frames through which lives are perceived as considerable or not are 'themselves operations of power'. Second, because life itself is 'constituted through selective means ... we cannot refer to this "being" outside of the operations of power' (ibid.). At the same time that moral frames facilitate our abilities to experience vulnerable others, they ontologically exclude those deemed inhuman and differentially distribute moral attention, concern and protection accordingly. As a result, the very conditions of being human that award moral status to some animate beings, 'deprive certain other individuals of the possibility of achieving that status, producing a differential between the human and the less-than-human' (Butler 2004: 2). The ontological invalidation of animal life enables the presumption that eating meat harms 'no one'.

As Butler explains, the 'human' functions as a normative, and not merely biological or morphological, concept. Humanity's high esteem for capacities believed to be unique to humans, such as rational thought, justify the superiority afforded to *homo sapiens* and define autonomy as freedom from nature. The figure of the inhuman animal separates civilised 'Man' from the assumed violent chaos and determinacy of the natural world. Consequently, our self-understanding and moral frameworks presuppose the devaluation of nonhuman animal bodies. As Butler points out, 'For the human to be human, it must relate to what is nonhuman, to what is outside itself but continuous with itself by virtue of an interimplication in life' (Butler 2004: 12). The power/knowledge apparatus of species classification naturalises anthropocentric evaluative criteria and invalidates bodily vulnerability and interdependency. Human exceptionalism, 'as a form of differential treatment...' frames the animal as '...the other against whom (or against which) the human is made. It is the inhuman, the beyond the human, the less than human, the border that secures the human in its ostensible reality' (Butler 2004: 30).[6]

In addition to exempting our commodification and instrumentalisation of non-human animals from moral sanctions against violence, the human-centric frames through which we recognise moral others significantly restricts human agency. Because subjectivity is produced by anthropocentric power networks, our receptivity to animal others is impeded at the level of perception. Moreover, for those individuals who are 'failures' of speciesist socialisation, global food production precludes the possibility of a non-violent relationship to nonhuman animals.

In *Frames of War*, Butler argues the 'tacit interpretive scheme': constituting ontological obstacles to the recognisability of certain lives impedes the autonomy of moral subjects to reflectively question and revise the boundary between moral and amoral others (2009: 51–2). A social ontology, which precedes the life of any human individual, frames the conditions for the possibility of moral response. As the affective, embodied capacity to respond to others, moral receptivity is conditioned, shaped and sustained by historically contingent frames that differentiate between others who require moral care and amoral beings who are available as instruments. Because speciesist relations of power, regimes of knowledge and modes of subjectivation invalidate animal lives, human superiority is incorporated into human self-understanding. The exclusion of animals from the moral community establishes the unconscious habits of perception, attention and response that constitute ethico-political experience and practice.

The unintelligibility of speciesist violence against animals, through the circumscription of moral frames, permits what Agamben terms the 'anthropological machine' (2002: 33) to function without detection, producing 'zones of indifference' (2005: 23). Existing in these amoral zones of indifference, nonhuman animals are not merely exclusions, but persist as the constitutive outside of humanity. Because the lives of animals are derealised and unrecognisable, they exemplify Agamben's concept of 'bare life' (1998), or biological resources that humans can use to pursue their own desires, goals and projects. Existing merely as equipment or raw material, the less-than-human world is an amoral realm that cannot be 'harmed' or 'violated' because such descriptive categories demand the ethico-ontological judgment that nonhuman creatures are worthy of moral consideration. The anthropological machine produces our moral frames. Because nonhuman animals are experienced as alien to moral community, their lives are available for biopolitical management, industrialisation and commodification.

The norm of food privacy assumes that food choice is an extension of personal autonomy. According to this liberal humanist conception of autonomy, food privacy permits individuals to intentionally select their diets without external influence. Individual dietary preferences, in this view, are distinct and private lifestyle choices. So long as the decision to abstain from animal products expresses only personal values, food privacy, at least in theory, should shelter vegans from public interrogation; at the same time, food privacy works to minimise public attention given to vegan practices. It should be noted; however, that in some instances, even the most apologetic veganism, as a departure from cultural food norms, violates social mores of politeness and civility. In any case, explicitly advocating veganism as an ethico-political practice of lesser- or non-violence disrupts the presumption of food privacy in nearly all food-related settings (see Cole, Chapter 10 this volume).

The presumptive norm of food privacy, expressed as the entitlement to consume animals without public judgment or regulation, camouflages compulsory anthropocentrism as freedom from interference. The public/private distinction grounding the norm of food privacy inaccurately assumes a liberal understanding of freedom as self-determination without outside influence. Because individuals are not discrete, atomistic units isolated from the social world, a simplistic appeal to the personal nature of diet operates according to an inaccurate conception of autonomy that neglects the productive role of normalisation. As a result, the deployment of food privacy mistakes participation in the hegemonic practice of meat eating as personal freedom, despite the fact that anthropocentric dietary preferences are always already incorporated into embodied selves long before we acquire conscious knowledge of what food is and where it comes from. The vast majority of humans' first interaction with animal others comes in the form of disembodied and deanimated parts (i.e. hamburgers or chicken wings) or products (i.e. eggs or milk). The taste and desire for animal flesh nourishes our bodies before we realise that our favourite foods require the death of another animated being.

By relegating food choice to the personal realm, food privacy depoliticises the exclusion of animals from the moral community. It is, however, through everyday personal habits that anthropocentrism is performed and perpetuated. Speciesist violence is justified through its banality and a 'might makes right' framework. In other words, the practice of eating animals justifies our entitlement to the pleasures of consuming animal flesh. As Chloe Taylor demonstrates:

> It is not the case that [humans] first determine that we are superior to non-human animals and then we conclude that we have the moral license to eat them. Rather, it is through our very eating of other animals that we constitute our superiority. According to this logic, we must be superior to other animals since we put them in cages and do horrible things to them. Human superiority is not a fact from which the permissibility of our practices is deduced; on the contrary, human superiority is something which we construct through our instrumentalization of other species.
>
> (2010: 75)

The lives of animals are ontologically invalidated by personal acts, in private spaces, by people we love. Yet, because such a micropolitics is imperceptible to liberal accounts of autonomy, food privacy norms disguise murder as (sometimes humane) slaughter with each meal, bite by bite (see Pachirat 2011).

The violent relation of 'non-relation' to our fellow creatures is not the result of a conscious, intentional individual choice but, rather, is given to us by the social world in which our selves emerge. As Butler explains, a 'constitutive sociality of self' precedes conscious self-awareness and autonomy (Butler 2004: 19). Because we become selves in constitutive relations to others, the self is thoroughly social in a way that problematises a clear division between the public and private self. In fact, the sociality of self challenges the liberal conception of subjectivity presumed by notions of food privacy. The individual is not the sole initiator of food choices and taste preferences but, rather, is ontologically constituted within a complex, relational network of unidirectional violence against nonhuman animals.

Social norms governing the allocation of moral concern and value fashion our existence before birth and beyond death. The frames of responsiveness facilitate our encounters with members of the moral community, while simultaneously obstructing this experience with animal others. Because these 'judgments' occur at the level of habituated bodily reactions, the craving of our moral cartography takes place 'below' conscious perception and prior to the self-awareness required for autonomous action. Consequently, participation in animal-centric eating practices is not an exercise of self-determination, but the path of least resistance. As author Melanie Joy points out, there is no free choice without awareness. 'When an invisible ideology guides our beliefs and our behaviors, we have become casualties of that system that has stolen our freedom to think for ourselves and to act accordingly' (Joy 2011: 64). Just as it makes little sense to say that one 'chooses' to be heterosexual in a heteronormative world, the continuance of speciesist dietary norms cannot be considered to be autonomous.

In a discussion of gender norms, Butler notes that, 'The staging and structuring of affect and desire is clearly one way in which norms work their way into what feels most properly to belong to me' (Butler 2004: 15). Similarly, anthropocentric techniques of power target, act upon, organise and produce the human body and its affective capacities, including perception and sensation. The speciesist moral hierarchy marking lives as considerable or not is transmitted through bodily capacities, responses and affects. According to Butler, this moral framing dividing:

> worthy from unworthy lives works fundamentally through the senses, differentiating the cries we can hear from those we cannot, the sights we can see from those we cannot, and likewise at the level of touch and smell. War sustains its practices through acting on the senses, crafting them to apprehend the world selectively, deadening affect in response to certain images and sounds, and enlivening affective responses to others.
>
> (2004: 51–2)

This kind of normalising power is more effective than juridical restrictions, because power must be experienced as a limitation before it can be resisted.

According to Taylor, 'Eating habits, like sexual habits, are affective, as well as a key part of our involuntary corporeal constitution by others' (2010: 78). Dietary discipline produces taste preferences through 'incitement and intensification' (Foucault 1990: 11) and, consequently, 'are highly difficult to get away from because they have become our habitual means of relating to our bodies, emotions, and selves' (Taylor 2010: 78). Pleasures of the body are inseparable from the constitutive sociality of self and its production within relations of power; the self is not only formed through its subjection to social norms, but also is affectively attached to its own subjection. Meat eaters will insist on their right to enjoy animal flesh without the interference of judgment or 'guilt-tripping'. Yet, even though they actively desire the taste, smell, visual presentation and texture of meat, we should not mistake this appearance of choice as autonomy; anthropocentrism acquires its dominance through the desirability of its effects. Power, as Foucault explained, becomes more efficient, mobile, continuous and pervasive when subjects voluntarily follow in accordance with norms. As a result, the loss of autonomy is more extensive than the restrictions of an external barrier, because it is 'hidden' from the self in its very constitution. In short, if a consumer craves a hamburger, voluntarily ordering one at a restaurant and enjoying its taste masks a more profound loss of freedom found in the inability to encounter the cow as a moral other and the unintelligibility of that animal's death. To paraphrase Foucault in a different context: we must not think that by saying yes to meat, one says no to power; on the contrary, one tracks along the course laid out by the deployment of anthropocentrism.[7]

Conclusions

Just as autonomy and privacy have pretensions toward being unambiguous promoters of freedom, the *freedom* to choose foods is firmly established as a norm in the context of first world nations in a globalised economy. However, we have argued that it is precisely opening up food practices to critical interrogation that can further a progressive (non-anthropocentric) notion of freedom. Assumptions amongst meat eaters that they are exercising something akin to free will ignores the political context of eating and the normative environment that steers people in the direction of particular hegemonic eating practices. Choice here is obviously anthropocentric, since the consumer is legally sanctioned and culturally encouraged to make a decision that pays for people to kill animals on their behalf. It cannot be a personal act when it is 'killing at a distance'. It is a choice that involves a whole series of decisions and includes many other people in the commodity chain and ultimately the practice of animal killing. It is then, literally, a highly social *and* impersonal practice with a whole series of consequences. What capitalism delivers so skilfully to the consumer is the disavowal of responsibility to reflect upon or even to fully know about these consequences.

We should not overstate the role of rational reflection and knowledge in such food practices. Instead we suggest that it is precisely their normative and habitual features that make them such successful practices, together with the cultural exclusion of alternative practices. As Grillo writes on this subject, *Nothing could be more public than the taking of a sentient life that cares about his own life, particularly when the act is not necessary and nor therefore morally defensible.*[8] Wrapping up food practices in discourses of autonomy, privacy and choice ultimately protects the consumer from thinking about how their practices are enmeshed within systems of violence against other animals and helps to secure the *habitual* life of these consumption practices. Eroding the power of this discursive framing must continue to be a target for CAS, both conceptually and empirically.

Notes

1 However, we do think feminist academics should be engaging more with food issues.
2 Rawlinson's use of Irigaray to point to intersections between gender and environmental exploitation is one exception (2001: 410).
3 The privacy of the economic sphere is undoubtedly a source of its power. CAS researchers certainly come up against this when trying to research the details and relationships of the animal-industrial complex (see Twine 2012).
4 The term 'nanny-state' could itself be seen as a misogynist jibe against the power of women. We thank Matthew Cole for pointing this out.
5 The term 'compulsory anthropocentrism' recalls Marti Kheel's discussion of compulsory meat eating (2008: 236). In a description of the "unprecedented proportions" of human violence against non-human animals, Derrida writes:

> [M]en do all they can in order to dissimulate this cruelty or to hide it from themselves; in order to organize on a global scale the forgetting or misunderstanding of this violence, which some would compare to the worst cases of genocide.
>
> (2008: 25–6)

6 It is important to note that while Butler's analysis of the process of derealisation and dehumanisation provides a rich conceptual tool for evaluating human exceptionalism, she has not expanded her arguments to their necessary, and logical, application to non-human animals. That being said, James Stanescu (2012) outlines the advantages and limitations of using Butler's ethical work to understand mechanisms of animalisation.
7 The original text states: 'We must not think that by saying yes to sex, one says no to power; on the contrary, one tracks along the course laid out by the general deployment of sexuality.' (Foucault 1990: 157).
8 See http://freefromharm.org/animal-products-and-psychology/five-reasons-why-meat-eating-cannot-be-considered-a-personal-choice/ (accessed 16 December 2013).

References

Adams, C. J. (1993) 'The feminist traffic in animals', in Gaard, G. (ed.) *Ecofeminism: Women, animals, nature*, Philadelphia: Temple University Press.

Agamben, G. (1998) *Homo sacer: Sovereign Power and Bare Life*, Palo Alto, CA: Stanford University Press.

Agamben, G. (2002) *The Open: Man and Animal*, Palo Alto, CA: Stanford University Press.

Agamben, G. (2005) *State of Exception*, Chicago: University of Chicago Press.

Beardsworth, A., Bryman, A., Keil, T., Goode, J., Haslam, C. and Lancashire, E. (2002) 'Women, men and food: the significance of gender for nutritional attitudes and choices', *British Food Journal*, 104(7): 470–91.

Benn, S. and Gaus, G. (eds) (1983) *Private and Public in Social Life*, London: Croom Helm.

Berners-Lee, M. and Clark, D. (2013) *The Burning Question – We Can't Burn Half the World's Oil, Coal and Gas. So how do we Quit?* London: Profile Books.

Berners-Lee, M., Hoolahan, C., Cammack, H. and Hewitt, C. N. (2012) 'The relative greenhouse gas impacts of realistic dietary choices', *Energy Policy*, 43: 184–90.

Butler, J. (1998) 'Performative acts and gender constitution: an essay in phenomenology and feminist theory', *Theatre Journal*, 40(7): 519–31.

Butler, J. (2004) *Undoing Gender*, New York: Routledge.

Butler, J. (2006) *Precarious Life: The Powers of Mourning and Violence*, Brooklyn, NY: Verso.

Butler, J. (2009) *Frames of War*, Brooklyn, NY: Verso.

Deckers, J. (2013) 'Obesity, public health, and the consumption of animal products – ethical concerns and political solutions', *Journal of Bioethical Inquiry*, 10: 29–38.

Derrida, J. (2008) *The Animal That Therefore I Am*, Bronx, NY: Fordham University Press.

Donchin, A. (2001) 'Understanding autonomy relationally: toward a reconfiguration of bioethical principles', *Journal of Medicine and Philosophy*, 26(4): 365–86.

Foucault, M. (1990) *The History of Sexuality, Vol. 1: An Introduction*, New York: Vintage.

Fraser, N. (1987) 'What's critical about critical theory?' in Benhabib. S. and Cornell, D. (eds) *Feminism as Critique*, Oxford: Polity Press.

Gaard, G. (ed.) (1993) *Ecofeminism: Women, Animals, Nature*, Philadelphia: Temple University Press.

Jenkins, S. (2012) 'Returning the ethical and political to animal studies', *Hypatia*, 27(3): 492.

Joy, M. (2011) *Why We Love Dogs, Eat Pigs, and Wear Cows*, Newburyport, MA: Conari Press.

Kheel, M. (2008) *Nature Ethics – An Ecofeminist perspective*, Lanham: Rowman and Littlefield.

Lynne Brown, J. and Miller, D. (2002) 'Couples' gender role preferences and management of family food preferences', *Journal of Nutrition Education and Behavior*, 34(4): 215–23.

Nordgren, A. (2012) 'Ethical issues in mitigation of climate change: the option of reduced meat production and consumption', *Journal of Agricultural and Environmental Ethics*, 25(4): 563–84.

Pachirat, T. (2011) *Every Twelve Seconds – Industrialized Slaughter and the Politics of Sight*, New Haven: Yale University Press.

Pateman, C. (1983) 'Feminist critiques of the public/private distinction' in Benn, S. and Gaus, G. (eds) *Private and Public in Social Life*, London: Croom Helm.

Plumwood, V. (1993) *Feminism and the Mastery of Nature*, London: Routledge.

Plumwood, V. (1995) 'Feminism, privacy and radical democracy', *Anarchist Studies*, 3(2): 97–120.

Rawlinson, M. (2001) 'Introduction', *Journal of Medicine and Philosophy*, 26(4): 339–41.

Rayner, G. and Lang, T. (2012) *Ecological Public Health – Reshaping the Conditions for Good Health*, London: Routledge.

Scully, J., Baldwin-Ragaven, L. and Fitzpatrick, P. (eds) (2010) *Feminist Bioethics: At the Centre, at the Margins*, Baltimore: Johns Hopkins University Press.

Shukin, N. (2009) *Animal Capital: Rending Life in Biopolitical Times*, Minneapolis: University of Minnesota Press.

Sobal, J. (2005) 'Men, meat, and marriage: models of masculinity', *Food and Foodways: Explorations in the History and Culture of Human Nourishment*, 13(1–2): 135–58.

Stanescu, J. (2012) 'Species trouble: Judith Butler, mourning, and the precarious lives of animals', *Hypatia*, 27(3): 567–82.

Stewart, K. and Cole, M. (2009) 'The conceptual separation of food and animals in childhood', *Food, Culture and Society*, 12(4): 457–76.

Taylor, C. (2010) 'Foucault and the ethics of eating', *Foucault Studies*, 9: 71–88.

Twine, R. (2001) 'Ma(r)king essence: ecofeminism and embodiment', *Ethics and the Environment*, 6(2): 31–58.

Twine, R. (2010a) 'Broadening the feminism in feminist bioethics' in *Feminist Bioethics: At the Centre, at the Margins*, edited by Scully, J., Baldwin-Ragaven, L. and Fitzpatrick, P. Baltimore: Johns Hopkins University Press.

Twine, R. (2010b) *Animals as Biotechnology – Ethics, Sustainability and Critical Animal Studies*, London: Routledge/Earthscan.

Twine, R. (2010c) 'Intersectional disgust? – Animals and (eco) feminism', *Feminism and Psychology*, 20(3): 397–406.

Twine, R. (2012) 'Revealing the "animal-industrial complex" – a concept and method for critical animal studies?' *Journal for Critical Animal Studies*, 10(1): 12–39.

Wirsenius, S., Hedenus, F. and Mohlin, K. (2011) 'Greenhouse gas taxes on animal food products: rationale, tax scheme and climate mitigation effects', *Climatic Change*, 108(1–2): 159–84.

12 The radical debate

A straw man in the movement?

Carol L. Glasser

Critical animal studies (CAS) is multidisciplinary in its theoretical and methodological foundation. However, not all disciplines are participating equally, with humanities at the forefront, the social sciences treating nonhuman animals as an occasional and often marginalized topic, and a near absence of respectful non-abusive treatment of animals as subjects in the hard sciences. As a sociologist, I am particularly interested in bringing the social sciences into the field of CAS. In Chapter one of this volume, Cudworth notes: "Whilst sociology has broadened its subjects, objects and processes of study, it has held fairly fast to the conception of the social as centered on the human" (p. 19).

Nonhuman animals are in our homes, both displaced by and living in our built environment, the subject of laws and policies, and the centre of fads and subcultures – and so they should be of greater interest to those who study society and culture. Cudsworth details a history of the study of animals in sociology, highlighting the treatment of animals in key sociological areas of inquiry, including a treatment of speciesism and an extension of theories of oppression and capitalism (e.g. see Clark, Chapter 7 this volume).

This work has paved the way for nonhuman animals to potentially become a serious focus of inquiry for sociology. As I was writing this chapter, a major social theory journal, *Sociological Theory*, published an article calling for a social theory that argues for incorporating Anthrozoology into a more comprehensive social theory (see York and Mancus 2013). While Anthrozoology lacks the critical subject-driven nature of CAS, it highlights the growing acceptance of the 'animal turn'.

This turn, however, has remained largely couched in the theoretical realm. What social science can uniquely add to the field of CAS is a set of methodologies and empirical evaluations to test theories and also to directly improve outcomes for animals. A few social movement organizations have embraced social science methodologies to improve work among activists and outcomes for animals. For example the US organization, ASPCA (American Society for the Prevention of Cruelty toward Animals), has a research branch that studies issues of relevance to adopting animals, including how names effect adoption rates (see ASPCA 2009) and effective ways to get people to place collars and tags on adopted animals (see Weiss *et al.* 2011). Sociologists working in academia have also employed empirical analyses of issues of relevance to animals and activists. Some examples include (but are not limited to) Irvine's work on desensitization to animals' death in 4H clubs (see Ellis and Irvine 2010) and the role of companion animals for homeless populations (see Irvine 2013) and the work of Fitzgerald *et al.* (2009) examining the impact that slaughterhouse work has on crime rates.

Though empirical investigation into animals and the movement that advocates for them is emerging, it remains scarce. This chapter seeks to add to empirical investigations that can benefit the field of CAS and, by extension, the animals who are its subject. This chapter also seeks to add to scholarship regarding the investigation of the social movement for animals.

An anthology addressing critical animal studies, particularly as it is unique from animal studies more broadly, would be remiss to neglect a discussion of the social movement that seeks to better the institutional and social status of non-human animals. For a core difference between the animal studies scholar and the critical animal studies scholar is an intended commitment to praxis. Praxis is the application of theory to action and vice versa. Social movements are a catalyst of social change and an avenue of praxis and so they must be better understood. The literature on critical animal studies often overlooks social movements and the existing literature addressing social movements rarely addresses the animal rights movement (ARM) (but see Jasper and Nelkin 1992; Wahlstrom and Peterson 2006). This chapter seeks to fill this void, as well as to add to the body of social science and empirical research within the field of CAS.

The animal rights movement contends with a number of crucial issues that are understudied within the social movements literature addressing democratic nations, including the effectiveness of activist tactics, the role of radical activism and state repression of activism, to name a few. Within the animal rights movement (ARM) and among activists there is much debate over the efficacy of radical activism. For some, these activists and actions represent a heroic ideal, for others they signify violence. At its best, such debate allows for growth and refinement, and helps to generate movement diversity. At its worst, it can damage the spirit and cohesion of a movement, rendering it powerless to confront the various factions' common enemies. Even so, there is a lack of empirical research that addresses the radical debate.

This chapter empirically examines some of the key debates over the role of radical and illegal activism on behalf of animals in the United States. The chapter first identifies problems with the debate as it is currently unfolding, including a lack of uniform definitions for controversial terms, incomplete historical knowledge and tendency to forward anecdotes as evidence. I then outline key claims by both sides of the debate that are empirically testable, taking seriously each perspective as potentially valid. I test the validity of these claims, using two primary datasets – one tracking illegal radical activism in the US from 1990–2010, another tracking media coverage of a sample of these actions.

Direct action and radicalism

Direct action (DA) is a term that has been used to describe the Boston Tea Party, Suffragists, conscientious objectors and various tactics utilized by sectors of the civil rights, anti-abortion, anti-nuclear, environmental and animal rights movements. Though the term is regularly used, it is rarely explicitly defined. For example, Tracy's (1996) book about radical pacifism is titled *Direct Action* but does not forward a concise definition of the term. Similarly, Epstein's (1991) research on the "direct action movement" deals exclusively with pacifist forms of protest, such as civil disobedience utilized in the anti-nuclear movement, but does not forward a cohesive definition of the term or address more controversial DA tactics embraced by the same movement.

In the case of the ARM, a direct action is any tactic that is explicitly aimed at stopping someone accused of animal exploitation (be it an industry, business or individual) or at aiding an exploited animal. Rather than attempting to change long-term cultural attitudes or the political environment (though these may be tertiary goals or latent outcomes), DA is aimed directly at the site of oppression, seeking to immediately impact the exploitation experienced by nonhuman animals. These actions may or may not be violent (regardless of how that is defined) and may or may not be legal (see Best and Nocella 2004: 17).

In the mainstream animal rights movement, the acceptability of DA tactics is often determined by whether actions are perceived as violent, if they are engaged in anonymously and if they seem likely to garner potential bystander sympathy. The tactics most typically considered radical include animal liberations, open rescues, electronic invasions (hactivism, black faxes, etc.), economic sabotage, property damage, tertiary targeting, bomb threats, personal threats, home demonstrations and office invasions.

The use of the word "violent" comes up frequently in these discussions but is rarely defined and is used inconsistently across academic and activist discussions over tactics. Often the term is used rhetorically to validate a point of view; actions acceptable to the author are defended as non-violent, while those disapproved of are labelled violent. However, the use of these terms does not always fall uniformly along ideological lines. Some who oppose the actions of the radical flank do not find them violent, while some may support the use of radical tactics even while characterizing them as violent.

A DA tactic becomes "radical" when it falls outside the range of what is socially accepted as legitimate means of protest. Every movement has fringe actors who engage in unpopular tactics and will not be pacified by normal means of control, including political, rhetorical or economic sanctions intended to prescribe a movement's boundaries through punishment, or via concessions that prescript the depth of claims that will be tolerated. Those who reject repression and do not accept concessions tend to also maintain a more dogmatic ideology about a movement's goals. They are also often willing to utilize tactics outside the rubric approved of by moderate organizations and activists. It is these actors and their tactics that are considered radical.

Importantly, whether or not a DA tactic is radical does not always correlate to whether it is legal. The popular acceptance of social movement tactics is historically, culturally and contextually dependent. *Sytyagraha* (pacifist resistance practiced by Mahatma Gandhi) and civil disobedience (practiced by Martin Luther King Jr), though typically illegal, are often addressed as the ultimate forms of DA and are not considered radical. Conversely, demonstrations outside the homes of animal researchers, while legal, are considered radical.

There is much debate in the ARM and other movements as to the impact of radicalism. In one of the first studies addressing the impact of radicalism, Gamson (1975) examined 30 different movements and found that radicalism had little impact. Since then, however, most scholars have found that radicals have a positive impact on movements, though the concept has remained understudied and under-theorized in social movements literature. Haines (1984, 1988) conducted the best-known research on the subject and coined the term "radical flank". He examined how the presence of a radical flank influenced the moderates of the US civil rights movement from 1954–1970. He introduced the concept of a "radical flank effect", identifying that the presence of a radical faction willing to engage in aggressive tactics directly affects the moderate faction of the movement (ibid.).

A negative radical flank effect occurs when the presence and actions of the radical flank "undermine[s] the position of moderates by discrediting movement activities and goals, and by threatening the ability of moderates to take advantage of the resources available from supportive third parties" (Haines 1984: 32). A positive radical flank effect occurs when the presence of radicals gives moderates more strength by increasing their bargaining power, either by bringing attention to their cause or by making them seem more reasonable in comparison.

Haines (1988) found a positive radical flank effect in the US civil rights movement; as militancy and radical actions increased, the moderates benefited. Compared to radicals, moderates of the movement had more tolerable goals and were easier to work with, thereby encouraging the state to make concessions to them. Haines (1984) also found that in times of increased radicalism, donations to all moderate groups also increased, particularly to the most moderate. Minkoff (1994) found a similar positive effect in her study of advocacy and service groups in the US women's movements of the 1950s and 1960s – advocacy oriented groups increased in number and support as protest organizations formed.

Haines established two possible effects of the radical flank – moderates benefit (positive) or moderates suffer (negative). Gupta (2002) built on this and identified four possible outcomes: (1) moderates benefit at the expense of radicals; (2) radicals benefit to the detriment of moderates; (3) both moderates and radicals gain and the overall social movement is strengthened; or (4) the movement is weakened overall, with both moderates and radicals losing.

Dillard's (2002) study of the rhetorical dimensions of radical flank effects in the animal rights and environmental movements found the third scenario (i.e. simultaneous benefit) to be most common. Not only did moderates enjoy resource-based positive radical flank effects as proposed by Haines, but radicals also introduced new rhetoric that brought deeper cultural and political challenge. Once this rhetoric was introduced by the radicals, it was eventually embraced and utilized by the moderates. Radicals also benefited because moderates, who had access to political processes and political elites, embraced some of the radicals' rhetoric.

Radicals make moderates appear more tolerable and provide a diversion from the moderate factions for state authority seeking to deter and control dissent. For example, the politics of Martin Luther King Jr, were initially detested by the government – King had an FBI-sanctioned wiretap on his phone, the Kennedy and Johnson administrations spoke ill of him and the Montgomery Bus Boycott was considered radical (Garrow 2002). However, in light of a growing radical faction of the civil rights movement (e.g. SNCC, Black Panther Party), King was begrudgingly endorsed and accepted, meeting personally with President Johnson and other political elites. Similarly, the environmental movement has been embraced by the government via 'green subsidies', while the most radical of activists, such as tree-sitters or those who physically or economically sabotage development, are actively repressed.

The debate over radicalism in the ARM exists on both a principled and a pragmatic level (Best and Nocella 2004). On a theoretical level, the debate centres on whether the actions of the radical flank fit within the ideological framework of an animal rights ethic. On a pragmatic level, the debate is whether radical flank actions accelerate or deter the goals of the movement.

The principled critique against radical direct action (RDA) maintains that illegal actions such as property damage, economic sabotage and arson are detrimental to the movement's goals and violate the moral imperative of the ARM:

> [W]hen a few people use violence and intimidation to achieve the desired goal, they undermine the animal movement's ethical basis. In a democratic society, change should come about through education and persuasion, not intimidation.
>
> (Singer 2004)

The pragmatic objection to such tactics is rooted in the concern that these actions may hurt or torment people or that they will alienate the public:

> Activities, for instance, such as break-ins, fires, things like that. Threats of violence – those are the types of activities that harm animals ... We throw out so much of the public who would otherwise be for animal protection if they see animal advocates as being, in their view, violent...
>
> (Josh Balk, Outreach Director of the HSUS Factory Farming Campaign, Interview 2010)

The principled argument in favour of RDA centres on the belief that if one truly believes nonhuman animals to be the moral equivalent of humans, to be morally consistent one should embrace any tactic necessary to stop their exploitation, particularly given the extremity of the abuse and the number affected. From this perspective, the ARM has been rhetorically and even theoretically positioned by some supporters of RDA within the movement as a "war" against animal exploiters and exploitation (see, for example, Best 2004). From the perspective that animal abusers are "combatants" (see, for example, Jonas 2004), the US ARM has shown restraint in that it has never harmed anyone physically. Though some who generally support RDA take issue with the war imagery (e.g. Kheel 2006), the point in support of RDA remains that, given the gravity of what nonhuman animals endure at the hands of humans, radical tactics are not only acceptable but ethically necessary.

On a pragmatic level, those in support of RDA argue that these actions directly and immediately save animals' lives. While more moderate paths might allow billions of animals to die in the process of seeking incremental gains, RDA tactics can immediately rescue an animal or economically cripple an animal exploitation business or industry. Some also argue that any media or other public coverage of such actions brings the issue of animal suffering into the limelight, no matter how the activists themselves may be perceived:

> Far from alienating our likely allies, I have found ALF[1] activities speak to people, regardless of their belief in animal rights. They provide an opportunity to discuss the gravity of the situation, the fact that animals suffer and die like we do, the fact that they are not less important than we are.
>
> (Friedrich 2004: 257)

Support of RDA is often also contextualized, with support for some tactics in some instances and not others at other times. Kheel (2006: 307) describes this point of view when she cautions readers that, "Some forms of direct action can be valuable tools for social change.... But when direct action is endorsed uncritically in any and all forms, or embraced as a universal norm, it can do more harm than good."

A straw man?

Though this debate over the radical flank could be productive for the animal rights movement, it often serves only as an avenue for activists to channel a priori notions about tactics. Though the radical debate has generated some theory (see, for example, Best and Nocella 2004), it has otherwise mostly served as a percolator for debate within the movement.

The radical debate is currently unproductive for several key reasons. First, the key terms are often left undefined. Ideologues wield undefined trigger words such as "violence" and "terrorism", thereby creating a "War of Words" (Best 2004). People who oppose radical actions generally label property damage or neighbourhood demonstrations "violent", while those who support radical tactics tend not to. A lack of uniform definitions leads to unproductive and contentious situations in which people engage in debate with different ideas about what is actually being discussed.

A second key issue is that animal activists and advocates often operate with incomplete or inaccurate historical knowledge. As Martin Luther King Jr and Mahatma Gandhi are invoked in support of pacifism or Che Guevara is heralded for taking aggressive action, the full story gets left behind. King was not the only voice in his movement and at times was criticized within his movement for not being aggressive enough; Gandhi was supported by activists who literally put their bodies on the line in an active way, a tactic not present in today's US pacifist movements; and Guevara's revolution came with a great deal of death and suffering for many Cubans.

This tendency to "cherry pick" from history only what fits a certain preordained ideology is exacerbated by discussing exemplar movements out of their social and historical context. The social movements to which opponents of RDA most readily turn are the US civil rights and the Indian independence movements. However, as Bruce Friedrich (2004) highlights, these movements are contextually different from the animal rights movement in a number of key ways: they were movements in which individuals were fighting for their *own* rights; there was global support for these movements; and a mass number of activists were willing to go to jail for the cause, among other factors.

More appropriate exemplars, Friedrich suggests, are the abolitionist movement to end slavery in the US and the anti-Nazi resistance in Europe. Unfortunately, these movements are rarely called upon as exemplars. While the abolitionist movement gets more attention as a parallel, it often occurs in an unhistorical manner. For example, some ARM activists label themselves "vegan abolitionists" and identify themselves as "radical" (see Wrenn 2012) but reject the use of most RDAs, labelling them as violent. This creates historical confusion about the movement their name evokes, as the US movement for the abolition of slavery used a wide variety of tactics, including RDAs (and even physical violence against individuals).

A further concern is that there is no distinction between the pragmatic and principled aspects of the debate. The theoretical ideal of what the animal rights movement represents should guide but not singularly determine on-the-ground decisions that are subject to shifting political, cultural and economic factors, as well as personal boundaries. For example, some activists assert that pushing for policy changes is not an ideal route to change, arguing that it asks for reform from an already broken system and further institutionalises government control over animal bodies. However, an activist who agrees with this *ideology* may still decide it is *pragmatic* in a specific socio-political context to work toward a specific legislative measure that s/he feels will end some animal exploitation.

Another common way in which theoretical ideal and activist practice don't align occurs when activists ideologically believe that RDA or illegal actions are acceptable or necessary, but are personally only willing to utilize legal tactics; they value the ideology but are unwilling to risk their own freedom. This boundary between what one values or believes and what one is willing to do is rarely discussed or distinguished in these debates. These distinctions need to be clearly made at the outset of debate so that principled ideals can then be discussed in terms of the social and political contexts in which the movement exists.

A final failing of many current debates is that they are based on anecdotes and lack systematic evidence. There is little systematic research on radical flank effects and the research that does exist is rarely referenced in the radical debate. There are a number of key assumptions and arguments in this debate that are empirically testable. Though these questions cannot all be tested to their full potential in one chapter, this is a first step toward a conversation about radicalism that is rooted in empirical evidence, not opinion or anecdote.

Many key arguments at the core of the radical debate are empirically testable. Testable arguments in support of RDA include:

1   Radical direct actions, such as animal liberations, save animals' lives.
2   Radical direct actions garner media attention.
3   Media coverage of radical activism creates a space for discussion about animal exploitation in mainstream media.

Testable arguments in opposition to RDA include:

1   Radical direct actions, such as property damage, may cause people or communities to be fearful and/or put the physical safety of human and nonhuman animals at risk.
2   Radical direct actions garner media attention.
3   Media coverage of radical activism opens up a discursive space to position the animal rights movement as violent and/or animal exploiters as victims.

Data and methods

To empirically investigate these claims I rely on two primary datasets. The first is a dataset of all known illegal radical actions from 1990–2010. *The Diary of Actions* (Young 2009) is a compilation of illegal radical direct actions in the US since 1979. After engaging in an illegal direct action, it is customary to issue a communiqué, a public statement describing the action and why it was initiated. Young compiled a listing of illegal direct actions from those communiqués, as well as from other sources, including reports in ARM "zines" and mainstream newspapers. The last year of data available in the *Diary of Actions* is 2008. For data from 2009 and 2010, I rely on the North American Animal Liberation Press Office (NAALPO), an organization that receives and publishes anonymous communiqués on their website.[2]

Information about each action is coded for various characteristics that might provide insight into the nature of the action or trends in actions over time including: the date of the action; industry targeted (e.g. fur, food, science, etc.); what was done (e.g. arson, vandalism, etc.); location of the action; if animals were liberated; the target of the action (e.g. retail store, personal property, etc.); if the target was a tertiary target.[3]

This is an important dataset, as it creates the ability to identify patterns in these actions. Even so, there are likely to be some gaps in this dataset. Since the proliferation of the Internet, communiqués have been easier to publish and collect, so the years before the mid 1990's are more likely to be missing actions than are later years. Further, not all activists who engage in these actions issue communiqués, and it is not clear that all animal liberations were done for animal rights causes, as a small number of actions were not "claimed" by any specific ARM group or did not leave any message stating that the action was taken for animals.

More importantly, this data set only covers *illegal* RDAs. Legal types of RDA are not included, as they do not customarily lead to communiqués. This includes home demonstrations, which are legal and more frequent, as well as less common events such as open rescues. Other events not captured are events in which no communiqué was written or was not included in the listing of actions used in this dataset. For this reason, the data must be approached more carefully after 2006, when the first major prosecutions under the Animal Enterprise Terrorism Act occurred. This law can enhance charges to the level of "domestic terrorism" for illegal activity done in the US with the ideological impetus of animal rights (see Lovitz 2010). The number of illegal radical actions does decrease in 2006. This decrease is likely to be a direct response to these prosecutions; however, it is also possible that activists are less likely to identify the ideological impetus of their actions so that, if caught, they will not suffer increased penalties.

Even with these limitations, this data set provides an amazing wealth of data that previous studies of the radical flank have been unable to analyse.[4] In total there are 1,679 illegal RDAs in this dataset, spanning a variety of industries.

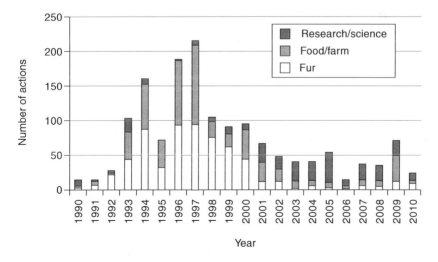

Figure 12.1 Illegal RDAs: most frequently targeted industries, by year.

Though industries targeted, frequency of actions, and the types of actions taken changed over time (see Figures 12.1 and 12.2), illegal RDAs have consistently been a part of the US ARM.

Media coverage

Media coverage of RDAs was also examined to gauge how these actions are represented to a public audience. A random 25 per cent sample of actions for each year from 1990 to 2010 was selected. For each of these actions, newspaper coverage of the event was tracked within the first week of the event. Newspaper coverage is used as a gauge of media response, as newspapers have wide circulations and there are databases available that accommodate systematic searches of local and national news coverage over this period of time. Newspaper searches were conducted in LexisNexis Academic using the "US Newspaper" and "Small Town Newspaper" listings, which include about 800 newspapers. Some actions that received coverage in smaller circulations may not be included. However, given the volume of actions that were searched for, this remains one of the most comprehensive article databases allowing for systematic searches.[5] Articles were only included if they associated the action with animal rights, as coverage of the actions that does not tie them to the animal rights cause will not have an impact on perceptions of the ARM.

In total there were 395 actions for which media coverage during the first week that the action occurred was searched. The news coverage of each action was coded by the author for the type of news outlet (local vs national circulation), which perspectives were represented and received direct quotes and the use of rhetorical words such as "violent" or "terrorism". Twenty-eight cases were dropped from the original sample ($N=423$) as the listings lacked sufficient searchable information (e.g. no specific location or date of the action). The final sample for which news coverage was searched is similar to the full sample in terms of the focus of the actions and the industries being targeted (See Table 12.1).

A matter of life and death

A key concern of the animal rights movement is a concern for life, and so much of the radical debate is centred on this very issue. Those in support of RDA make an argument about lives saved, while those opposed make arguments about lives potentially lost or damaged. While these arguments fly rhetorically, there is often little more than anecdote to support either side.

Supporters of RDA argue that these actions immediately save lives, so it is important to know if animals are actually being permanently liberated (How often? How many?) and what proportion of RDAs are actually saving animals' lives. Animal liberations occurred every year from 1990 to 2010, though they were not frequent. On average, they accounted for 13 per cent of all actions during this 21-year period, with an average of nine reported liberations per year.

Table 12.1 Descriptive statistics

	Full sample	*Media sample*
INDUSTRY TARGETED		
Entertainment	21 (1.3%)	6 (1.0%)
Farm	34 (2.0%)	12 (2.8%)
Food	550 (32.8%)	128 (32.7%)
Fur	635 (37.8%)	154 (38.2%)
Hunting/Sports	49 (2.9%)	11 (2.5%)
Companion animals	19 (1.1%)	5 (1.0%)
Science/research	294 (17.5%)	70 (17.2%)
Wildlife	15 (0.9%)	6 (1.8%)
Other	22 (1.3%)	5 (1.3%)
Unknown	57 (3.4%)	11 (2.8%)
ACTION TAKEN		
Violent(?)	307 (18.3%)	68 (17.2%)
Liberation	184 (11.0%)	49 (12.4%)
Vandalism	1,132 (67.4%)	262 (66.3%)
TOTAL	1,679 (100%)	395 (100%)

Notably, as the focus of RDAs shifted from the fur industry to science and research in the 2000s (see Figure 12.1), the number of animal liberations and the number of animals liberated decreased. Functionally, this is likely to be because laboratory rescues require activists to physically remove and re-home animals, whereas with fur farm liberations (e.g. mink, fox, etc.), activists may only need to open the cages as the animals can survive in the wild on their own. In only two years (2006 and 2008) did animal liberations account for more than one-fifth of recorded illegal RDA's (see Figure 12.2).

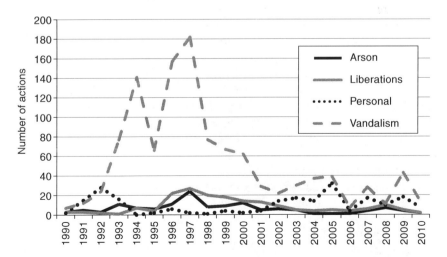

Figure 12.2 Number of illegal RDAs, by year and type.

Though animal liberations were relatively infrequent, there is no doubt that animals' lives were saved. From 1990 to 2010, there were at least 181 liberations in which at least 141,670 nonhuman animals were rescued.[6] The fate of these animals after rescue is generally unknown or unreported in communiqués. The media often reports that released mink are hit by cars or eventually re-caught (e.g. Bachman 2007; Wangstad 1998), but studies have also shown that farm-raised mink can survive in the wild (e.g. Mason *et al.* 2011). In either case, there are undoubtedly at least some who lived full, safe lives as a result of their rescue.

Only a minority of RDAs liberated animals. Given the extent of animal exploitation the ARM is combating, animal liberations are not making a large impact. For example, in the period from 2001 to 2010, the number of land animals killed in the US for food alone was about 92.7 billion (USDA 2012), while the number of nonhuman animals liberated by RDAs in the US during that same time was around 19,000 (and some of these animals were re-caught and returned to their former lives of abuse). The number of animals liberated did not even total one-tenth of 1 per cent of the number of animals killed for food alone.

Liberations may be an ineffective strategy statistically but if you are one of the animals being liberated, such numerical parsing is likely to be unpersuasive. The answer to whether RDAs save lives is a solid "yes". Ideologically, to the ARM, animals are not a "mass group" of "things" (Adams 2007), they are individuals and each individual should always matter. As such, there is weight to the argument that RDAs are useful and important because they do save lives, regardless of the number.

For those who oppose RDAs, the question of life focuses not on lives saved, but on lives inconvenienced, damaged, risked or lost. Actions that would put the lives of others in harm's way can be considered violent. Since the definition of "violence" and even of "harm" differs, I measure it in a variety of ways. At the most basic level, there has been no known human loss of life or physical injury at the hands of the US ARM. However, there is a spectrum of actions taken that are considered violent by some. Arsons and bomb threats are typically classified as violent; once a fire is set it can be difficult to control, human and nonhuman animals may unexpectedly be in the building, bomb threats can cause chaos and fear, fire-fighters may be injured putting out a blaze, etc. Of the 1,679 instances of RDAs in this dataset, 135 actions (7 per cent) involved arson, threat of arson, a bomb threat or an incendiary device (regardless of whether it went off).

Another controversial tactic is personal threats, whether made directly to a person or by doing damage to one's personal property. During this 21-year period, 11 per cent of illegal RDAs (215 actions) included some sort of personal threat or attack of an individual's personal property. These actions involved everything from vandalizing people's homes with spray paint, to puncturing car tyres in work parking lots, to sending razor blades in the mail along with claims that they were HIV infected.

The type of threats changed over time, with arsons and incendiary devices losing popularity to personal threats in the 2000s. Further, these threats were increasingly aimed not directly at the activists' target, but at tertiary targets.[7] All told, actions that included personal threats or arson accounted for less than one-fifth of all RDAs in the US from 1990 to 2010. Like animal liberations, these actions constitute only a minority of RDAs. Even so, they definitely do occur.

Overall, there is validity to the arguments on both sides of the radical debate regarding harm to life. Yes, animals' lives are saved and yes, some threatening and potentially harmful actions do occur. The weight of each on a balance scale remains in the purview of ethical and theoretical arguments but, either way, the gravity of these actions is being overplayed considering how infrequently they occur in relation to radical actions alone, much less the more frequent legal and moderate forms of animal rights activism. Given the infrequency of such actions, what becomes a more relevant is how they are reported by the media.

Media matters

Movements often judge their success by the amount of attention they receive from the mass media. The moderate majority of a movement is typically interested not only in media coverage, but also in having that coverage convey the desired message, or frame, of the movement's point of view:

> Persuasion works through increasing the prominence of one's preferred frame in the mass media. By changing the way various publics and bystanders understand an issue, those who are opposed or neutral may redefine how their interests and values are affected.
>
> (Gamson 2007: 254)

Media coverage is not a major goal of all factions of a social movement because, as Koopmans (2007: 389) highlights: "Reliance on mass media makes movements dependent on what others define as important and legitimate." Some in the radical faction of the ARM have been vocally critical of utilizing the mass media as a strategy (Dawn 2004), but many radicals and the moderate majority of the ARM rely on the mass media as a tool. With few monetary or political resources, mass media becomes an important means by which a movement's activities and messages are carried to a wider public. Therefore, the debate over the effect that the radical flank has on media coverage remains crucial. But again, there is a paucity of empirical research on the subject.

Those in support of RDA argue that they generate more media coverage for animal rights causes, which they claim is positive regardless of framing. In a survey of US adults, over one-half of respondents indicated that in the past three months they rarely (29 per cent) or never (26 per cent) engaged in or heard discussions of animal protection or welfare issues (HRC 2012); considering this reality, supporters of RDAs for the ARM argue that any coverage is good coverage.

Arguments opposing RDAs and their impact on media coverage focus on the type of coverage garnered, with claims that the ARM is painted in a negative light when RDAs are covered. The concern is that ideologically-charged words like "terrorist" and "violent" can be used to describe the ARM, and that the targets of radical actions, the animal abusers themselves, are positioned as victims.

Arguments both for and against RDA in the ARM assume that coverage of ARM RDAs occurs in the first place. However, no research has sought to validate such an assumption. In a review of literature addressing media bias in the coverage of other social movements, Earl *et al.* (2004) found that there is selection bias in what newspapers report on (see also Koopmans 2007), such that conservative organizing is omitted from news coverage, with a preference for reporting on protest activity instead. I find an additional bias against reporting on radical activism as well.

In my sample, only 8.9 per cent ($N=36$) of actions received any mainstream newspaper coverage. Most of these actions had only one (46 per cent) or two (20 per cent) articles written about them. There were scant exceptions; one action involving an incendiary device at the home of an animal experimenter in 2008 was covered in 29 separate articles, and the release of 10,000 mink from a fur farm in 2003 was covered in 15 different articles. Removing these outliers, each action that received newspaper coverage within the week it occurred had an average of 2.7 articles written about it.

Not all illegal RDAs are equally as controversial, so it is important to understand if the type of action affects the chances that an RDA will be covered. Because so few actions received coverage, tests of statistical significance need to be approached cautiously. With that caveat in mind, findings suggest that acts of vandalism are significantly less likely to receive newspaper coverage compared to all other actions while arsons, attempted arsons or bomb threats are significantly more likely to receive newspaper coverage.[8]

Not only do illegal RDAs represent a minority of actions taken on behalf of animals, they are not widely covered by the news media. It is still useful, however, to understand the type of coverage they receive when they are covered, as this is a central part of the radical debate. If arguments in support of RDAs are valid, the animal rights point of view will be presented in newspaper articles; there will be even more support for this perspective if animal rights activists or organizations are directly quoted. If arguments against RDA are valid, coverage will include charged words like "violent" and "terrorist", and the targets of illegal RDAs will be interviewed, creating a discursive space to position themselves as victims. These arguments are not mutually exclusive and there is some evidence to support both sides of the debate.

In about one-third (36 per cent) of articles, both the targets of the RDA and an animal rights activist or organization were consulted. In about one-quarter of the articles (27 per cent) neither side was specifically consulted and just as often, only the target of the action was consulted. In a minority of articles (10 per cent) only the ARM was consulted. Overall, about two-thirds (63 per cent) of articles consulted with the targets of actions, while about half (46 per cent) consulted with the ARM (See Figure 12.3).

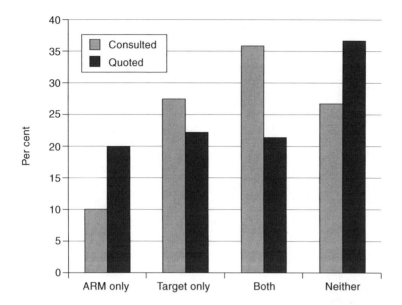

Figure 12.3 Which perspective is presented?

Though the non-ARM perspective is more likely to be given space in these articles, most articles (90 per cent) that consult someone from the ARM also directly quote a movement actor or printed material, such as a communiqué. As such, the ARM is just as likely to have someone quoted as is the target of the action. In 22 per cent of the articles, only the target of the actions was quoted. Similarly, in 20 per cent of articles, only an ARM activist or organization was quoted and in 21 per cent of the articles both sides were quoted.

Only about 9 per cent of the illegal RDAs in this sample received any coverage in 800 local and nationally circulated newspapers, including the widely circulated *Associated Press (AP)*. RDAs are already a minority of animal rights activity, and even when they do occur they rarely receive mainstream newspaper coverage. The few actions that are covered are slightly more likely to present the perspective of the target of the action. This does open up more opportunity for animal abusers to present themselves as victims, supporting arguments against RDAs. However, at the same time, the ARM is just as likely to be directly quoted as is the target of an action, supporting arguments in favour of RDAs.

There is some credibility to the concern of the ARM being discussed with charged words, though it occurred in fewer than half of the articles. Of the 38 actions that received newspaper coverage, the words "terrorist", "extreme", "violent", "radical", and/or "militant" were present in at least one article covering 16 (42 per cent) of the actions. In all, 40 per cent of all articles covering illegal RDAs used this language. While it cannot be said that this does not occur, it is so infrequent that it may be negligible, at least in terms of illegal RDAs, as these terms may also appear in coverage of less radical forms of animal rights activism as well.

Overall, there is little support for any of the arguments over RDA and media coverage because there is so little coverage of these events. Going forward, debates of public perception (as generated by the media) would be more fruitful from a pragmatic perspective if they were targeted at better understanding non-radical and legal protest activities and how media portrays them, for they are the actions that are actually getting covered and impacting public perceptions.

Conclusion

The radical debate in the ARM currently suffers from a lack of empirical evidence. Much of the debate is centred on public perception, even without evidence that the public receives information about RDAs and has opinions about radical animal rights activity, or that public opinion directly impacts outcomes for non-human animals. This study begins this discussion in an important way.

Those who support RDA claim that lives are saved and these actions open a discursive space for discussing animal rights ideology. I find that lives *are* saved, and each life saved is a huge victory. However, only a fraction of 1 per cent of animals killed and exploited is rescued by these actions. In terms of public relations, animal rights activists and organizations are just as likely to be quoted in news media as the targets of these actions. At the same time however, illegal RDAs rarely receive newspaper coverage.

Those who oppose RDA argue that these actions are often violent. The debate is nuanced at this point, as the question becomes, 'What is violent?' If any type of property damage is violent then they are correct, as most illegal RDAs involve vandalism, such as damaging windows and doors or leaving messages with spray paint. However, actions that are more likely to scare or hurt someone, such as vandalism at a private residence, arson or bomb threats, occur infrequently and have never caused known physical harm. Even so, they do happen. Those who oppose RDA also worry about negative media coverage. Negatively charged words are common, but appear in fewer than half of the articles covering illegal RDAs.

This study finds that core arguments on both sides of the radical debate have little empirical evidence to refute them and only scant evidence to support them. Academic audiences all too often categorize findings like this as a "non-finding". However, in the context of praxis-centred research, this non-finding speaks volumes. The core findings of this study are that these actions happen infrequently, when they do they are rarely discussed in mainstream media and if they are, their impact is not clearly positive or negative for the movement. Inserted into the context of a movement that spends energy, draws schisms and invests resources into the outcome of this debate, this finding provides a directive for the movement – the time, energy and resources spent to combat illegal and radical actions is misdirected.

The radical debate strikes deeply at the core of the animal rights community. Movements are strongest when there is a variety of organizations spanning a breadth of goals and utilizing a variety of tactics (Downey and Rohlinger 2008) and they make the most gains when these factions work together (Koopmans 2007; Staggenborg 1998). For this reason the need to avoid schism is especially important. It is worthwhile and important for scholars to continue understanding the radical flank effect of the ARM, but it is ineffective at this point, outside of systematic empirical investigation, for activists concerned with pragmatic outcomes of radicalism in the ARM to continue pouring energy into these debates.

The radical debate is a straw man, hastily thrown up and torn down by each side. The debate is divisive for the ARM, as each side makes sweeping claims about the role of radicals that are based in opinion and supported selectively by anecdotes and historical examples that are often out of context. Importantly, it is not the radicals who catalyze this debate but their opponents. For example, in 2005 a major animal rights organization, the Humane Society of the United States (HSUS), withdrew its participation in the largest US animal rights conference because the programme included speakers from an animal rights campaign, Stop Huntington Animal Cruelty USA (SHAC USA), which utilized legal radical tactics and also vocally supported illegal radical actions. Such schisms are also displayed publicly, such as when, in 2008, HSUS donated $2,500 in reward money for information leading to the capture of activists who placed an incendiary device outside the home of an animal experimenter.

RDA at this point in the ARM is rarely utilized and its consequences are unclear. Therefore, from a pragmatic perspective, energy from within the movement to demonize such tactics is unproductive and doing the ARM a disservice. Those in the ARM who are generating public and divisive debate over radicalism must be careful to avoid the tendency to engage in a "war of words", to cherry pick from historical movement exemplars and to replace evidence with anecdote. These arguments are at best exaggerated and at worst embrace the repressive rhetoric of animal activists as violent.

Suggestions for future research

A more fruitful way forward in this debate is to continue empirical examination into the outcome of radicalism. In this chapter I have focused on testing some of the premises underlying pragmatic questions in the debate over radicalism in the ARM. Future research needs to continue excavating these questions rather than engaging in the debate rhetorically for the sake of driving home one's a priori point of view.

Media

This study finds a lack of coverage of RDAs in print media. Other studies can seek to determine what *is* being covered – is it protest activity, certain organizations, certain issues, etc? To the degree that study of media coverage of RDAs is pursued, it will be useful to examine not only the coverage of a specific RDA, but to determine if a single action spurs more articles about a specific topic. One of the few studies examining the outcome of RDAs suggests that this may be the case. Dawn (2004) studied media coverage of three RDAs and argues that in some cases these actions lead to separate coverage of the animal rights issue being protested. For example, she describes that, following the vandalism of a foie gras chef's home, a local news channel carried a story of the treatment of ducks in the foie gras industry.

Media coverage ensuing from RDA may also be examined to determine if the ARM is portrayed in a more positive or negative light than it is in coverage of moderate tactics used by the ARM, or if RDA influences how moderate organizations and tactics are covered. Also, future study can examine what types of actions get what type of press. Some involved in the radical debate argue that open rescues receive more positive media attention than covert liberations. Claims such as this should be systematically examined.

Attitudes

Because RDAs cannot be predicted, it is difficult to survey public attitudes about an animal issue before and after a major event. However, when major actions do occur, surveys can be conducted to determine knowledge of and attitudes toward these actions. General population surveys may also gauge social attitudes toward animal rights. According to a 2012 survey (HRC), "animal protection" (along with "workers' rights") is viewed favourably by more US adults than other political movements or social causes. Sixty-eight per cent of respondents had a favourable opinion of "animal protection" compared to 49 per cent for homeless advocacy and immigration reform and 37 per cent for gay rights. It is important to understand the nuances of this support and how ideological and tactical variation in the movement is perceived and supported.

Measuring outcomes

While public opinions are important to explore, it is not necessarily the case that public perception engenders outcomes for animals. It is a common mistake to assume that winning a PR battle will win more freedom for animals. Even as veganism in the US has doubled in the past five years, it is not clear that this has resulted in shifts in the rates of slaughter or in the treatment of animals. More research should systematically track legislative success and basic statistical information on the rates of animal exploitation and animal deaths in various industries. There is also debate over the utility of economic sabotage. Future research assessing the utility of economic sabotage may also study whether acts of economic sabotage have caused enough of a financial burden to halt the exploitation of animals or shut down businesses.

Donations

One key way to understanding radical flank effects is by tracking donations during periods of high and low radical flank activity. Giving Foundation USA (2011) tracks donations to environmental and animal organizations; between 1990 and 2010 donations have increased, but by very little. In 1990, they comprised 1.54 per cent of all donations made in the US and in 2010 they comprised 2.29 per cent, with the peak being a total of 2.39 per cent of all US donations in 2009. Haines' (1984) study of radical flank effects tracked donations in a more nuanced manner, evaluating shifts in giving between more moderate and more radical organizations within the same movement following periods of increased radicalism. A similar study could provide a better understanding of the radical debate in the ARM.

Role of radical objectors

Finally, the impact of those that denounce radicalism should also be studied in more depth. For example, does the media report widely on ARM organizations denouncing RDA? Do mainstream organizations see an increase or decrease in donations or membership when they choose to denounce or support RDA? Or do these acts go widely unnoticed? These questions are key because it is the rejection of RDA that drives schism in the movement and it is important to understand the full cost of these decisions.

Notes

1 ALF refers to the Animal Liberation Front, a group of anonymous activists often claiming or being attributed to illegal covert actions for animals.
2 See https://animalliberationpressoffice.org/NAALPO/category/communiques/, accessed 21 March 2011.
3 All coding was conducted by the author. A second coder coded a sample of cases to test for accuracy when developing the coding scheme, with a final subset tested by another coder at the end of the coding to determine inter-coder reliability and ensure the accuracy of these findings. For more information about the coding process please contact the author.
4 There is only one similar dataset I know of, created by Michael Loadenthal (2010), which tracks actions that might be labelled "ecoterrorism".
5 For a more detailed description on methods, including a listing of newspapers covered in the database and the search terms used to search for news coverage please contact the author.
6 The estimate of animals liberated is a baseline; in some cases no number was mentioned and therefore was not calculated into the total number of animals, and in cases when a range was estimated the lowest number given was recorded.
7 The shift to tertiary targeting was a tactical innovation in the US that began with the popularity and success of the Stop Huntingdon Animal Cruelty (SHAC) campaign (see Glasser 2011: 106–108).
8 Significance was tested using the Chi-squared test of significance. Both of these findings were significant ($p \leq 0.001$).

References

Adams, C. J. (2007) "The war on compassion", in J. Donovan and C. Adams (eds) *The Feminist Care Tradition in Animal Ethics*, New York: Columbia University Press.

American Society for the Prevention of Cruelty to Animals (ASPCA) (2009) "ASPCA Research: Do Cat Names Affect Adoptability? ASPCA", available at: www.aspcapro. org/resource/saving-lives-adoption-marketing-research-data/aspca-research-do-cat-names-affect. Last accessed 20 July 2013.

Bachman, J. (2007) "Farm missing mink", *The Berkshire Eagle*, 14 August.

Best, S. (2004) "It's war! The escalating battle between activists and the corporate-state complex", in S. Best and A. J. Nocella, II (eds) *Terrorists or Freedom Fighters? Reflections of the Liberation of Animals*, New York: Lantern Books.

Best, S. and Nocella, II, A. J. (2004) "Introduction', in S. Best and A. J. Nocella, II (eds) *Terrorists or Freedom Fighters? Reflections of the Liberation of Animals*, New York: Lantern Books.

Dawn, K. (2004) "From the Front line to the front page – an analysis of ALF media coverage", in S. Best and A. J. Nocella, II (eds) *Terrorists or Freedom Fighters? Reflections of the Liberation of Animals*, New York: Lantern Books.

Downey, D. J. and Rohlinger, D. (2008) "Linking strategic choice with macro-organizational dynamics: strategy and social movement articulation", *Research in Social Movements, Conflicts and Change*, 28: 3–38.

Dillard, C. L. (2002) "The rhetorical dimensions of radical flank effects: investigations into the influence of emerging radical voices on the rhetoric of long-standing moderate organizations in two social movements", Ph.D dissertation, Austin: University of Texas, Scientific Commons.

Earl, J., Martin, A., McCarthy, J. D. and Soule, S. (2004) "The use of newspaper data in the study of collective action", *Annual Review of Sociology*, 30: 65–80.

Ellis, C. and Irvine, L. (2010) "Reproducing dominion: emotional apprenticeship in the 4h youth livestock program", *Society & Animals*, 18: 21–39.

Epstein, B. (1993) *Political Protest & Cultural Revolution: Nonviolent Direct Action in the 1970s and 1980s*, Los Angeles: University of California Press.

Fitzgerald, A. J., Kalof, L. and Deitz, T. (2009) "Slaughterhouses and increased crime rates: an empirical analysis of the spill over from 'The Jungle' into the surrounding community", *Organization and Environment*, 22(2): 158–184.

Friedrich, B. G. (2004) "Defending agitation and the ALF", in S. Best and A. J. Nocella, II (eds) *Terrorists or Freedom Fighters? Reflections of the Liberation of Animals*, New York: Lantern Books.

Gamson, W. (1975) *The Strategy of Social Protest*, Homewood, IL: Dorsey.

Gamson, W. (2007) "Bystanders, public opinion, and the media", in D. Snow, S. A. Soule and Hanspeter Kriesi (eds) *The Blackwell Companion to Social Movements*, Malden: Blackwell Publishing Ltd.

Garrow, D. J. (2002) "The FBI and Martin Luther King", *The Atlantic*. July/August 2002, available at: www.theatlantic.com/past/docs/issues/2002/07/garrow.htm. Last accessed 21 March 2012.

Giving USA Foundation (2011) "Giving data", in "Giving USA 2011: The Annual Report on Philanthropy for the Year 2010", Chicago: Giving USA Foundation.

Glasser, C. L. (2011) "Moderates and radicals under repression: the U.S. Animal Rights Movement, 1990–2010", Ph.D dissertation, Irvine: University of California.

Gupta, D. (2002) "Radical flank effects: the effect of radical-moderate splits in regional nationalist movements", Conference of Europeanists, Chicago, IL.

Haines, H. H. (1984) "Black radicalization and the funding of civil rights: 1957–1970", *Social Problems*, 76: 31–43.

Haines, H. H. (1988) *Black Radicals and the Civil Rights Movement, 1954–1970*, Knoxville: University of Tennessee Press.

Humane Research Council (HRC) (2011) "Animal tracker–wave 4", Olympia, Washington, DC: Humane Research Council.

Humane Research Council (HRC) (2012) "Animal tracker–wave 4", Olympia: Washington, DC: Humane Research Council.

Irvine, L. (2013) "Animals as life changers and lifesavers: pets in the redemption narratives of homeless people", *Journal of Contemporary Ethnography*, 42(1): 3–30.

Jasper, J. M. and Nelkin, D. (1992) *The Animal Rights Crusade: The Growth of a Moral Protest*, New York: The Free Press.

Jonas, K. (2004) "Bricks and bullhorns", in S. Best and A. J. Nocella, II (eds) *Terrorists or Freedom Fighters? Reflections of the Liberation of Animals*, New York: Lantern Books.

Kheel, M. (2006) "Direct action and the heroic ideal: an ecofeminist critique", in S. Best and A. J. Nocella, II (eds) *Igniting a Revolution: Voices in Defense of the Earth*, Oakland: AK Press.

Koopmans, R. (2007) "Protest in time and space: the evolution of waves of contention", in D. A. Snow, S. A. Soule and Hanspeter Kriesi (eds) *The Blackwell Companion to Social Movements*, Malden: Blackwell Publishing Ltd.

Lovitz, D. (2010) *Muzzling a Movement: The Effects of Anti-terrorism Law, Money, and Politics on Animal Activism*, Brooklyn, NY: Lantern Books.

Mason, G. J., Cooper, J. and Clarebrough, C. (2011) "Frustrations of fur-farmed mink", *Nature*, 410: 35–36.

Minkoff, D. C. (1994) "From service provision to institutional advocacy: the shifting legitimacy of organizational forms", *Social Forces*, 72(4): 943–969.

Singer, P. (2004) "Humans are sentient too", *The Guardian*, 30 July, available at: www.utilitarian.net/singer/by/20040730.htm. Last accessed 20 July 2013.

Staggenborg, S. (1998) "Social movement communities and cycles of protest: the emergence and maintenance of a local women's movement", *Social Problems*, 45(2): 180–204.

Tracy, J. (1996) *Direct Action: Radical Pacifism from the Union Eight to the Chicago Seven*, Chicago: University of Chicago Press.

United States Department of Agriculture (USDA) (2012) "Livestock and slaughter annual summary", National Agricultural Statistics Service, available at: http://usda.mannlib.cornell.edu/MannUsda/viewDocumentInfo.do?documentID=1097. Last accessed 20 November 2012,

Wahlstrom, M. and Peterson, A. (2006) "Between the state and the market: expanding the concept of 'political opportunity structure'", *Acta Sociologica*, 40(4): 363–377.

Wangstad, W. (1998) "Group takes responsibility for fur farm raids, releases; ranch near Rochester Reports 2,800 mink freed", *St. Paul Pioneer Press*, Local: 2C, 29 August.

Weiss, E., Slater, M. R. and Lord, L. K. (2011) "Retention of provided identification for dogs and cats seen in veterinary clinics and adopted from shelters in Oklahoma City, OK", *Preventative Veterinary Medicine*, 101(2/3): 265–269.

Wrenn, C. L. (2012) "Applying social movement theory to nonhuman rights mobilization and the importance of faction hierarchies", *Peace Studies Journal*, 5(3): 27–44.

Young, P. (ed.) (2009) *Diary of Actions: The First 30 Years*, Voice of The Voiceless Communications (a publish-on-demand book).

Conclusion
Future directions for critical animal studies

Helena Pedersen and Vasile Stanescu

Animal studies[1] is expanding rapidly throughout academia. In our editors' introduction to Kim Socha's book *Women, Destruction, and the Avant-Garde: A Paradigm for Animal Liberation* (Pedersen and Stanescu 2012), we ask into what circuits of the knowledge economy animal studies scholarship actually feeds, when global institutionalized animal use and abuse shows no sign of decline but rather continues to take on new and creatively oppressive forms.

As this volume shows, there is a strong and committed counter-movement to these developments in and outside academia. In a certain sense it is possible to say that CAS goes back to the roots of animal studies, to the animal advocacy movements, where the ideas of creating academic platforms for work against animal exploitation originated. On the other hand, as a field for academic-activist intervention and transformation, CAS also points distinctly toward the future; a future free from animal and human oppression, which we refuse to see as utopian, but as a possibility to begin to form here and now (see Watson [1947], quoted in Cole, Chapter 10 this volume).

As Steven Best anticipated in 2009, we see an increasing tendency in the animal studies literature to co-opt CAS and incorporate the field as one of several sub-fields of "animal studies". We oppose this tendency, which we see as a move toward a "domestication" of CAS that seeks to depoliticize and render the field devoid of its radically transformative force. With our concluding chapter, we join Jessica Gröling (Chapter 5) and Nathan Stephens Griffin (Chapter 6) in positioning CAS not as part of, but in profound *opposition* to, animal studies and situating CAS among its theoretical, historical, and ethical allies in critical theory. Seeing CAS as a strand of critical theory (broadly defined), rather than as a sub-field of animal studies, radically shifts the power dynamics of the animal studies landscape and highlights its inherent tensions, contradictions and conflicts. By "critical", we mean the application of critical theory towards actual liberation. Max Horkheimer's famous definition of critical theory as that which tries "to liberate human beings from the circumstances that enslave them" (Horkheimer 1982: 244) is correct as far as it goes, but wrong in that it places the limits of liberation at only "the human". We would say that critical theory and, therefore, *critical* animal studies, is that which seeks to liberate the animal from the circumstances that seek to enslave her.

The purpose of this chapter, however, goes beyond a summing up of what CAS "is". Instead we would like to use this space to reflect on possible future directions for the development of the field. We know that this is a risky undertaking, and also a contradiction in terms: The future(s) of CAS can never be enclosed or concluded in a final chapter. Rather, the "final chapter" should, together with the other contributions in this volume, be viewed as *one of many* possible beginnings, continuations, interventions or catalyzers for action. We will start with charting our three guiding principles – (1) animals as research subjects and stakeholders in knowledge production; (2) intersectional approaches; and (3) the rejection of animal welfarism as a form of commodity fetish that we feel should underpin any future research and other activities in the field of CAS. Roughly following the content and organization of this volume, these principles address, in a succinct and overlapping manner, CAS methodology, the theory/practice connection and anti-capitalism. We end with a few personal reflections on possible CAS futures (4), and on activist-based scholarship (5). These are claims that we have made before (Pedersen and Stanescu 2012; Stanescu 2013; Stanescu and Jenkins forthcoming). Since, in large part, people do not yet know what CAS "is", we believe there is a real intellectual value in repeating, and *re*-repeating, our key ideas behind CAS. In fact, to be frank, we don't plan to *stop* repeating them until we can witness a widespread adoption of these ideas and principals (of which we believe this volume represents an important part).

Animals in CAS knowledge production

The key and wholly motivating "centre" for CAS is a firm, unwavering normative commitment to ending the exploitation of nonhuman animals for human consumption and pleasure. An animal is not merely a concept or a metaphor but, instead, a real, living and embodied *person* who requires our respect, support and solidarity. We do not believe that nonhuman animals should be property. We do not believe that they should be bought and sold at their "owners" whims. We do not believe that they should be raised, and killed, for human consumption as meat, as leather, as fur or in any other way whatsoever. We do not believe that they should be experimented on for cosmetics, for household products or for supposed "scientific" or "medical" benefits. We also oppose all the "harmless" daily humiliations that are visited on nonhuman animals – being forced to perform in circuses, used to sell products, forced into movies for "our" amusement, and being subject to speciesist[2] terminology and jokes. In short, we reject and live opposed to all examples of human chauvinism and anthropocentric privilege.[3] Our rejection of this thus makes no essential or a priori difference between ostensibly "necessary" (such as medical science research) or "trivial" (such as the representation of animals in visual culture) forms of animal exploitation. Not only because these forms and practices are frequently entangled and interconnected in the complex production networks of capitalist society, but because they are based on the same ontological assumption of a total accessibility of nonhuman animals for human use. Our rejection of anthropocentric privilege extends to our own academic practice and scholarship as we would suggest that any text, no matter how well written, how insightful, or compelling, that violates these core ideas, while it may (or may not) be classified as an example of "animal studies", "animality studies", or "human–animal studies", is not, nor even could become, *critical* animal studies. *For us,* critical *refers not only to an engagement with critical theory, but equally a commitment to be critical of anything that purports to study animals and at the same time fails to engage, support, protect and stand with the animal herself.* For us, our goal is always and already to become "scholar-activists". Not in the sense of scholars who write to, or about, activists but in the sense that as scholars we must also learn to write *as* activists. This is, for us, the most basic and necessary pre-requisite for any work that attempts to reside in the field of CAS.

We therefore agree with Jessica Gröling (Chapter 5), Nathan Stephens Griffin (Chapter 6) and Lynda Birke (Chapter 4) that CAS knowledge production must critically reflect on the performativity of its own work and the power relations it produces, reconfigures or challenges. Above all, CAS knowledge production must be accountable to its nonhuman animal subjects by striving to contribute to the improvement of the situation of animals and by considering the broader political consequences of our research. We need to approach the critical study of human–animal relationships, and the systems of exploitation in which these relations are enmeshed, from multiple methods and levels of analysis, and bring a diversity of methodologies into productive work. Such methodological diversity may include critical ethnography, interviewing, action research and critical discourse analysis, as well as innovative hybrid methodologies, such as ethno-mimesis, which combine the principles of biographical and visual research (Stephens Griffin, Chapter 6 this volume). As CAS scholars, we strongly believe that we must *think* and *do* methodology and analyses radically and differently, seeing them as integrated parts of a larger political strategy. As Sara Salih (Chapter 3) points out, knowledge production in CAS enacts critical intervention into our existing knowledge structures, breaking down and breaking with, rather than "producing ', knowledge; a form of knowing in unknowing. With the following sections, we hope to contribute to opening up this form of critical intervention.

Intersectional approaches and the theory/practice connection

Critical animal studies, and its intrinsic praxis, veganism, is not (only) about not eating animals, but is instead representative of an entire social justice worldview of which individual dietary choice represents the visible tip of the iceberg. To make an analogy, the boycott of segregated business in the civil rights movement (lunch counters, buses) was not only about segregation-based practices at lunch counters and buses, but about a systematic rejection of racism, Jim Crow and segregation-based politics. So too, the economic boycott of the meat, dairy and egg industries is not (only) about the violence and exploitation inherent in the production of meat, dairy and eggs, but also about the entire rejection of speciesism and anthropocentric ethics. In fact one could, at least in theory, refrain from eating animal products, for example for health reasons (a wholly humanist-centred motivation), and not constitute an "engaged vegan" in this larger ontological articulation of the term (Stanescu and Jenkins, forthcoming). However, it is this shared rejection of a worldview that is premised on the flawed idea that "might makes right", that, we believe, is a unique and valuable contribution to the field of "critical" animal studies. Moreover, while we find the commonly used term "intersectionality" useful in helping to 'connect' various interlocking forms of violence and oppression, at the same time, we worry that it misses the reality that they are not "separate" in the first place. We will now discuss two examples of intersectionality issues in CAS: "disability" rights and the issue of forced mating, that not only show the *inherent* entanglement of forms of subordination and oppression, but also illuminate the entanglement of theory and practice as two inseparable parts of the same struggle.

Don LePan envisions, in his novel *Animals* (2010), a futurist dystopia in which, because of mass species extinction, the human population now raises, breeds, and eats cognitively "less developed" humans, who are no longer even included within the realm of the "human" at all because of their presumed lower IQs. We use this story as an example of why, as critical animal scholars, we find the question of the cognitive ability of animals irrelevant in terms of the ethical practice of veganism. While the idea is factually false (as Peter Singer, Tom Regan, Marc Bekoff and many others have pointed out), *even if* every animal possessed lower cognitive ability than every "human", it would be irrelevant in terms of the basis of their ethical inclusion and care, since we reject the micro-fascist notion[4] that a difference in cognitive, physical or linguistic capacity can ever translate into a permissibility of torture or death (Stanescu 2012).

While deeply mindful and grateful of the work done by such scholars as Peter Singer, we reject the inherent anthropocentrism in his move to invoke "marginal cases" as the basis for ethical care. For Singer it is "speciesist" to exclude ethical care for animals because they have the same or similar cognitive capacity as certain humans – the young, the coma patient, the Alzheimer patient and, above all, to quote Singer "the retarded" (1977: 21). It is therefore, in his worldview, unfair, or if you will "speciesist", to treat two living entities with the same IQ differently when the only relevant distinction is one of species. However, as critical animal scholars, we reject not only the "speciesism" of treating two living entities differently when they possess the same level of sentience, but also the inherent anthropocentrism of believing that cognitive ability itself should be the basis of inclusion in the sphere of ethics. As critical animal scholars, we are concerned with "disability" rights and "animal" rights (both of which equally need to be in quotation marks) not again as separate issues with some type of intersection, but as the same issue expressed in two related areas of care. In both cases we reject the "micro-fascism" inherent in a worldview that cognitive awareness should be a prerequisite for inclusion in the moral community.

This then represents, we believe, key areas of "intersectionality". We must endorse a worldview that accords dignity and worth to nonhuman animals, but we must do so in a way that critiques the anthropocentric worldview and value system. Nonhuman animals must matter, but they must matter based on their own worth and not based on the value system of reason created by our own species. For throughout the critical tradition, from Jacob von Uexküll, to Heidegger, to Deleuze and Guattari, to Derrida there is an awareness that animals may, in fact, inhabit entirely different "worlds" from humans and indeed from each other. For Heidegger (building on the work by Uexküll), this difference represents a lack and a deprivation generating Heidegger's claim that animals' lives are world "poor". For Deleuze and Guattari, it represents a benefit, generating in part their idea of "becoming animal". But, in both cases, there is an acknowledgment that nonhuman animals may experience, interact, enjoy or dislike their worlds in vastly different ways from either humans or each other. When Thomas Nagel (1979) asks if we can imagine what it is like to be a bat, he is only beginning to scratch the surface of what is possible to consider (bats are mammals; they age and die), but can we ever truly know or understand the worlds that the vastness of different nonhuman animals experience? Continental philosophy, as well as critical race studies, queer theory, feminism, postcolonial studies and many other areas have helped us see how even good spirited attempts to deal with others that ignore issues of difference can, at the same time, enact a certain type of ontological violence. We would suggest that this represents the importance of critical animal studies. We must be in shared solidarity with the suffering of nonhuman animals but we must, at the same time, do so in ways that are attentive to uniqueness and difference of the subjecthood of their very experience of the world (see Birke, Chapter 4). As CAS scholars we must continue to develop an ontology of care (as Carol Adams, Josephine Donovan, and many others have written about) that not only prohibits suffering but also supports the need of animals for freedom of movement, companionship with loved ones and even joy, within a framework where these emotions are understood within the animal's own different worldview.

We would argue that one cannot be a meat eating feminist or queer advocate. Extending Cole's (Chapter 10) critical analyses of the sexual exploitation of animals in food production, let us provide what to us has become a touchstone of the absolute incompatibility of any form of ownership and consumption of animals and any articulation of feminism or queer theory. The following episode is a veritable mirror image of the exploitative practice described by Barlow (1945) and quoted in Cole (Chapter 10). Catherine Friend and her partner are both committed feminists, who entered into the farming practice as a way to combat both traditional gender roles and hetero-normativity. And yet Friend and her partner, Melissa, hold down a female goat and force her to have what is clearly undesired sex in order to effectively match their breeding programme. In a chapter entitled "Let's just forget this ever happened" (Friend 2006) this is how Friend describes the practice:

At Mary's, we led Ambrosia [Friend's female sheep] to the converted chicken house, into a building about twenty feet by ten feet, with bare board door and Bozeman [the male sheep] came flying in, eyes wild, lip curled at the scent of Ambrosia. Our goat took one look at this creature and began running. I couldn't blame her. Not only do intact bucks reek with an indescribable scent, but this guy's head and neck were oily, greasy, and matted with something foul ... Ambrosia wasn't buying it. Who could blame her? We watched Bozeman chase in a circle for five minutes. "Is this how goat sex usually goes?" I finally asked. "No", Mary said. "Usually the doe stands still. Ambrosia must be near the end of her cycle. She can still get pregnant, but isn't willing to stand still". She sighed. "I'm afraid we have to hold her". Groaning, we stepped forward. Melissa grabbed Ambrosia's collar but she twisted away. Mary and I cornered her but she slipped past us. Finally it took all three of us to catch Ambrosia. Then, unbelievably, we restrained her head and torso while Bozeman, loopy with lust, flung himself on her and began thrusting his hips. No one said a word as Bozeman concentrated on the task at hand, and Ambrosia grunted indignantly. I held my breath to avoid Bozeman's aroma. Finally I muttered, "Can I still call myself a feminist after this?"

(Friend 2006: 146)

The answer to this rhetorical question is, we fear, "no" (Stanescu 2013). Carol Adams (1990) and many other scholars have helped us to see how the system of the production of meat is part of a shared particular system of mutual exploitation of nonhuman animals and female animals. We agree. However we wish to offer that this analysis is, in a way, secondary and that eating meat would in and of itself be wholly rejectable even if it had not the slightest impact on female humans. Since *all* forms of animal consumption from "farmed animals" represent acts of sexual violence, virtually all meat and dairy are intrinsically incompatible with the core ideas of feminism, queer theory or articulation of sexual freedom or identification. As such, every purchase is *always and already* a support for sexual violence and likewise every rejection of purchasing of meat is in and of itself already a feminist act. That is, of course, not to suggest that we are so naïve as to be unaware of the fact that one can be "vegan" (in the restrictive sense of not consuming animal products) and still grossly patriarchal. (One need merely peruse any of the advertisements of People for the Ethical Treatment of Animals to witness this reality.) But the act of not consuming meat is *itself* always an anti-sexist action (Stanescu and Jenkins, forthcoming).

Capitalism, "happy meat" and the commodity fetish

As CAS scholars we oppose commodity culture, commodity fetishism and capitalism because we are vegan, and we are vegan because we are ethically opposed to the notion that life (human or otherwise) can, or should, ever be rendered as a buyable or sellable commodity. Hence, for us, the difference between the structural violence to the human slaughterhouse worker engaged in the dis-assembly lines is a difference in degree (but not of kind) to the structural violence rendered around the bodies of the animals who represent the slaughterhouse end "product". As Noëlie Vialles has argued in *Animal to Edible* (1994) and Nicole Shukin has argued (explicitly building on Vialles' work) in *Animal Capital* (2009), both represent a commodification of life: a shared commodity fetishism where the body and suffering of both worker and animal vanish into a single consumable product.[5] As an expression of our own vegan practice, we boycott even non-animal products that we know exploit human animals, such as certain types of sugar, coffee, chocolates, bananas etc. These are not separate boycotts from a critical animal studies perspective, but a world-view that the pain and death of any animal (human or otherwise) should not be rendered invisible in a market place. To rephrase these same ideas in Marxist terminology, our goal is a rejection of the invisibility of the various oppressions at play in a capitalist world, where exchange value comes to dominate all other forms of value. As critical animal scholars, we are vegan because we are anti-capitalist, we are anti-capitalist because we are vegan. These are two aspects of one original, and shared, ethical and ontological engagement in the world (see, for instance, Jenkins and Twine, Chapter 11 this volume).

Consequently, we also hold that any idea or articulation of "humanely" raised meat is an inherent and impossible oxymoron. And it makes no difference to us if this slaughter is marketed as "locavorism" (a geographic distinction); "slow food" (a communal eating experience); done on behalf of a "compassionate" or "conscientious" carnivore (ethical distinctions); "free range", "organic", or "pasture raised" (distinctions based on specific practices of animal husbandry). Such practices are offensive because fundamentally at their core, they still treat animals as buyable and sellable commodities.

Admittedly, on the one hand, in terms of pure numbers, a focus on the loca-vore movement may seem ill-placed, since 99 per cent of all animals killed for human consumption within the United States are killed in the factory farm system (Farm Forward 2008).[6] So to focus on the locavore movement is to focus on the 1 per cent exception that, merely, "proves the rule". And, moreover, it seems misplaced in that no matter how much actual suffering occurs on these supposedly "humane farms" no one, including ourselves, could argue that the animals suffer "as much" as those housed in factory farm conditions. However our worry is, in part, that drawing increasing focus to these statistically unrepresentative examples of the theoretical "ideal" farm serves primarily to hide from the average consumer the reality of the life of animals and our species relation to them. These locavore farms represent a manner in which the inherent power relations of anthropocentrism become masked in a, now literal, rhetoric of "pastoral care" and supposed benevolence. The reality is that the locavore movement could never function on any scale beyond its current tokenistic existence, as the United Nations (Steinfeld 2006) has pointed out, since it utilizes more land per pound of meat than the current factory farm system. So the locavore movement both represents a movement that *does not* currently help virtually any animals at all, and *cannot* do so in the future. But this is not the essence of our critique. It is instead to suggest that even if, via the suspension of the laws of ecology and physics, the entire 60 (and soon to be 120) billion land animals currently raised and killed each year (Steinfeld 2006) could be transferred from CAFOs to local, free-range, and "humane" farms, such a practice would only serve to help render the staggering level of speciesist violence even more naturalized and therefore "invisible". In other words, we believe that rather than being a critique of the current system, the "locavore" or "free-range" movement and the "factory farm" system work in tandem. The locavore movement helps to set up a false (but viewable) proxy for what is supposedly occurring throughout all meat production, and, at the exact same time, the statistical reality that a "humane farming" system is wholly impractical and would in practice render 99 per cent of all meat consumption unviable is, in turn, rendered invisible via the universal nature of the factory farm system.

Therefore all that seems to occur, in reality, is that consumers of higher socio-economic status purchase overpriced "humane meat", which they consume to "atone" for the factory farmed animals that they continue to consume in the vast majority of their diets. And, indeed, this is exactly the case we see with virtually all advocates of "local humane farming" such as Michael Pollan, who repeatedly assures his readers that there are no strict rules and that small steps constitute actual change (see Salih, Chapter 3, for a critique of Pollan), or Catherine Friend (author of *The Compassionate Carnivore*) who, in exactly parallel fashion not only assures her readers that they are wholly free to continue to purchase factory farm meat (as long as they also purchase some "happy meat" as well) but even assures us that she herself continues to consume factory farm meat for at least 25 per cent of all her meals because it is so "convenient" (2008: 197; Stanescu 2013; Stanescu and Jenkins forthcoming).

Our point is that the locavore and humane meat movement is not a different system, or a critique of the factory farm system; it is instead a function of the *same system*. The factory farm cannot be rendered visible; however, via the lens of the "humane" farm the killing of animals can, falsely, be shown – in other words, people can come and "see" what it is like to kill animals on a farm even if, in reality, what they are seeing is nothing like actually witnessing the killing of 99 per cent of all animals, which they are never allowed to see.

To phrase the same idea in a Marxist register,[7] in a post-Fordist economy consumers increasingly seem to want to "feel" like they are piercing the commodity fetish and the standardization of the assembly line, even if, in reality, this belief is wholly false. We know we are missing something even if we don't want to actually know what it is that we are missing. There are stores for children where they can "make" their own teddy bears by assembling pre-fabricated components into the bear. We can see this trend also in the rise of "open" kitchens in restaurants in which a small section of the food production will be visible to the public so that people can "see" the food being prepared (even if it is only a small part of the kitchen that we ever see).

But perhaps the best example is the move towards the creation of berry farms, where people can pay a fee to pick their own berries. This move in which labour is re-classified as a hobby (and a novelty) is, itself, a shift and reflection on the current state of post-Fordist society structure. But what we want to focus on here is the manner in which a supposedly experimental critique of the commodity fetish actually serves to reinforce both its reality and invisibility, i.e. children "assembling" a teddy bear at a mall bears little similarity to the mass production of teddy bears (perhaps also manufactured by children in far different conditions) in other countries, or the pleasure of paying to "pick your own berries" bears little resemblance to the reality of exploited farm labour in an industrialized agriculture system. These experiences do not represent a critique of the system, but instead, we wish to suggest, are part of the same system in which a certain anxiety of exploitative labour practices can, in part, be alleviated via a wholly token, expensive and recreational experience of playing at "labour".

So too, we wish to suggest that both the locavore movement and the associated DIY slaughter movement represent a similar false, and wholly token, indict of the commodity fetish. For example both Michael Pollan and Joel Salatin (Pollan 2006: 2010) speak, at length, for the need to go "beyond the bar code" (their term for, in essence, transcending the commodity fetish) by personally interacting with the farmer, personally raising one's "own" animals, and even personally killing these animals. Indeed Pollan (in the two most disturbing scenes of *The Omnivore's Dilemma*) personally kills "his" own chickens on Polyface farms and personally kills "feral" pigs in Santa Cruz (Pollan 2006). And, in turn, these practices have spawned "cottage industries" of individuals raising and then killing their own animals (primarily chickens and rabbits). Our point is that while these descriptions are filled with a rhetoric of transcending the commodity fetish, like paying to leisurely pick berries, it represents a way to deal with the actuality of not knowing where our food comes from – while at the exact same time still managing to not elucidate where our food comes from.

In the first place the pastoral romanticisation of only eating locally, of meeting the farmers, of "piercing the bar code"; these examples simply hide the reality of both the violence of these supposedly "happy" farms and their retrenchment of commodity culture (Stanescu and Jenkins forthcoming). For example, as Foer (2010) and others have pointed out, Salatin (Pollan 2006; Wood 2010) uses the exact same selectively bred birds on his farms (who live abnormally short and painful lives) as factory farms do – and does so purely because it is more profitable. Likewise Catherine Friend, on her supposedly "humane" farm engages in castration, tail docking and forced sexual violence upon her lambs (who she then sends to the exact same industrial slaughter houses) (Friend 2008; Stanescu 2013). For all the rhetoric of "knowing" how these animals live, the consumers of these products actually do not know how these animals are actually raised (or killed). However, since they now believe that they know, the reality that they are still wholly engaged in the politics of the commodity fetish is not revealed but, instead, becomes even more hidden. So, like the "maker" of teddy bears, what these consumers are purchasing, at a premium, is an illusion of authentic knowledge and relationships, which is carefully constructed to produce the appearance of better care (regardless of the reality of that situation, or complete lack thereof). Perhaps the best example of this "marketing of care" (or the retrenchment of the commodity fetish under the guise of piercing it) is Niman Ranch, which purports to be a humane beef producer. However it is no longer even owned or operated by the founder, Bill Niman, who was forced out by share holders who wanted to lower animal standards while still keeping the same name (and price). And in fact the former owner and founder, Bill Niman, will no longer eat "Niman" beef because of the current treatment of the animals (Finz 2009; Stanescu 2013). What we are trying to suggest is that what is actually being purchased, the true "product" is, in fact, less the meat (or for that matter the teddy bears) but instead the *product itself is the idea of transcending the commodity fetish*, which since it is always and already still itself a "product", does exactly the opposite.

We see in all such cases the reification of the commodity fetish under the guise of piercing it. But the simple reality is that one cannot "buy" one's way out of the commodity fetish no matter how much one pays for organic or pasture raised meat. And, likewise, we would argue that it is equally impossible to kill one's way out of anthropocentrism and human chauvinism, no matter how well the animals are supposedly treated before their murder. We therefore reject all calls for animal "welfare", which still keep the reality of human supremacy left unchecked and unchallenged. For the issue of being a "compassionate carnivore" inherently alters the issue of social justice captured in the terms of animal rights, or animal liberation, to one of mere charity, which consequently can be ignored, compromised or abandoned at will (Stanescu 2010; 2013).

We believe that CAS must concern itself with the lives, joys and suffering of real animals. But it must also do so in a manner that refutes as inherently ironic, paradoxical and impossible any reformist method of animal welfarism, locavorism, or "humane" slaughter. We would argue that the goal of CAS is, and can only be, that of actual and true animal liberation. Or, as Alice Walker (1988) more eloquently phrased the same sentiment:

> [W]e are used to drinking milk from containers showing "contented" cows, whose real lives we want to know nothing about, eating eggs and drumsticks from "happy" hens, and munching hamburgers advertised by bulls of integrity who seem to command their fate. As we talked of freedom and justice one day for all, we sat down to steaks. I am eating misery, I thought, as I took the first bite. And spit it out.
>
> (Walker 1988: 6)

Reflections on CAS futures

There is an enormous and radical potential in CAS as a field and movement of activist scholarship, politics and critical intervention. At the same time, we are also concerned that the word *studies* in CAS takes us to an impasse, denoting a sense of academic detachment and passivity. We follow Carol L. Glasser's analysis (Chapter 12, this volume), suggesting that the animal liberation movement needs in-depth understanding of the historical context of other social justice movements (such as the anti-colonial, anti-apartheid or anti-capitalist movements). We feel the need to resist the academic mainstreaming of our field, but also to include those possible forms of present or future activity that do not readily fall under either of the categories of academic studies and activist practice, but find a place in between, or entirely beyond, those familiar labels. Like Nathan Stephens Griffin (Chapter 6), we think that all new forms of "doing" CAS will require a repertoire of novel and creative methodological approaches.

We believe that moving CAS beyond familiar labels is a necessary step to take, as new forms of biotechnology, commodity culture and ecosystem destruction require new ways of doing radical theory as well as practice. Therefore, another possible future development we see for CAS is to mutate and proliferate into new territories, such as those indicated by the Anthropocene. As we see it, the notion of humanity as a destructive geophysical force that is on its way to irreversibly destabilizing and destroying ecological life-support systems and living conditions on the planet, as it potentially moves toward a global warming of 4°C within the next 50 years (The World Bank 2012), presents a deep conflict perspective on humans and the environment, articulating a humanity at war not only with other species, but also with our own.[8] We believe that this radically negative future renders the consensus, co-evolutionary, and "mutual" symbiosis perspective on human–animal relations that presently guides much HAS, posthumanist and feminist-materialist scholarship, to be deeply problematic, parasitic and genuinely flawed. In this light, CAS might find a new role in articulating human–animal relations, not as an ideal or romanticized form of "partnership" that feeds into human desires for narcissistic self-confirmation through cross-species "interaction", but in a negative space that asks us to *leave them alone*.

Acknowledging that the most acute and serious research problem for CAS may not be the animals but the humans, CAS might change its object of analysis and morph into what Richard Twine and Vasile Stanescu have termed "post animal studies" (Stanescu and Twine 2012). We refer to this move towards post animal studies in two ways. In the first sense, the major response by the animal agribusiness to all of the critiques levelled against them has been to try and remove the "animal" from the "meat" in the form of *in vitro* meat production (Miller 2012), genetic manipulation to offset environmental effects (Clark 2012) or the removing of pain receptors from animals in confinement (Terhaar 2012). Such a move has silenced the traditional theorists of animal studies, such as Peter Singer who, for example, endorses *in vitro* meat on his website (Singer 2013).

So to think in terms of "post animal studies" is to think deeply about the questions yet to come – how can we think of human commodification of life and anthropocentric privilege beyond issues of pain or suffering? Beyond even what has traditionally been viewed as an "animal" at all? The first sense of "post animal studies" is, therefore, to try and think the technofuture that is to come that is, in some sense, post "the animal" (in terms of cultured cell tissue or mothers so genetically manipulated that they no longer respond to the cries of their own children) (Miller 2012; Terhaar 2012). The second sense of "post animal studies" is to try and change the focus. Considering the scale of vivisection and other forms of animal experimentation, one could argue that the "animal" has indeed been "studied" quite enough. What if we no longer always tried to understand or solve the "animal question", since the problems are not of the animals' making at all, but of our own? What if we began to ask the "human" question? How can we shift the focus from always seeing the *animals* as the topic of inquiry when, in reality, the problem is the *human* animals' mistreatment of all other animals? Following the argument of Carmen Dell'Aversano (2010), such a leap would indicate a shift of focus from our relationships with animals to a critical intervention with our own species' practices, including taken-for-granted assumptions, such as the continued expansion of the human species with its disastrous effects on animals, ecosystems and the planet as a whole. Revisiting the first section of this chapter, it would be one example of (in the words of Sara Salih, Chapter 3 this volume) a breaking with existing knowledge structures, a move toward creating a space for something other-than-human to actually emerge and prosper, not on human-defined conditions, but on their own.

Direct action

Most importantly, we would argue that CAS must acknowledge, support and sustain direct action taken for the purpose of animal liberation. We see this as a continuation of a long tradition of civil disobedience for issues of social justice and we support such actions, and those who engage in them. We are mindful of feminist critiques forwarded by Marti Kheel (2006) against a hyper-masculinity of direct action that at least some advocates for direct action would seem to support. Not to mention the entire issues of class, race and citizenship, which both format the stakes and underpin the ability to engage in any direct action at all (to be arrested when one is an undocumented worker produces very different results than to be arrested when one is a "full" citizen). It is certainly not our claim that everyone must, or should, personally engage in direct action or even that direct action should be seen as a "higher" form of action. Nothing could be further or more offensive to the intersectional notion at the core of CAS. However, as scholars and activists we support direct action even when controversial (although always open to critique). We must support real, concrete and direct action (however that term is understood) in order to help actual animals and end their suffering. In other words we must not only write "to" activists; we must also write *as activists*. Our overarching goal would be for CAS to echo Foucault's (1974) call for his theory to serve as "a kind of tool-box others can rummage through to find a tool they can use however they wish in their own area ... I don't write for an audience, I write for users, not readers" (Foucault 1974). We, too, must write not only for readers but also for "users", by which we mean activists, supporters and comrades. Moreover, we would expand the Foucauldian "tool-box" view on theory above and let our critical theories *work through* our scholarship and become active agents in the shaping of subversive knowledge, as well as of scientific and political debates. Or, to paraphrase Marx, the reason that CAS matters at all is because we believe that it is not only our goal to interpret the world (human–animal relations, speciesism, global destruction etc.). The point is for us: academics and activists in joint collaboration, to actually *change it.* This means that for us, the theory/practice divide is profoundly false. Like anti-capitalism and veganism, or feminism and veganism, they constitute two dimensions of the same phenomenon, and only by working in close conjunction can they transform each other and transform the situation of actual animals.

Acknowledgements

We wish specifically to acknowledge, and thank, all our previous co-authors, reviewers, conference organizers and fellow attendees, whose helpful and thoughtful feedback underpinned this chapter. We would also like to thank the Culture and Animals Foundation, the Woods Institute for the Environment, Minding Animals International, The Institute for Critical Animal Studies, the programme in Modern Thought and Literature at Stanford University and a cultural grant from the European Union (provided for the "Occupy Species" conference), which helped to provide support for this previous research. Most importantly, we have to thank our excellent editors, Nik Taylor and Richard Twine.

Notes

1 This chapter draws directly from our previous talks and symposiums. Specifically, in terms of presentations, we took from "The Future of Critical Animal Studies" symposium and the official launch for the *Critical Animal Studies* book series in Utrecht, the Netherlands; The "Occupy Species" Conference in Hamburg, Germany; the "Conscious Eating" conference and the "Funny Kinds of Love" conference in Berkeley, United States; and the "Chi" conference at Stanford University, USA.
2 See Cudworth, Chapter 1 this volume, for a critical discussion on the notion of speciesism.
3 See Cole, Chapter 10 this volume, for a historical analysis of human species privilege.
4 It is not unrelated to note that the fascists first developed the "IQ" test as a way to determine at which point life became killable – a point most clearly articulated by Paola Cavalieri in *The Death of the Animal* (2009).
5 In contemporary globalized commodity chains, this phenomenon needs some modification, since the end product may contain the labour, suffering and body tissue of not only one, but multiple humans and animals.
6 As Farm Forward, an organization which works to support "humane" farms, explains:

> [T]he reality of meat is unambiguous. And at Farm Forward we don't pull any punches when we face inconvenient realities: Most of the animals raised and killed for food (more than 99 percent, to be precise) come from unsustainable and cruel factory farms or, in the case of sea animals, other industrial operations.... Every person who adopts a vegetarian diet reduces suffering and environmental

degradation and helps stretch the small supply of non-factory meat, dairy, and eggs currently available for those who choose to eat meat. As long as the demand for non-factory animal products exceeds the supply to this degree, it is best to avoid even these products.

(Farm Forward 2008)

7 See Clark, Chapter 9 this volume, for a Marxist analysis of animals in the capitalist labour process.
8 We want to emphasize here that Anthropogenic causes of climate change, ecosystem destruction and natural resource depletion are not evenly distributed among humanity.

References

Adams, C. J. (1990) *The Sexual Politics of Meat: A Feminist-Vegetarian Critical Theory*, New York: Continuum.

Barlow, E. G. (1945) "Dairy farming", *The Vegan News*, 5 September.

Best, S. (2009) "The rise of critical animal studies: putting theory into action and animal liberation into higher education", *Journal for Critical Animal Studies*, 7(1): 9–52.

Cavalieri, P. (2009) *The Death of The Animal: A Dialogue*, New York: Columbia University Press.

Clark, J. (2012) "Ecological biopower, environmental violence against animals, and the 'greening' of the factory farm", *Journal for Critical Animal Studies*, 10(4): 100–129.

Dell'Aversano, C. (2010) "The love whose name cannot be spoken: queering the human-animal bond", *Journal for Critical Animal Studies*, 8(1/2): 9–53.

Farm Forward (2008) Farm Forward website, available at: www.farmforward.com/ (accessed 28 April 2013).

Finz, S. (2009) "Niman ranch founder challenges new owners", *San Francisco Chronicle* 22 February.

Foer, J. S. (2010) *Eating Animals*, New York: Little, Brown and Company.

Foucault, M. (1994) [1974] *Prisons Et Asiles Dans Le Mécanisme Du Pouvoir. Dits et Ecrits*, vol. 11, Paris: Gallimard.

Friend, C. (2006) *Hit By A Farm: How I Learned To Stop Worrying And Love The Barn*, New York: Marlowe & Co.

Friend, C. (2008) *The Compassionate Carnivore: Or, How To Keep Animals Happy, Save Old Macdonald's Farm, Reduce Your Hoofprint, And Still Eat Meat*, Philadelphia, PA: Da Capo Lifelong.

Horkheimer, M. (1982) *Critical Theory*, New York: Seabury Press.

Kheel, M. (2006) "Direct action and the heroic ideal: an ecofeminist critique", in Best, S. and Nocella, A. (eds) *Igniting A Revolution: Voices In Defense Of The Earth*. Oakland, CA: AK Press.

LePan, D. (2010) *Animals: A Novel*, New York: Soft Skull Press.

Miller, J. (2012) "In vitro meat: power, authenticity, and vegetarianism", *Journal for Critical Animal Studies*, 10(4): 41–63.

Nagel, T. (1979) *Mortal questions*, Cambridge [Eng.]: Cambridge University Press.

Noske, B. (1997) *Beyond Boundaries: Humans and Animals*, Buffalo, NY: Blackrose Books.

Pedersen, H. and Stanescu, V. (2012) "What is 'critical' about animal studies? From the animal 'question' to the animal 'condition'", in Socha, K., *Women, Destruction, and the Avant-Garde: A Paradigm for Animal Liberation*. Amsterdam and New York: Rodopi.

Pollan, M. (2006) *The Omnivore's Dilemma: A Natural History Of Four Meals*, New York: Penguin Press.

Pollan, M. (2010) "The food movement, rising", *The New York Times Book Review*, 10 June.

Salatin, J. (2007) *Everything I want to do is Illegal*, Swoope, VA: Polyface.

Shukin, N. (2009) *Animal Capital: Rendering Life In Biopolitical Times*, Minneapolis: University of Minnesota Press.

Singer, P. (1977) *Animal Liberation: A New Ethics For Our Treatment Of Animals*, New York: Avon Books.

Singer, P. (2013) Peter Singer's personal website, available at: www.princeton.edu/~psinger/faq.html (accessed 28 April 2013).

Stanescu, J. (2012) "Species trouble: Judith Butler, mourning, and the precarious lives of animals", *Hypatia*, 27(3): 567–582.

Stanescu, V. (2010) "'Green' eggs and ham? The myth of sustainable meat and the danger of the local", *Journal for Critical Animal Studies*, 8(1/2): 8–32.

Stanescu, V. (2013) "Why 'loving' animals is not enough: a response to Kathy Rudy, locavorism, and the marketing of 'humane' meat", *The Journal of American Culture*, 36(2): 100–110.

Stanescu, V. and Jenkins, S. (forthcoming) "Principal six: one struggle", In *Critical Animal Studies Reader*, Atsuko Matsuoka And Anthony Nocella (ed.).

Stanescu, V. and Twine. R. (2012) "Post animal studies: the future(s) of critical animal studies", *Journal for Critical Animal Studies*, 10(4): 4–19.

Steinfeld, H., Food and Agriculture Organization of the United Nations and Livestock, Environment and Development (2006) *Livestock's Long Shadow: Environmental Issues And Options*, Rome: Food and Agriculture Organization of the United Nations.

Terhaar, T. (2012) "The animal in the age of its technological reducibility", *Journal for Critical Animal Studies*, 10(4): 64–77.

The World Bank (2012) *Turn Down the Heat: Why a 4°C Warmer World Must be Avoided*, Washington, DC: The World Bank.

Vialles, N. (1994) *Animal to edible*, Cambridge: Cambridge University Press.

Walker, A. (1988) "'Am I blue?' through other eyes: animal stories by women", Edior Zahava, Freedom, CA: Crossing Press.

Watson, D. (1947) "An address on veganism", Ambleside, Westmorland: The Vegan Society.

Wood, G. (2010) "Interview: Joel Salatin", *The Guardian*, 31 January.

Index

Page numbers in **bold** denote figures.